Lecture Notes in Mobility

The book series Lecture Notes in Mobility (LNMOB) reports on innovative, peer-reviewed research and developments in intelligent, connected and sustainable transportation systems of the future. It covers technological advances, research, developments and applications, as well as business models, management systems and policy implementation relating to: zero-emission, electric and energy-efficient vehicles; alternative and optimized powertrains; vehicle automation and cooperation; clean, user-centric and on-demand transport systems; shared mobility services and intermodal hubs; energy, data and communication infrastructure for transportation; and micromobility and soft urban modes, among other topics. The series gives a special emphasis to sustainable, seamless and inclusive transformation strategies and covers both traditional and any new transportation modes for passengers and goods. Cutting-edge findings from public research funding programs in Europe, America and Asia do represent an important source of content for this series. PhD thesis of exceptional value may also be considered for publication. Supervised by a scientific advisory board of world-leading scholars and professionals, the Lecture Notes in Mobility are intended to offer an authoritative and comprehensive source of information on the latest transportation technology and mobility trends to an audience of researchers, practitioners, policymakers, and advanced-level students, and a multidisciplinary platform fostering the exchange of ideas and collaboration between the different groups.

George A. Giannopoulos · Yidong Li

Human Mobility, Artificial Intelligence and Climate Change

Springer

George A. Giannopoulos
Academy of Athens
Athens, Greece

Aristotle University of Thessaloniki
(Emeritus)
Thessaloniki, Greece

Yidong Li
School of Computer Science
and Technology
Beijing Jiaotong University
Beijing, China

ISSN 2196-5544 ISSN 2196-5552 (electronic)
Lecture Notes in Mobility
ISBN 978-3-032-08170-4 ISBN 978-3-032-08171-1 (eBook)
https://doi.org/10.1007/978-3-032-08171-1

This Springer imprint is published by the registered company Springer Nature Switzerland AG
The registered company address is: Gewerbestrasse 11, 6330 Cham, Switzerland

Preface

This book was greatly motivated by two key developments that became evident during the period immediately after the COVID pandemic. The first was the realisation of what it is like to live in emission-free cities as because of the greatly reduced traffic flows during the COVID mobility restrictions all our big cities lost their usual congested and polluted look and became an "idyllic" place with clear atmosphere and easy access everywhere. This situation was most welcome by most urban residents and served as a "window of opportunity" to a carbon-free future with less traffic on the roads. The second development was the application of Artificial Intelligence (AI) in the transport and mobility sector which had been applied for the first time some two decades ago but now emerged more visibly by the developments in driverless technologies and automated mobility demonstrations. The first realisation brought home the importance and necessity of decarbonisation in the transport sector while the second revealed the great potential of AI as an enabler for a fully optimised and user-centric mobility system. Investigating the potential of AI in speeding up implementation of the various decarbonisation plans in the transport sector for an environmentally and climate friendly mobility and at the same time optimising the operation of the (transport) system, became a very pertinent intellectual and research activity nowadays. Exposing this intellectual and research activity further and promoting the cause for a carbon-free transport sector became the main justification and impetus for writing this book.

As it turned out, the timing of the publication of this book constituted another justification for its making. The year 2025 can very well be termed a "turning point" in the road towards the future of transportation and human mobility. This is so, for technological, socioeconomic, and environmental/climate change reasons. In terms of the technology of transportation, many new technologies that have been on the research and development stage for the past 10 years or so, are now at the brink of taking off on the path to commercial application and viability i.e., the "slope of enlightenment" to use the terminology of the famous Gartner Hype Cycle for emerging technologies. Take for example the technologies of electric, connected, intelligent, automated vehicles, or the various Artificial Intelligence (AI) techniques and algorithms that feed practical mobility Apps, the positioning and guidance technologies, and many others. In terms of the socioeconomic aspects, most users of the transport system—especially in the urban areas—are now becoming aware and accept the need for using their private cars less in favour of a more shared type of mobility, and they are more keen to protect a clean and healthy environment. In terms of the environmental and climate change aspects of transportation, the year 2025 is a turning point as it is the deadline year by which critical decisions should have been made vis-à-vis achieving the intermediate climate mitigation goals set for the milestone year of 2030. This is mentioned in most countries' climate change mitigation plans as a critical milestone year on the road to full decarbonisation by 2050. The year 2025 is also a critical turning point year on account of the complete reversal of the environmental and climate change policies of one leading world economy, which is also one of the most polluting, i.e., the US. Instead of strengthening the stance and quickening the pace to decarbonisation and leading the world by example, the new US government that took office in January 2025, reversed the previously held US climate and environmental policies and abandoned all efforts at decarbonisation even denying that climate change exists (as a human induced phenomenon). At the time of writing, the full impact of this reversal of policies or its long term results are not yet known but it is evident that there will constitute a big blow to the efforts of the UN to promote a common stance globally towards averting a full-blown climate crisis.

By all accounts, the year 2025 may stand out in the world's archives as a negative turning point as regards environmental and climate change policies. Besides the negative developments in the US, the implementation so far of decarbonisation plans has been poor and insufficient to meet the planned intermediate targets for 2030 and beyond. Few years ago, the mood was different. Many countries had formulated long-term strategies—the so-called *Nationally Determined Contributions* (NDCs)—for fully decarbonising their economies by mid-century. The number of countries that submitted such strategies to the UN at the end of 2023, were 75 representing 87% of the world's GDP, 68% of the global population (in 2019), and around 77% of global greenhouse gas emissions as they were estimated in 2019. At the same time, businesses started to reallocate resources from high-emissions to low-emissions products, climate-related venture capital investments have been increasing, and the global financial sector was also adapting its practices to assist actions combating climate change. The annual global investment in transition technologies doubled, from $660 billion in 2015 to more than $1 trillion in 2023 while at the same time technological innovation kept making new climate mitigation technologies more viable. Solar and wind power generation kept getting cheaper and, in many countries, these energy sources accounted for more than 40% of the electricity generation capacity (75% of all new electricity-generating capacity).

All these developments of previous years, though in the right direction, they did not produce sizeable or visible results as regards the targets set for 2025 and 2030. Whereas, in order to limit the global warming to an average of + 1.5 °C by 2050, as set by the 2015 United Nations Framework Convention on Climate Change in Paris, global greenhouse gas emissions should start falling by 2025 at the latest and reach a – 43% reduction by 2030 (over 1990 levels), nothing of the sort is happening. As documented in the *GHG emissions of all world countries* report of the EU's EDGAR (Emissions Database for Global Atmospheric Research), the global greenhouse gas emissions in the last few years up to 2024 were increasing by an average of + 2% per year rather than falling. Compared to 1990 levels these emissions were + 62% in 2023 and even more in 2024, so reducing them by 2030 to – 43% is a rather impossible task. Most alarming of all, at the beginning of 2025 the global trade and geopolitical environment seems set to derail whatever progress has been achieved in the past and burden the planet with multiple economic and geopolitical crises, additional carbon emissions, and a further setback in the pace of international cooperation to mitigate climate change.

Transport, though one of the most important contributors to a national and regional economy, it is also one of the most polluting sectors of human activities and for some countries *the* most polluting sector. Its relationship with climate change is a two-way affair. On the one hand, climate change brings about a series of primary (direct) and secondary (indirect) impacts on the transport system's infrastructure and operation and on the other, the operation of the Transport system is one of the main contributors to the greenhouse gas emissions that are responsible for the development and evolution of the phenomenon of global warming. AI technologies come as the tool of choice in treating this two-way interaction and breaking the vicious circle of "transport system operation/further increases of carbon emissions/further global warming/impacts of climate change on transport operation", and so on. What makes AI appropriate for the multitude of applications in the transport sector is its superior ability for quantitative and qualitative data analyses, its suitability for traveller behavioural studies, the management of travel demand, and its superior handling of optimisation problems for achieving safety, efficiency, and sustainability in the operation of the transport system. AI's ability to understand, train, and model the complex nature of transportation and mobility systems operation and the interactions among their many components, is unique.

These capabilities make the use of AI an important and indispensable tool for intelligent, individually tailored and efficient mobility services coupled with lower car ownership and use, and favouring the development and operation of shared services. AI-powered mobility platforms will soon be offering multiple services such as comprehensive intermodal travel information, real-time traffic and other forecasts, bike and car sharing, direct booking and payment, and several others. Many of the foreseen future types of mobility services are already happening today, but they are still in their initial, premature, stages. For example, shared on demand mobility is already here, with the car-sharing community experiencing a steady growth from around 0.35 million registered members in 2006, globally, to 7 million in 2015, and more than 35 million in 2025 (Asia leads the car-sharing community with 58% of the total users, followed by Europe with 29%.). Despite the current rising car-sharing membership, only a fraction of car-sharers integrates regularly car sharing into their everyday trips and car ownership remains high. This shows that there is a long way to go, and AI will play a crucial role in maturing this trend.

Using AI related algorithms in the transportation field is not new. Their first use can be traced to approximately 25 years ago, at the beginning of the 2000s, when AI related algorithms and techniques were used for the study of travel demand and traffic management phenomena (e.g., fuzzy sets theory, artificial neural networks, genetic algorithms and agent-based models). After the COVID pandemic, however, their use and application in the transport and mobility sector became much more visible and started expanding exponentially, especially for the operation of automated intelligent vehicles. As we make clear in this book, the use of AI in mobility applications is booming but caution is needed for them to be reliable, transparent, accountable, overseen by humans and capable of being shut down if necessary. They must also adhere to some form of global oversight, agreed by all implementing parties and governments so that it can be expressed in the form of technical as well as ethical legally binding guidelines. To supervise all these and provide coordination and harmonisation at regional and global level, there is a need for an internationally backed organisational structure that is inclusive, equitable and properly staffed to exercise such oversight. One of the first duties of such government would be to set and oversee that AI is developed and used within the above frame and—even more importantly—to see that AI research does not produce machines and software capable of acting independently with an intelligence that exceeds that of human beings.

Coming back to the climate change mitigation issues the pertinent question is, can the world make the current negative trends change and comply with the provisions of the 2015 Paris COP21 Agreement? We remind that these provisions involved limiting global warming to + 1.5 to + 2.0 °C by 2050 while maintaining and even improving the quality of life for citizens around the world. Unfortunately, there is no "silver bullet" for achieving these UN-set objectives. Success can only come through a concerted action using a multitude of policies and measures to be implemented by governments around the world over a wide range of economic sectors. For the transport and mobility sector, we refer to most of them, in Chap. 3. In view of the diversity, complexity, and sheer magnitude of the necessary actions, the full utilisation of Artificial Intelligence (AI) is, as already pointed out, the most promising expediting factor and tool of choice for the planning, implementation, and monitoring of these actions. An honest and transparent international cooperation is also needed to provide the necessary coordination in prioritisation of these actions as well as to secure joint financing and technical assistance to the lesser developed economies. Instead of this, the year 2025 sees the world's largest economy set to inaction and complete denial of such cooperation instead of accelerated action and convergence. Despite these negative developments these authors argue that in the not-too-distant future the unfolding climate crisis and the extreme weather events that will result, will eventually convince even the most extreme and stubborn political minds that climate change is real and that setting a plan of action is mandatory and merits top priority in the national agendas. Human induced climate change is fast becoming a climate crisis, and its impacts are no longer something that *will* happen at some time in the future but something that is happening right now. Many ecosystems and people around the world are affected with dozens of victims and billions of damages to infrastructures. The international cooperation effort against climate change must therefore be kept alive and going strong in spite recent setbacks.

This book addresses all issues related to the application of AI for a decarbonised and operationally efficient transport and mobility sector in the face of the incoming climate crisis. Its material is intended for a readership that includes students, academic and research personnel, as well as transport sector administrators at various levels, involved in the study or research or administration of the transport and mobility sector. It is also hoped that it will make an interesting reading for the climate or environmental professional as well as for the public at large. It is written in a way that simplifies the notions involved in both AI and transport and presents various relevant applications in developing and implementing a future mobility that is sustainable, climate friendly, and resilient. After analysing some key AI applications in the transport field, the book ends with a 70-page chapter presenting the views and projections for future mobility under the form of a *futures thinking* exercise outlining the 2050+ mobility scene, globally.

The book is also aimed to familiarise its readers with the AI applications in the transport and mobility sector, i.e., to let them know what they are likely to be, what they will consist off and what their true capabilities are. Also, on ways to maximise the AI's usefulness and utilisation in a benign and positive way to assist in making the climate crisis better understood and accepted by all as a human-caused phenomenon.

We hope to assist in sharpening the current differences and bring closer together the various national interests and political viewpoints as regards making the transport and mobility sector cleaner and more climate friendly. Existing differences can be reconciled and countries be more and more aligned with the findings and recommendations of science as well as with the majority of countries as expressed by their positions in international organisations like the United Nations.

It is finally hoped that the material in the 10 chapters of this book, will contribute to a better understanding of the benefits and potential uses of AI technologies and methodologies in the field of transport and mobility as planners strive to protect this system and prepare it to face better the impacts of climate change.

Athens, Greece
Beijing, China
May 2025

Prof. George A. Giannopoulos
Prof. Yidong Li

Acknowledgments We are most grateful to those who have helped us, directly or indirectly, to complete this book. The idea for looking at the future of mobility under the positive influence of AI and the negative one of climate change was discussed with several colleagues in Europe and China who encouraged us to attempt looking at this complex and all too important interaction. These colleagues are many and with some of them we've talked only on the phone, so we won't even try to be inclusive. Yet—somehow—talking to these colleagues, whom we choose to acknowledge here anonymously, gave us the confidence to proceed and the perspective to see that there is common sense in tying all three elements together. Ultimately, the credit for the positive attitude and encouragement for this book must go to our editor and project coordinator at Springer Nature, Leontina Di Cecco who has been supportive and positive all the way to the end of the project. Many thanks are also due for a similar reason, to Prof. Christos Zerefos Secretary General of the Academy of Athens and world-known expert on climate change, with whom Prof. Giannopoulos has a long-time cooperation originally at the Aristotle University of Thessaloniki and in the last 12 years at the Academy of Athens and the EMEKA—Climate Change Study Committee of the Bank of Greece.

For the technical and other general assistance that we received in making this book, many thanks are due to the staff and faculty of the School of Computer Science and Technology of Beijing Jiaotong University and more particularly to the following individuals:

- Dr. Shenyuan Ren, for her general coordination assistance and for contributing to the use cases mentioned in Chap. 5.
- Dr. Hui Zhang, for her contribution to Chap. 5.
- Dr. Xiaqing Li, for his contribution to Chap. 6.
- Dr. Yuanzhouhan Cao, for his contribution to Chap. 7.
- Ms. Yushan Han, for technical assistance in the various diagrams and Figures, especially those of Chap. 10.

We are also indebted to Prof. Constantinos (Costas) Antoniou, Chair of Transportation Systems Engineering, TUM School of Engineering and Design, Technical University of Munich, for authoring Chap. 4 together with Dr. Qing-Long Lu, Department of Civil and Environmental Engineering, National University of Singapore.

Professor Giannopoulos acknowledges the important inspiration and support from his wife Diana, his sons Anastasis and Thanos, and his long-time colleagues at the Hellenic Institute of Transport Drs. Angelos Bekiaris and Georgia Aifadopoulou.

The writing is ours; the voice is ours; and the responsibility for any mistakes or oversights is ours too…

Competing Interests The authors have no competing interests to declare that are relevant to the content of this manuscript.

Contents

About the Authors

Professor George A. Giannopoulos is a transportation planner (Ph.D. University of London—Imperial College), professor emeritus of the Aristotle University of Thessaloniki, correspondent member of the Academy of Athens, and Advisory Professor of Beijing Jiaotong University. He is also member of the steering panel—energy sector—of the European Academies Science Advisory Council (EASAC), and member of the US/TRB's standing Committee on Research Implementation and Innovation. For 20 years (1981–2001), he was the director of the Transport Engineering Laboratory of the Civil Engineering Department of the Aristotle University of Thessaloniki and the founder and director, for 7 years (2003–2010), of its Postgraduate course on "*Organisation, Operation and Management of Transport Systems*". For 16 years (2001–2016), he was the director of the Hellenic Institute of Transport (HIT) of the Centre for Research and Technology Hellas—CERTH. He has extensive experience in Transportation planning and Transport Policy issues as well as on research governance, implementation, and international transport research cooperation issues. For the last 15 years, he has been a member of the high-level advisory panel on climate change for the Greek government and the Bank of Greece (the EMEKA—Climate Change Study Committee). He has authored the strategic transport adaptation plan to climate change for Greece that is used as input to the *Life-IP AdaptInGr* project for climate change adaptation, of the Greek Ministry of the Environment. He has participated in more than 200 studies and/or research projects, over the last 35 years, in most of which as coordinator. He has published 260 scientific papers and articles in magazines and conferences, as well as 18 books, six of which in English. e-mail: ggian@academyofathens.gr

Professor Yidong Li is a computer scientist, professor of Computer Science and Artificial Intelligence at Beijing Jiaotong University and Dean of the School of Computer Science and Technology of same university. He is also the director of the Key Laboratory of Big Data and Artificial Intelligence in Transportation sponsored by the Ministry of Education of China. He received his B.Eng. degree in Electrical and Electronic Engineering from Beijing Jiaotong University, and the M.Sc. and Ph.D. degrees in Computer Science from the University of Adelaide, in 2006 and 2010, respectively. His research interests include big data, AI, advanced computing, and intelligent transportation systems. He has published more than 200 research papers in various journals and refereed conferences. He has also co-authored or co-edited 5 books and contributed several book chapters. e-mail: ydli@bjtu.edu.cn

Chapter 1
The Climate Crisis Is Here and "Do Nothing" Is not an Option

Abstract This chapter presents in a concise and documented way the story of the climate change phenomenon and its anthropogenic origin. It presents the arguments put forward by those who deny that climate change is caused by human activities as well as the arguments of those who accept this notion and support the need for immediate action. Both arguments are presented with a summary of supporting data and documentation. Key data of CO_2 emissions per capita and per GDP are presented for some key polluting countries for a period of 30 years while data of actual quantities of CO_2 emitted in Gt/year are given for the six top polluting economies (China, USA, EU, Russia, India, Japan). The trends in the quantities of CO_2 emitted each year by the six most polluting countries, show a marked increase in the emissions after the year 2000 for some of these countries. The concise and accurate presentation and comments on emissions data and the reference to key reports and documents by various relevant organisations, is complimented by presentation of some of the most pronounced impacts of climate change that are already here. These are the major natural disasters that occurred around the world over the last decade which bring the most convincing evidence of the dire consequences that can be caused by climate change and global warming. The concluding sections of this chapter defend the notion that climate change is here, it is getting accelerated by human activities and their related GHG emissions, and that "doing nothing" is not an option. Already, the negative socioeconomic impacts of extreme weather events are felt in an increasing number of countries, and this should provide people and governments with enough evidence as to the painful consequences of a "do nothing" situation. It is argued that the consequences and impacts of "doing nothing" will be much bigger and more costly to humanity than the cost of the actions necessary to mitigate climate change and achieving full decarbonisation in the foreseeable future. The fight to combat climate change is seen as the only path in front of humanity to save the planet from the catastrophic consequences of a full-blown climate crisis.

Keywords Sustainability · Climate crisis · Global warming · IPCC · GHG emissions · Emissions data · GHG emitters · Climate Change

G. A. Giannopoulos and Y. Li, *Human Mobility, Artificial Intelligence and Climate Change*, Lecture Notes in Mobility, https://doi.org/10.1007/978-3-032-08171-1_1

1.1 Introduction—Sustainability and Climate Change

The first environmental conservation and preservation movements started in Europe in the late 1800s aiming to protect the countryside, the wilderness, and the wildlife from industrial development and urbanization. In the modern age, the worldwide concern for the environment can be said that started in 1970 when, on April 22, the first Earth Day was celebrated in the United States (Back, 2013). This event marked the start of the modern environmental movement that advocates for political action, public awareness, and environmental justice. In 1992, the Earth Summit in Rio de Janeiro produced the "*Rio Declaration*", the "*Agenda 21*", and the "*Convention on Biological Diversity*" reports, setting the global agenda for environmental governance and cooperation. Since then, the concern for the environment is evolving and growing by the day. Concurrent with the development of the environmental awareness in the 70s came the energy crises that started in 1973 and continued almost regularly ever since. Following the awareness of the need for protecting the environment, another notion, that of *sustainability* came to the fore. This started in 1987 when the UN's Brundtland Report defined the concept of "sustainable development" as the development that meets the needs of the present without compromising the ability of future generations to meet their own needs (WCED, 1987).

Later definitions distinguished three dimensions, or pillars, of sustainability namely *environmental*, *economic*, and *social* sustainability (Purvis et al., 2019). Environmental sustainability focuses on environmental resources such as the atmosphere, land, and water and all the relevant plant or animal ecosystems. The economic sustainability refers to the need to achieve economic development and growth without exhausting resources for future generations and without detriment to the environment. This may sound somehow controversial as economic development is most often connected to an extensive use of resources and inevitably harmful effects on the environment but nevertheless if proper actions are taken to decouple economic development from such negative impacts, economic sustainability has meaning and can indeed be achieved. Social sustainability can be expressed as the free and unimpeded use of social services without compromising long term economic or environmental resources. The "free and unimpeded use" of social services refers to the removal of structural obstacles for people in accessing and utilising key social areas such as health, education, residential provisions, cultural services, and similar (Missimer et al., 2017). In this sense, social sustainability can be also thought off as a standalone dimension that includes the other two (Boyer et al., 2016). The use of AI will support all kinds of sustainability (Abduljabbar et al, 2019; IEA, 2021; Chan et al, 2022).

Sustainability is now the prevailing notion for many aspects of our lives including "sustainable transportation and mobility". In the case of transportation, achieving sustainability usually means to develop a transportation system that meets the mobility needs of a given society without exhausting or endangering the physical or environmental resources for future generations to meet their own mobility needs (OECD-ITF, 2016; EASAC, 2019). Thus, a sustainable transportation system is the one that uses renewable resources at below their rates of regeneration, or non-renewable resources at below the rates of development of renewable substitutes. The notion of sustainability is a core notion that has shaped the environmental movement to date and has prevailed in transportation policies worldwide. Of course, there are many more factors and perspectives to consider as regards a *sustainable* operation or system but the overall concern for the environment is evolving and growing and will result in more theories and aspects of sustainability in the future. Climate change brings into the environmental debate a different dimension and several new parameters to consider. It refers to the phenomenon of warming of the planet that has happened in previous epochs too, but which this time is connected to the rise of CO_2 in the atmosphere, caused by human activities and the resulting "greenhouse effect". A sustainable human activity may or may not be compatible with the climate change related need for reduction of the emissions of CO_2 and other greenhouse gases (GHG). It may, for example, be an activity that is compatible with the sustainability requirement of not exhausting natural resources, thus preserving them for future generations, but emitting nevertheless considerable amounts of GHG and thus being detrimental to climate change abatement requirements.

The story of the climate and its anthropogenic change started quite early last century, much earlier than most people think and much before the media took it up as an almost daily subject of reporting. The alarm that a man-made climate warming due to the greenhouse effect is occurring, was raised as early as 1938 when a British engineer, Guy Callendar who, using records from 147 weather stations around the world, showed that temperatures had risen over the previous century while at the same period CO_2 concentrations had also increased. He then suggested that these CO_2 concentrations were causing the warming. After the 2nd world war, in 1958, C. D. Keeling (an American researcher at the University of California, San Diego) begun systematic measurements of atmospheric CO_2 at Mauna Loa in Hawaii and in Antarctica. Within four years, those measurements showed the first unequivocal proof that CO_2 concentrations in the earth's atmosphere were rising. These measurements continue until today while new ones have piled up during the years giving rise to more and more voices sounding the alarm for a manmade climate warming due to rising CO_2 concentrations. The term "global warming" was coined by Wallace Broecker in a 1975 paper titled "Climatic Change: Are we on the brink of a pronounced global warming?" (Broecker, 1975).

1.2 Is the Current Climate Change Real and is It Influenced by Human Activities?

1.2.1 The Arguments Against

The view of those that refuse to accept that there is a real climate change phenomenon and more so that it is due to human activities, can be summarised as follows. Climate is influenced by a multitude of factors that operate at timescales ranging from hours to hundreds of millions of years. Many of the causes of climate change are external to the earth's system while others are part of the earth system but external to its atmosphere. Among the mentioned, by the climate change deniers, non-human related causes of climate change are the orbital variations of the earth, tectonic and volcanic activity, solar eruptions, interactions among the various components of the earth system, and others. Other arguments posed by the deniers of human induced climate change, are the following:

- The observed changes in temperature and weather patterns are just normal fluctuations.
- There is no consensus among scientists that climate change is real (i.e., there are experts that question the validity of the climate warming models and projections).
- The policies for mitigation, i.e. reducing greenhouse gas emissions to minimise the growth in temperatures, are too costly and impractical and they will harm the global economy and the quality of life more than the assumed negative climate crisis impacts.
- Any of the suggested mitigation actions are insignificant and futile and they will have no power to make a real difference on the climate.

Furthermore, the climate change deniers, argue that global warming is not having always a negative impact on the planet, after all. It can, for example, lead to more crop growth, longer growing seasons, and fewer deaths from cold. In short, the deniers hold the thesis that climate change is a natural phenomenon not connected to human activity, as it has also happened at other points in history. It also will not be so destructive as to justify taking costly mitigation measures.

1.2.2 The Arguments for

That the climate change is occurring and much faster than in any other period in the history of the earth due to human activities, is the documented view of most relevant scientists and organisations globally. Detailed climatic records have been analysed since the early 1900s and changes in the climate have been confirmed by instrumental records and other means. These records, in some locations go back to a few hundred years but globally—by examining the atmospheric samples contained in ice cores that are taken at the two poles—they go back to many thousands of years as shown in Fig. 1.1. These records show that the fluctuations of CO_2 in the earth's atmosphere (in parts per million—ppm) over the last 800,000 years remained well below the mark of 300 ppm fluctuating between this level as maximum and approximately 200 ppm as minimum. In the last 100 years, however, the amount of CO_2 content in the atmosphere has risen almost vertically to a level of more than 430 ppm today as compared to just over 300 ppm in 1911 and is still rising (Fig. 1.1).

Based on such data, there are thousands of scientific publications, articles, blogs and other reports that present hard evidence and proving that climate change is happening and happening at a far speedier pace than in the past. Furthermore, that this is related to human activities and the CO_2 and other GHG that they emit in the

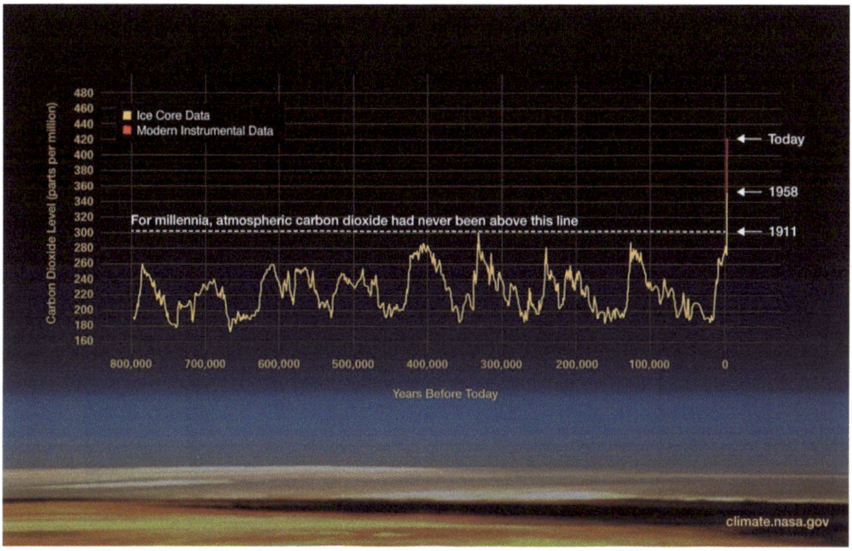

Fig. 1.1 The fluctuations of CO_2 (parts per million) in the earth's atmosphere over the last 800,000 years before today. Since 1911 the rise is almost vertical. *Source* NASA (2023)

earth's atmosphere (Hawkins et al., 2017; IPCC, 2023; NASA, 2023; NOAA, 2022; UNEP, 2024). The following is a simplified summary of the evidence to that effect:

1. There is a documented rise in the average surface temperature of the earth by about 1.2 °C (2.2 °F) since the late nineteenth century and most of this increase has occurred in the past 40 years. This is shown in the earth's maps of Fig. 1.2 which was compiled by the US National Oceanic and Atmospheric Administration (NOAA) based on data of the US National Centres for Environmental Information (NCEI).
2. The rise in global surface temperatures has gone almost in parallel with the rise in the quantities of CO_2 in the atmosphere. Figure 1.3 shows these trends. The increased quantities of CO_2 in the atmosphere are causing the greenhouse effect that warms up the planet and raises global temperatures.
3. Most of the CO_2 in the atmosphere, as well as other Green House Gases (GHG), are due to human activities such as burning fossil fuels for energy or transport, and deforestation. It is estimated that since the onset of industrial times in the eighteenth century, human activities have raised atmospheric CO_2 by 50% (NASA, 2023). In 2022, the three most CO_2 emitting sectors globally, were transportation, electric power generation, and industry in that order (IEA, 2023b).
4. The ice sheets and glaciers in the poles (the Arctic, Greenland, and Antarctica) as well as in other parts of the world are melting at ever increasing rates (NASA, 2023).
5. The global mean sea level has risen by about 20 cm (8 inches) since 1880 with about half of it occurring since 1993 (National Geographic, 2023). This is due to

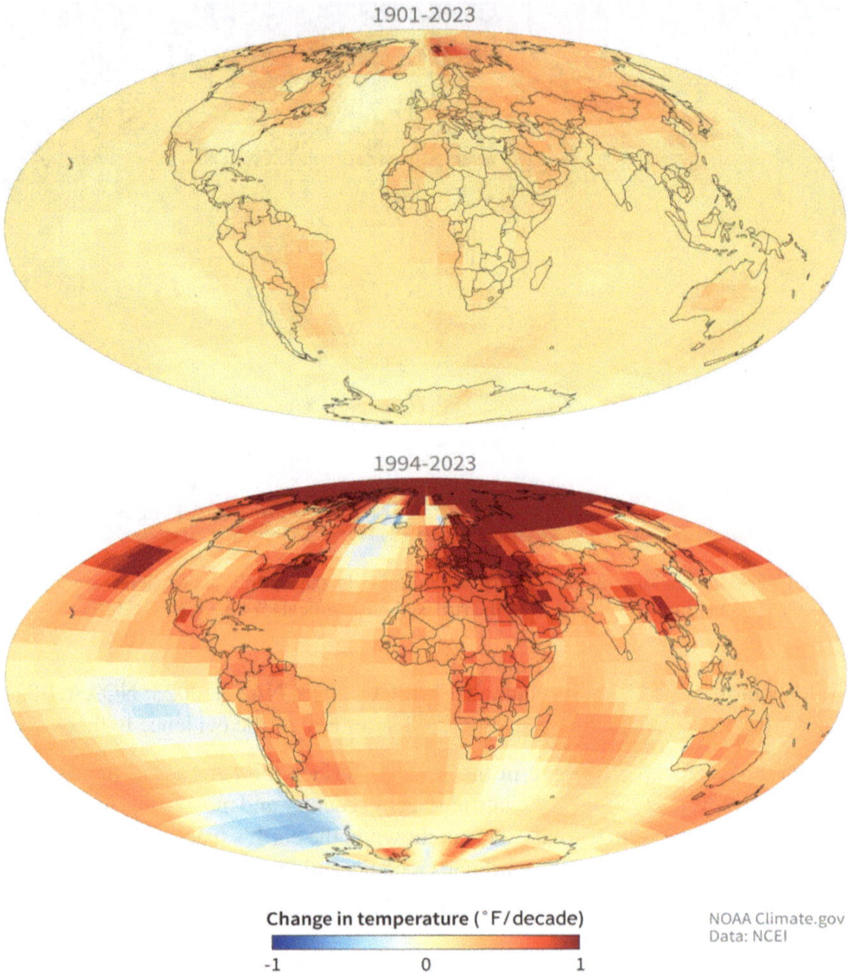

Fig. 1.2 Warming of earth's surface between 1901 and 2023 (above) and in the last 30 years (below). *Source* US National Oceanic and Atmospheric Administration (NOAA), Data of National Centres for Environmental Information (NCEI), at https://www.climate.gov/

the expansion of seawater as it warms and the addition of meltwater from melting ice sheets and glaciers.

6. The amount and distribution of rainfall and snowfall have changed in many regions of the world. Some areas have become drier while others wetter with many floods and storms occurring more frequently in many areas.
7. Many plants and animals have altered their ranges, phenology, and behaviour. Some species have expanded their habitats, while others have contracted or have become extinct. These changes can only be explained as a response to the changing climate patterns.

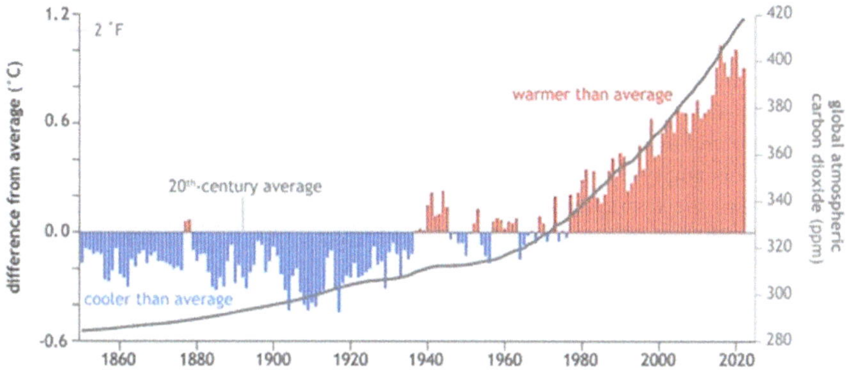

Fig. 1.3 Yearly mean global surface temperature (histogram lines) and the content of CO_2 in the atmosphere in parts per million—ppm (solid line) for the period 1850–2022: the correlation of the two trends is clear. *Source* US National Oceanic and Atmospheric Administration (NOAA), with data from Ozone-Earth System Research Laboratory/Ethereum 2/National Centres for Environmental Information—ESRL/ETH2/NCEI (in US climate.gov)

A lucid account of these trends and their due documentation can be found in the relevant entry at the *Encyclopaedia Britannica* under the title "Evidence for climate change". In the same entry the reader can also find some iconic BBC videos on this subject, by Sir David Attenborough (see https://www.britannica.com/science/climate-change/Evidence-for-climate-change—accessed December 2024). Further documentation and evidence can be found in the relevant reports and announcements of many leading scientific organizations around the world including science academies, universities, research organisations, and also in scientific papers by individual experts. They all document the existence of climate change and the fact that it is advancing much faster than in previous epochs due to the GHG emissions from human activities that create the "greenhouse effect". It is now appropriate to look in more detail at the levels of these emissions and how they are distributed among the various countries of the world.

Data presented by the *Global Carbon Project—GCP* (www.globalcarbonproject.org/) in the December 2024 meeting of the COP29, showed that global carbon emissions from fossil fuels had reached a record high in 2024 and increased 0.8% over 2023 whereas it should have been reduced by more than 7% if the world was to achieve the UN set target of a global decrease, in global carbon emissions, of – 40% by 2030 over 2019, and – 100% by 2050! In 2024 the Mediterranean Sea reached for the second consecutive year record temperatures of its waters. According to Copernicus (the earth observation component of the European Union's space programme), the Mediterranean Sea waters reached in 2024 an average temperature of 28.71 °C. This is 2.5 °C higher than its average temperature for the period 1991–2020 and has several negative impacts for the fauna and flora of the land and maritime areas. Overall, the earth's average temperature has already increased between 1.1 and 1.9 °C over the pre-industrial era according to the World Meteorological Organisation (WMO,

2025). According to the "pessimistic" hypothesis of the Intergovernmental Panel on Climate Change (IPCC) if the average global temperatures increase by more than 2 °C, then the negative impacts of climate change will be very severe and devastating. *The hard evidence from the field, is therefore showing that climate change is on-going, and that humanity is not doing enough to stop it.* The power of "doing something" is amply demonstrated by the severe reduction of GHG emissions that resulted from the restrictions to human activities during the Covid pandemic (Li et al, 2022; Shaik and Ahmed, 2022; Li and Lasenby, 2023; Oestreich et al, 2023; Vagliasindi, 2023).

1.2.3 The UN's Intervention

Following the convincing evidence that an anthropogenic [human-caused] climate change is real and fast converting to a crisis, the United Nations (UN) has been mobilised since the late 1980s. In 1988 the UN formed the *Intergovernmental Panel on Climate Change* (IPCC). The panel consisted of more than 2000 scientists from many countries, and its mission was to collect, collate and assess evidence on climate change. The first IPCC report (for which the Panel shared the 2007 Nobel Peace Prize together with US former vice president Al Gore), was definitive in stating that climate change is real and largely attributable to human CO_2 and other GHG emissions. The Fifth Assessment Report of the IPCC named the "do nothing" situation, as the *Representative Concentration Pathway*—RCP scenario 8.5 (RCP 8.5). A successful and rigorously applied mitigation effort throughout the world, would limit global warming (global average temperature) up to + 2 °C and this was termed the "optimistic" RCP hypothesis, known as RCP 2.6.

Another most important action by the UN, which aimed at mobilizing the governments of all countries and coordinating them towards common action, was the establishment of the permanent *United Nations Framework Convention on Climate Change* (UNFCCC). This Convention met for the first time in Rio de Janeiro in 1992 and was initially signed by 154 countries. By 2022 the signatory countries, or parties, had risen to 198 which is almost the totality of countries in the planet. The supreme decision-making body of the UNFCCC is the *Conference of the Parties* (COP), which meets annually to assess global progress in dealing with climate change. On its third meeting the Conference of the Parties (COP3) met in Kyoto, Japan, on 11 December 1997 and adopted the Protocol of the United Nations Framework Convention on Climate Change (the "Kyoto Protocol"). This protocol came to force in 2005 and was kept in force until 2020 when it was superseded by the Paris Agreement, which was signed during COP21 in December 2015 in Paris, France and entered into force in 2016. In this COP21 meeting, the unanimous decision was to take measures to limit the global warming to + 1.5 °C (preferably) up to + 2 °C (maximum) by the end of the century. This level of global warming was set, in article 2, as the level at which the stabilization of greenhouse gas concentrations in the atmosphere would prevent dangerous interference with the climate system. In the latest COP28 and COP29 UN

Climate meetings in Dubai (UAE) and Azerbaijan correspondingly, some further decisions were taken including a deal that was reached (at the COP28 Conference in Dubai) to accelerate emission reductions, towards net zero (0% emissions) by 2050, with urgent actions scheduled in the critical decade 2020–2030. This includes an agreement to transition away from fossil fuels and to reduce global emissions by − 43% by 2030. Additionally, world leaders have committed to tripling renewable energy capacity and doubling the rate of energy efficiency improvements by 2030.

1.3 Data on GHG Emissions and GHG Emitters

1.3.1 Key Data Bases

Data on CO_2 and other greenhouse gas (GHG) emissions at world level are regularly collected and stored by a small number of international organizations and some scientific or research bodies. The most well-known of these are the following:

1. **EDGAR** (Emissions Database for Global Atmospheric Research): Managed by the European Commission's Joint Research Centre, EDGAR provides comprehensive data on greenhouse gas emissions and air pollutants globally (see: https://edgar.jrc.ec.europa.eu).
2. **UNFCCC** (Database by the United Nations Framework Convention on Climate Change): The UNFCCC Annex I signatory countries (i.e., industrialized, developed, and "economies in transition"—EITs) collect and publish (under the UNFCCC) national greenhouse gas inventories (see: https://unfccc.int/topics/mitigation/resources/registry-and-data/ghg-data-from-unfccc). These inventories are used as crucial evidence for tracking progress towards the implementation and effectiveness of international climate agreements.
3. **IEA** (Database of International Energy Agency): The IEA provides extensive data on energy-related CO_2 emissions, including detailed sectoral breakdowns and country-specific information (see: https://www.iea.org/data-and-statistics/data-explorers). Their data is related to the energy sector and is widely used for formulating or changing energy policies but nevertheless they are very indicative of the overall global trends as the energy producing sector is the main GHG emitter globally.
4. **World Resources Institute—WRI.** The WRI has created the Climate Watch Platform (CWP) which offers a wealth of information and data on global GHG emissions including historical data, projections, and national climate commitments (see: https://www.wri.org/research/climate-watch-country-greenhouse-gas-emissions-data-and-methodology).
5. **Global Carbon Project (GCP)**: The GCP focuses on quantifying global carbon emissions and their sources and provides annual updates on global carbon budgets which are essential for understanding trends in carbon emissions (see: https://www.globalcarbonproject.org/).

These organizations play a critical role in providing reliable (and accessible) GHG emissions data, which is essential for climate research, policymaking, and tracking progress towards emission reduction targets. Perhaps, the most influential and wide-ranging databases are the ones maintained by the United Nations Framework Convention on Climate Change (UFCCC). Following the UN's Climate Change Conference (COP21) in Paris, France, on 12 December 2015, which entered into force and became legally binding for the signatories on 4 November 2016, each participating country undertook the obligation to develop and implement a regularly updated plan containing its targets, policies and measures to be implemented for climate change mitigation and adaptation. These plans are the *Nationally Determined Contributions* (NDCs) which are the official documents with the climate action plans that each country has put together to cut GHG emissions and at the same time prepare for the future climate change impacts. The NDCs are updated every five years (at least this is what the countries are supposed to do). An interesting monitor of the progress made in this domain is the *Climate Action Tracker*—CAT (https://climateactiontracker.org/). This is a web-based platform developed and maintained by an independent scientific project supported by some countries (basically European) and it tracks government climate actions in countries around the world. In its most recent version, it includes the NDCs for some 50 countries which, together, cover around 85% of global GHG emissions.

The other most versatile and wide-ranging global emissions database is the *Emissions Database for Global Atmospheric Research* (EDGAR) developed by the Joint Research Center (JRC) of the European Commission. The EDGAR database is a multipurpose, independent, global database of anthropogenic emissions of greenhouse gases and air pollution that in effect complements the UNFCCC database (which is based on what is reported by the various Parties of the UNFCCC through their periodic NDCs). A long-standing collaboration of EDGAR with the International Energy Agency (IEA) produced the *Global Emissions Initiative*—GEIA, a community effort dedicated to emissions information exchange and competence building (see: https://www.geiacenter.org). GEIA is dedicated to quantification of chemical emissions (primarily CO_2 emission calculations from fossil fuel combustion). The EDGAR database, in its 2023 report states clearly that CO_2 emissions, which are the main contributors to global warming, are still increasing at world level despite climate change mitigation agreements (Crippa et al., 2023; EC-EDGAR, 2023).

1.3.2 Key GHG Emitters

The six largest GHG emitters globally, at least for the time being, are China, the United States, India, the EU27, Russia, and Brazil in that order. In 2022, they collectively accounted for 61.6% of global GHG emissions. At the same year they accounted for 50.1% of the global population, 61.2% of the global Gross Domestic Product (GDP), and 63.4% of the global fossil fuel consumption. The trends in CO_2

emissions for a select number of countries (including the above six top emitters) for the period 1990–2022 are shown in Table 1.1, the CO_2 emissions per capita, and Table 1.2, for the CO_2 emissions per GDP. From these Tables the following can be noted:

- The per capita CO_2 emissions at global level (i.e. for the world as a whole) show a steady increase of approximately 0.5%/year or 13% in the 32-year period 1990–2022. In the EU of 27 member countries, there is the opposite trend i.e., one of decreasing emissions per capita of almost – 30% in the 30-year period.
- At country level, the per capita CO_2 emissions data show that the USA remained the highest polluting country in the world until approximately 2015 when the top polluter became Saudi Arabia with Australia second and the USA third with a close fourth South Korea.
- As regards the trends in the per capita CO_2 emissions, a varying picture emerges with several countries showing a marked decline while others show a marked increase. The following cases show a marked decrease: the EU (overall decrease over the considered period approx. – 30%), Germany (– 36%), Sweden (– 44%), the UK (– 51%), and USA (– 27%). The following cases show a marked increase: China (+ 331%), S. Arabia (+ 60%), S. Korea (+ 93%), Türkiye (+ 97%), Brazil

Table 1.1 List of CO_2 emissions per capita for a select group of countries (alphabetically)

Country or region	1990 tCO_2/cap	2000 tCO_2/cap	2005 tCO_2/cap	2010 tCO_2/cap	2015 tCO_2/cap	2020 tCO_2/cap	2021 tCO_2/cap	2022 tCO_2/cap
Global	4.22	4.17	4.58	4.87	4.90	4.61	4.84	4.84
EU27	9.03	8.28	8.44	7.72	6.96	5.94	6.37	6.32
Australia	16.27	18.55	18.96	18.76	16.92	15.65	15.10	15.12
Brazil	1.47	1.97	1.95	2.22	2.50	2.09	2.34	2.15
China	2.05	2.85	4.74	6.71	7.71	8.45	8.90	8.85
Germany	12.74	10.63	10.23	10.00	9.54	7.72	8.22	8.16
Greece	7.60	8.62	9.17	7.78	6.19	4.71	4.89	5.14
India	0.38	0.37	0.33	0.33	0.32	0.27	0.27	0.27
Japan	9.36	9.78	9.99	9.49	9.65	8.40	8.53	8.61
Norway	0.20	0.16	0.15	0.16	0.14	0.13	0.12	0.11
Russia	15.95	11.33	11.97	12.01	12.00	12.44	13.46	13.31
S. Arabia	10.63	12.76	14.47	17.74	19.36	16.46	16.75	16.98
S. Korea	6.34	10.01	10.58	12.06	12.71	11.95	12.44	12.26
Sweden	6.60	6.50	6.09	5.70	4.52	3.86	3.73	3.69
Türkiye	2.87	3.59	3.62	4.32	4.71	5.07	5.52	5.66
UK	10.16	9.28	9.23	7.86	6.33	4.76	5.01	5.00
USA	19.74	21.03	19.95	17.89	16.05	13.48	14.29	14.44

Source EC-EDGAR (2023)

(+ 46%). The rest of the countries in Table 1.1, show rather stable trends with either a small increase or a small decrease in their per capita CO_2 emissions.

- By contrast, the CO_2 emissions per GDP dollar, i.e. the emissions per unit of the gross domestic product (Table 1.2) show a remarkable decreasing trend for all cases. At global level a decrease, for the period considered, of – 39% is noted. For the EU, this decrease is – 56%, for Australia – 50%, for China – 75%, and so on. However, this general decreasing trend is not because the actual emissions are reduced but mainly because in this period all countries have experienced steep increases in their economic output and development which caused the GDPs to increase faster (IEA, 2023a).

Of interest too, is the total quantity of CO_2 emitted each year by each country. This is of interest because it is this, total quantity of CO_2 in the atmosphere, that is causing climate change and global warming. The trends in the quantity of CO_2 emitted each year for the six most polluting countries is shown in Fig. 1.4. As shown there, of the six highest CO_2 emitters in the world (China, USA, EU, Russia, India, Japan) for the period 1990–2019, China and India show a marked increase in their emissions after the year 2000 with the increases for China being the highest. Between 2001 and 2019 the yearly CO_2 emissions for China increased threefold from approximately 4 Gt/year to 12 Gt/year while for India the increase was twofold from 2 Gt/year to just over 4 Gt/year.

Table 1.2 List of CO_2 emissions per GDP for a select group of countries (alphabetically)

Country	1990	2000	2005	2010	2015	2020	2021	2022
Units	tCO_2/GDP	tCO_2/GDP	tCO_2/GDP	tCO_2/GDP	tCO_2/GDP	tCO_2/GDP	tCO_2/GDP	tCO_2/GDP
Global	0.64	0.54	0.51	0.48	0.44	0.40	0.39	0.39
EU27	0.41	0.30	0.28	0.25	0.22	0.18	0.18	0.18
Australia	0.86	0.74	0.67	0.61	0.52	0.46	0.44	0.43
Brazil	0.44	0.46	0.46	0.40	0.42	0.42	0.43	0.40
China	2.52	1.24	1.21	0.97	0.77	0.65	0.63	0.61
Germany	0.42	0.29	0.27	0.24	0.22	0.17	0.18	0.17
Greece	0.40	0.39	0.34	0.30	0.30	0.25	0.23	0.23
India	0.91	0.70	0.59	0.54	0.47	0.41	0.40	0.39
Japan	0.33	0.30	0.29	0.27	0.26	0.23	0.23	0.23
Norway	0.35	0.28	0.26	0.25	0.23	0.20	0.20	0.19
Russia	0.96	1.00	0.77	0.66	0.61	0.62	0.63	0.64
S. Arabia	0.35	0.41	0.44	0.51	0.49	0.47	0.47	0.45
S. Korea	0.61	0.51	0.42	0.40	0.36	0.32	0.32	0.31
Sweden	0.27	0.22	0.20	0.18	0.13	0.12	0.11	0.11
Türkiye	0.33	0.33	0.27	0.29	0.26	0.26	0.25	0.24
UK	0.44	0.31	0.26	0.23	0.17	0.15	0.14	0.14
USA	0.61	0.51	0.44	0.40	0.34	0.28	0.28	0.28

Source EC-EDGAR (2023)

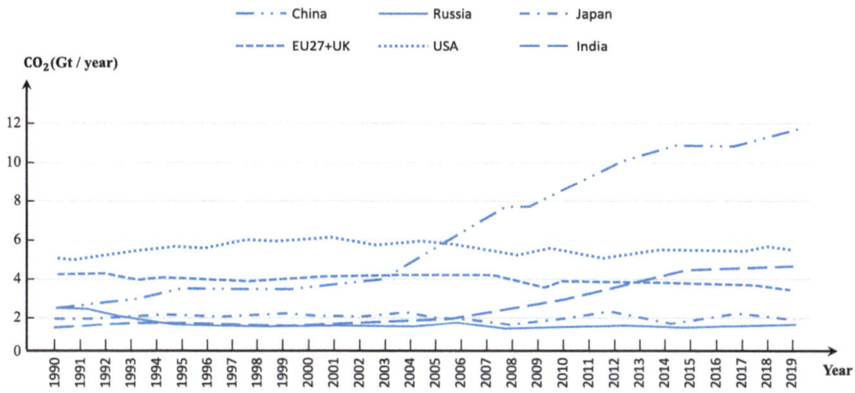

Fig. 1.4 Fossil CO_2 emissions in Gt/year of the six major emitting economies. *Source* JRC (2020)

1.4 Why "Do Nothing" Is not an Option

Those who deny the notion that climate change is accelerated and—to a large extent—caused by human activities-related GHG emissions, put forward the view that "doing nothing" is better than "doing something" because any such action will be too costly and impractical, and it will harm the global economy and our quality of life. This argument ignores the reality of the consequences and impacts of "doing nothing" which will be much bigger and more costly to humanity that the actions necessary to prevent or at least mitigate climate change. Doing nothing would lead to disastrous consequences for the planet and humanity and the economic impacts of a full-blown climate crisis will far outweigh the costs of the actions to avoid it. There is now a large body of research and scientific evidence that documents the economic, social, and other cataclysmic consequences of leaving climate change unchecked. Already, the negative socioeconomic impacts of extreme weather events are felt in an increasing number of countries, and this should provide people and governments with enough evidence as to the painful consequences of a "do nothing" situation.

Perhaps the most inclusive and objective report documenting beyond any doubt the severe and catastrophic impacts of climate change, if left unchecked, is the World Bank's "*Shock Waves*" report that consists of 14 background papers in collaboration with research institutions across the globe (World Bank, 2015). A quantified estimate of the potential impacts of climate change is given in a report by the Swiss Re Institute (Swiss Re, 2021). In this report, the estimates are that the global economy stands to lose between 11 and 14% of global GDP by mid-century if today's climate trends continue, compared to a world without climate change. Based on the current trajectory of a 2.0–2.6 °C temperature rise by the end of the century, China's economy could be 18% smaller, the US one 7% smaller and the Euro-area 8%, by mid-century. Under a more severe scenario, i.e. with higher temperature increases and no action on climate change, the loss figures for the global economy could reach as much as 18% of the overall global GDP. According to World Bank data, the world's GDP was standing in 2022 at US $101,325,686 million (approx. US $101 trillion). This brings the estimate of the Swiss Re Institute for the climate change, in the "do nothing" option, by mid-century, to between US $11 and 18 trillion in 2022 prices. The GDP of the EU in 2022 was $16,746,223 million (16.7 trillion). For China it was $17,963,171 million (17.9 trillion), and for the US $25,439,700 million (25.4 trillion). Calculating the 11–18% of these figures, the interested reader can have the estimate of the impact of "do nothing" in each of these economies. Another report, from the world's largest humanitarian network, the Geneva based International Federation of Red Cross and Red Crescent Societies, states that because of a "do nothing" situation in climate change there will be an increase of 100% of the people needing humanitarian assistance, as early as by 2030 (IFRC, 2019). This report further suggests that the rising human toll would come with a huge financial price tag, with climate-related humanitarian costs reaching $20 billion per year by 2030, in the most pessimistic scenario.

For China, a 2020 report by China's National Climate Centre of the China Meteorological Administration, states the potential impacts for China over the next few decades from the effects of climate change (https://www.cma.gov.cn/en/). These will include sea-level rise, stronger storms, and more intense heat waves. Already China's average temperatures and sea levels have risen faster than the global average and some of China's coastal cities, such as Shanghai, could see large parts submerged if the global average temperature continues to rise. An estimated forty-three million people in China live on land that could be underwater by the end of the century if the global average temperature rises by approximately 2 °C. If it increases by more than 2 °C the number of people affected could be as high as one hundred million. Additionally, as temperatures rise, China will experience more frequent extreme weather events, such as heavy rainfall and flooding. Already, extreme natural disasters kill thousands of Chinese people every year and destroy millions of acres of crops.

For the EU, a 2024 report by the *European Environment Agency* assessing the climate risk in the EU, noted (EEA, 2024):

- Europe is the fastest-warming continent in the world. When applying the scales of severity used in the European climate risk assessment, several climate risks have already reached critical levels.
- If decisive action is not taken now, most climate risks identified could reach critical or catastrophic levels by the end of this century. Hundreds of thousands of people would die from heatwaves, and economic losses from coastal floods alone could exceed EUR 1 trillion per year.
- The extent to which Europe (the EU) can avoid large scale damages will depend on how quickly it can reduce global greenhouse gas (GHG) emissions, and how fast and effectively it can prepare its societies and adapt to the unavoidable impacts of climate change.
- Europe's climate change strategies should prioritise emission reductions without over-reliance on forest carbon sinks. Forest-based mitigation should only play a complementary role in bridging the transition to a low-carbon economy and offsetting remaining emissions while providing other co-benefits.

On a more global scale, the National Geographic foundation sums up the impacts of "do nothing" in climate change as follows (National Geographic, 2023):

- Global temperatures would rise by over 4 °C by the end of the century, causing extreme heatwaves, droughts, wildfires, and floods.
- The Arctic would be ice-free every summer, and the Greenland and Antarctic ice sheets would melt, raising sea levels by several meters and displacing millions of people.
- The world's coral reefs would die, along with many other species and ecosystems that are vital for biodiversity and human well-being.
- Water scarcity, food insecurity, air pollution, and health problems would affect billions of people, especially in the most vulnerable and poorest regions.
- Social and economic instability, conflicts, migration, and humanitarian crises would increase, posing serious threats to human rights, peace, and security.

The most convincing evidence of the dire consequences of climate change, which makes a "do nothing" policy not an option, are the disastrous effects of extreme weather phenomena that are already happening. The *Emergency Events Database* EM-DAT (https://www.emdat.be), was created in 1988 as a joint initiative between the Centre for Research on the Epidemiology of Disasters (CRED) and the World Health Organization (WHO), to gather data on disaster records worldwide from 1900 to the present day (EM-DAT, 2024). Already 26,000 of such records have been classified and these are grouped into two groups of hazards: *natural* and *technological*. From this data we can extract information on the real-life impacts from extreme weather-related events that have happened in the past. In 2022, for example, there were 387 *natural* disasters recorded worldwide, and these have resulted in the loss of 30,704 lives while the total number of people affected were 185 million. The total economic losses due to natural disasters have reached, in 2022, an estimated US $223.8 billion globally (EM-DAT report, "*2022 disasters in numbers: climate in action*"). Heat waves have caused an estimated 16,000 excess deaths in Europe alone with at least five record-breaking heat waves in Europe in 2022, with summer temperatures reaching 47 °C.

In the Americas too, major natural disasters took place in the same year with hurricane *Ian* single-handedly causing damage costing US $100 billion. When this book was at its final stages of preparation, in January 2025, the great fire of Los Angeles occurred (Palisades, Eaton, Kenneth, and other areas). This terrible event was immediately transmitted live in the news around the world as a prime example of climate crisis happening. The fire, which took place in mid-winter starting on January 6th and lasting for almost 2 weeks, was attributed to a very dry weather with no or minimal rainfall for more than 10 months, low humidity (below 10%), high temperatures in the ocean water, and the resulting high winds due to all these climatic conditions in connection to the topography of the area. This event, together with the other major fire in Maui (Hawaii) that began on August 8, 2023, with 102 people killed and thousands of houses burned, is quite indicative of the dire impacts of climate change. The 2025 fire in Los Angeles burned hundreds of thousands of acres of land and tens of thousands of houses and can eventually prove to be one of the most costly and difficult to overcome climate change events in US history.

In Africa, the human and economic impacts of natural disasters is also very high. Over the twenty-year period (2002–2021), 793 severe floods and 137 droughts were recorded representing 55% of all types of natural hazards in Africa (that totalled 1693 incidents), with 14,053 deaths by flooding and 20,821 deaths by droughts, respectively. Another EM-DAT report, *The interplay of drought-flood extreme events in Africa over the last twenty years 2002–2021*, states that "climate change is expected to increase the frequency and magnitude of hydro-climatic extreme events, such as floods and droughts, impacting economies, livelihoods, and the environment, particularly in vulnerable economies such as in some African countries where the occurrence of multi-hazard events is likely to amplify disaster impacts".

In the EU, the European Environment Agency's *Climate Risk Assessment* (EUCRA) report has substantiated in full detail the impacts of a "do nothing" situation for the EU member countries. It identified 36 climate risks with potentially catastrophic consequences and these risks were grouped in 5 clusters i.e.: *ecosystems, food supply, health systems, infrastructures*, and *economy and finance systems* (EEA, 2024). A summary of the conclusions of the EUCRA report is given below:

- Human-induced climate change is already affecting the planet and 2023 was globally the warmest year on record.
- Of all continents in the world, Europe is the fastest-warming continent with extreme heat waves becoming more frequent while precipitation patterns are changing with catastrophic floods interchanged with long periods of low overall rainfall and severe droughts.
- Climate change is felt as a risk multiplier that can exacerbate existing risks and crises that can cascade from one system or region to another, including from the outside world to Europe. Examples include mega-droughts leading to water and food insecurity, disruptions of critical infrastructure, and threats to financial markets and stability.
- Several climate risks are classified, in the EUCRA report, as having reached already critical levels while all of them will reach critical or catastrophic levels by the middle or end of this century, if no decisive action is taken, with hundreds of thousands of people in risk to die.
- Climate change related economic losses from coastal floods alone could exceed EUR 1 trillion per year by the end of the century.
- While the governments of the EU-member countries have made considerable progress in understanding the climate risks, their societal preparedness is still low as policy implementation is lagging substantially behind, quickly increasing risk levels.
- Adaptation policies can both support and conflict with other environmental, social and economic policy objectives. Thus, an integrated policy approach considering multiple policy objectives is essential for ensuring efficient adaptation.

The case of leaving the climate change unchecked and continuing our GHG emitting activities as in the past, is clearly not an option. This has, fortunately, now been accepted by most governments across the world with a major diversion from this trend being the new (in 2025) US federal government. With the US being one of the largest polluters globally, such diversions from the necessary global approach to facing the coming climate crisis, is cause for serious concerns and threatens global action for mitigation as well as adaptation efforts. In spite the US's diversion from the global efforts, the rest of the world seems committed to the fight and hopefully the decarbonisation efforts will continue with the share of solar, wind, hydro and nuclear energy to continue increasing. The fight to combat climate change is the only path in front of humanity to save the planet from the catastrophic consequences of a full-blown climate crisis, and this can be done while saving existing and creating new jobs and improving the health and well-being of citizens around the world.

References

Abduljabbar, R., Dia, H., Liyanage, S., & Asadi Bagloee, S. (2019). Applications of artificial intelligence in transport: An overview. *Sustainability, 11*(1), 189. https://doi.org/10.3390/su11010189

Back, R. (2013). A brief history of climate change. *BBC News Blog,* September 20, 2013. In: https://www.bbc.com/news/science-environment-15874560. Accessed December 2024.

Boyer, R. H. W., Peterson, N. D., Poonam, A., & Caldwell, K. (2016). Five approaches to social sustainability and an integrated way forward. *Sustainability, 8*(9), 878. https://doi.org/10.3390/su8090878

Broecker, W. S. (1975). Climatic change: Are we on the brink of a pronounced global warming? *Science, 189*(4201), 460–463. https://doi.org/10.1126/science.189.4201.460

Chan, L., Hogaboam, L., & Cao, R. (2022). Artificial intelligence in transportation. In *Applied artificial intelligence in business: Concepts and cases* (pp. 231–247). Springer Textbook. ISBN: 9783031057397.

Crippa, M., Guizzardi, D., Pagani, F., Banja, M., Muntean, M., Schaaf, E., Becker, W., Monforti-Ferrario, F., Quadrelli, R., Risquez Martin, A., Taghavi-Moharamli, P., Köykkä, J., Grassi, G., Rossi, S., Brandao De Melo, J., Oom, D., Branco, A., San-Miguel, J., & Vignati, E. (2023). GHG emissions of all world countries. *JRC report* JRC134504. Publications Office of the European Union. https://doi.org/10.2760/953322

EASAC. (2019). Decarbonisation of transport: Options and challenges. *European Science Advisory Council,* policy report 37, March 2019. https://easac.eu/projects/details/decarbonisation-of-transport/

EC-EDGAR. (2023). GHG emissions of all world countries. *European Commission—Emissions Database for Global Atmospheric Research (EDGAR),* 2023 report. https://edgar.jrc.ec.europa.eu/report_2023. Accessed December 2024.

EEA. (2024). European climate risk assessment—Executive summary. *European Environment Agency,* report no. 01/2024. ISBN: 978-92-9480-627-7

EM-DAT. (2024). The international disaster database. *Centre for Research on the Epidemiology of Disasters (CRED).* https://www.emdat.be/

Hawkins, E., Ortega, P., Suckling, E., Schurer, A., Hegerl, G., Joshi, M., Osborn, T. J., Masson-Delmotte, V., Mignot, J., Thorne, P., & van Oldenborgh, G. J. (2017). Estimating changes in global temperature since the preindustrial period. *Bulletin of the American Meteorological Society, 98*(9), 1841–1856. https://doi.org/10.1175/BAMS-D-16-0007.1

IEA. (2021). Trends and developments in electric vehicle markets. *International Energy Agency.* https://www.iea.org/reports/global-ev-outlook-2021/trends-and-developments-in-electric-vehicle-markets
IEA. (2023a). World energy investment 2023. *International Energy Agency.* https://www.iea.org/reports/world-energy-investment-2023/overview-and-key-findings
IEA. (2023b). CO_2 emissions in 2022. *International Energy Agency* report, No. 1, the Global Energy Transitions Stocktake series, Paris, March 2023. https://www.iea.org/reports/co2-emissions-in-2022. Accessed December 2024.
IFRC. (2019). The cost of doing nothing: The humanitarian price of climate change and how it can be avoided. *International Federation of Red Cross and Red Crescent Societies*, Geneva, September 2019.
IPCC. (2023). AR6 synthesis report: Climate change 2023. *The Intergovernmental Panel on Climate Change.* 6th Assessment Report of the Panel's 58th Session held in Interlaken, Switzerland, 13–19 March 2023. https://www.ipcc.ch/report/sixth-assessment-report-cycle/
JRC. (2020). Fossil CO_2 emissions of all world countries 2020. M. Crippa, D. Guizzardi, M. Muntean, E. Schaaf, E. Solazzo, F. Monforti-Ferrario, J. G. J. Olivier, & E. Vignati (Eds.), EU *Joint Research Centre*—JRC, Science for policy report no. JRC 121460 (EUR 30358 EN). ISBN: 978-92-76-21514-1.
Li, D., & Lasenby, J. (2023). Investigating impacts of COVID-19 on urban mobility and emissions. *Cities, 135.* https://doi.org/10.1016/j.cities.2023.104246
Li, Q., Bai, Q., Hu, A., Yu, Z., & Yan, S. (2022). How does COVID-19 affect traffic on highway network: Evidence from Yunnan Province, China. *Journal of Advanced Transportation, 2022*, Article ID 7379334, 23. https://doi.org/10.1155/2022/7379334
Missimer, M., Robèrt, K.-H., & Gör, B. (2017). A strategic approach to social sustainability—Part 2: A principle-based definition. *Journal of Cleaner Production, 140*, 42–52. https://doi.org/10.1016/j.jclepro.2016.04.059
NASA. (2023). Evidence: There is unequivocal evidence that Earth is warming at an unprecedented rate, Human activity is the principal cause. *NASA Global Climate Change.* https://climate.nasa.gov/evidence/. Accessed, January 2025.
National Geographic. (2023). Sea levels are rising at an extraordinary pace—Here's what to know. *National Geographic* report by Christina Nunez, April 10, 2023. In: https://www.nationalgeographic.com/environment/article/sea-level-rise-1. Accessed December 2024.
NOAA. (2022). What evidence exists that Earth is warming and that humans are the main cause? *National Oceanic and Atmospheric Administration*, US Dept of lCommerce, a report by D. Herring, R. Lindsey, October 12th, 2022. climate.gov/news-features/climate-qa/what-evidence-exists-that-Earth-is-warming-and-that-humans-are-the-main-cause. Accessed December 2024.
OECD-ITF. (2016). Shared mobility: Innovation for liveable cities. Organisation for Economic Cooperation and Development—*International Transport Forum*, a Corporate Partnership Board Report, November 2016. www.itf.oecd.org
Oestreich, L., Sandri, R. P., da Silva Vieira, J., & Ruiz-Padillo, A. (2023). Impacts of the COVID-19 pandemic on the profile and preferences of urban mobility in Brazil: Challenges and opportunities. *Travel Behaviour and Society Journal, 31*, 312–322. https://doi.org/10.1016/j.tbs.2023.01.002
Purvis, B., Mao, Y., & Robinson, D. (2019). Three pillars of sustainability: In search of conceptual origins. *Sustainability Science, 14*(3), 681–695. https://doi.org/10.1007/s11625-018-0627-5
Shaik, M. E., & Ahmed S. (2022). An overview of the impact of COVID-19 on road traffic safety and travel behaviour. *Transportation Engineering*, May 21, 2022. https://doi.org/10.1016/j.treng.2022.100119
Swiss Re. (2021). The economics of climate change: No action, is not an option. *Swiss Re Management Ltd*—Swiss Reinsurance Company—Swiss Re Institute. April 2021.
UNEP. (2024). Debunking eight common myths about climate change. *UN Environment Program*, 04 Jun 2024. https://www.unep.org/news-and-stories/story/debunking-eight-common-myths-about-climate-change. Accessed December 2024.

Vagliasindi, M. (2023). *The impact of COVID-19 on mobility and congestion.* World Bank, Policy Research Working Papers; 10351. Washington DC. http://hdl.handle.net/10986/39533

World Bank. (2015). Shock waves: Managing the impacts of climate change on poverty—Background papers. *The World Bank*, November 8, 2015.

WCED. (1987). Our common future. *World Commission on Environment and Development*, United Nations, Oxford University Press, October 1987, pp 383. ISBN: 019282080X. (This is also known as the «Brundtland Report» by the name of the Commission's first chairperson Gro Harlem Brundtland).

WMO. (2025). State of the Global Climate. Annual statement 2024. *World Meteorological Organisation*, 19 March 2025.

Chapter 2
Actions and Reactions in a World of Different Geopolitical Priorities

Abstract This chapter looks at what the world is doing so far to mitigate the climate change and if this is enough. It also examines the role of technology in mitigating climate change as well as the role of the various geopolitical events in delaying governmental efforts to decarbonise human activities. At first the chapter presents, the plans and declared actions of six major economies for decarbonisation and climate change mitigation as well as for adaptation to face the potential impacts. The six economies examined are, China, the European Union, the USA (under the Biden administration), India, Russian Federation, and Brazil. For the first three and especially for the European Union, the presented data are given in sufficient detail which the reader will find useful as a tentative blueprint for use in other potential cases. For the other three the presented information is a short summary of the information provided by these countries in their *Nationally Determined Contribution* (NDCs) reports which they submitted to the UN's Framework Convention on Climate Change (UNFCCC). Despite the ambitious plans and policies for decarbonisation announced by national or international governments, the world so far is slow in implementation actions with local wars and other political and geopolitical upheavals usually delaying or cancelling all together plans and priorities for enforcing policies for decarbonisation. They destabilise whole regions causing diversion of funds to military or other expenditures that otherwise could have been devoted to climate mitigation and adaptation actions. Further to the assessments made by the authors, the assessments made by various independent international organisations are presented and their results are assessed and discussed. The overriding conclusion is that the world is not on track to achieve the goals of the Paris Agreement, and this conclusion is accentuated by the complete reversal of US climate policies of the new (in 2025) US government. In concluding, this chapter attempts to answer the question: In the wake of failure can technology help? It notes that the era of "climate AI", as well as of "mobility AI", is already here and excitement over these technologies is great. Early applications are almost everywhere, and AI software is now in the core of practically all "soft" technologies used for climate change mitigation and adaptation applications as well as for control and optimisation of transport and mobility operational processes. The arguments in favour of the application of new technologies and AI in the field of transport and mobility are presented and at the same time the reader is cautioned

G. A. Giannopoulos and Y. Li, *Human Mobility, Artificial Intelligence and Climate Change*, Lecture Notes in Mobility, https://doi.org/10.1007/978-3-032-08171-1_2

of the challenges that exist which include managing the risks inherent in financing the development and application of new technologies, using ethically the novel technologies of AI, rethinking core business processes and retraining or developing new skills and capabilities in the workforce. Last but not least, the most important issue of securing robust data in terms of quality and quantity is brought up as necessary in order to build and train the new AI algorithms and models.

Keywords Decarbonisation plans · Decarbonisation policies · Decarbonisation · Climate policy · Nationally determined contributions · Climate AI · Mobility AI · Climate mitigation · Net-zero targets

2.1 Is the World Doing Enough to Handle the Climate Crisis?

Human induced climate change is here and is fast becoming a "climate crisis". The evidence given in the preceding chapter can convince any good intentioned reader that "doing nothing" is not an option. The contrary view that climate change does not exist and whatever part of the climate is changing is not due to human carbon emissions but due to natural influencing factors and causes that cannot be affected, is dismissed and not supported by scientific evidence as it is largely based on common myths and unsupported beliefs. Urgent action is now being planned or undertaken by governments around the world and international organisations to reduce carbon footprints of human activities and support sustainable solutions for most human activities in the future.

The target years for ultimate results are set in the middle of this century (2050 or 2060) with the year 2030 being a crucial milestone in the progress to zero the carbon and other Green House Gas (GHG) emissions (i.e., CO_2, methane, nitrous oxide, fluorocarbons, and similar), by 2050. These targets are basically the ones that have been agreed in 2015 by many world governments during the COP21 UN Conference on the climate that took place in Paris, France. At the time of publication of this book, we are more than 10 years after that Conference and less than 5 years before the milestone year 2030. So, the crucial question is whether national and international governments are on track in taking and implementing actions that will bring the desired results. In other words, is the world doing enough? In the following, we attempt a well-documented answer to this question by looking at the steps taken by major national and international governments and assessing the progress so far.

2.1.1 Plans and Proposed Actions by the European Union

2.1.1.1 Overall

The plans and provisions by the European Union (EU) and its member countries for climate change mitigation foresee full decarbonization of all economic sectors by 2050. They include reducing the net GHG emissions in the EU by at least 55% by 2030 and reaching carbon neutrality by 2050 mainly through enhancing energy efficiency and use of renewable energy sources. Also, by promoting a circular economy and green infrastructures and strengthening carbon pricing and phasing out any fossil fuel subsidies (more details on later section).

As soon as the 2019–2024 EU Commission (the executive body of government in the EU) took office in 2019, it announced its strategic goal for making the European Union climate neutral by 2050 and declared the so called *European Green Deal* as its primary set of policy initiatives that would be taken to meet that goal. In July 2021 an "umbrella" Commission announcement (Communication) was made introducing a vast package of 17 Commission proposals for the revision of climate and energy legislation in the EU. This became known as the "Fit for 55 Package" because it placed at the centre of its attention the adoption of legislation that would result in reducing EU GHG emissions by at least 55% by 2030. That was perhaps the most comprehensive set of legislation ever put together in response to climate change. It covered various sectors of the economy such as energy, transport, industry, and agriculture. Table 2.1 shows the main elements of the "fit for 55" legislature which addresses the 4 main sectors of GHG emissions, i.e. electricity generation, industry, transport, and buildings.

As it turned out, the COVID pandemic and the Russian Ukrainian conflict consequences caused new priorities to be set by most European governments and resulted in some revisions and considerable delays in the implementation of the "Fit for 55 Package". It is estimated that the "Fit for 55" package will be fully adopted and implemented by the end of 2025 after the new European Commission (EC) that resulted from the European elections of June 2024, takes office on January 2025. In the process of implementation there will be, most probably, quite a few revisions and member states will have to transfer this EU legislation to their own legal systems before starting to implement it. The result will be that the provisions of the "Fit for 55" package will start being implemented in the year 2030 or thereabouts. So, the target for a 55% reduction (compared to 2019) of EU emissions by 2030, will be impossible to attain. This is a typical situation in which plans and policies that were originally well formulated and coordinated, misfire due to a change in policy priorities and options as well as delayed application caused by external factors (e.g., major geopolitical events) or internal ones (e.g., lack of proper coordination or change in government philosophy).

The "Fit for 55" legislative package, as it was initially assembled, is perhaps one of the most comprehensive, all inclusive, and ambitious legislative packages in the world to combat climate change. It therefore merits a more detailed presentation as

Table 2.1 Main contents of the EU's "fit for 55" legislative package

Name of "fit for 55" element	Main contents and objectives
Renewable energy directive (RED)	Increased use of renewables in the EU energy mix from 32 to 40% by 2030
Energy efficiency directive (EED)	Mandates energy efficiency improvements within the European Union.
Energy performance of buildings directive (EPBD) (Q4 2021)	Aims to promote the improvement of the energy performance of buildings within the European Union
Emissions trading system directive (ETS)	Makes polluters pay for their greenhouse gas emissions on the 'cap and trade' principle. In 2023, a new emissions trading system was created, the *Emissions Trading System* 2 (ETS 2)
Climate action regulation	The regulation that obliges member countries to reduce by 55% their GHG emissions by 2030. It replaces the *Effort Sharing Regulation* (*ESR*) and establishes binding annual greenhouse gas emission targets for Member States from 2021 to 2030
Carbon border adjustment mechanism (CBAM)	Establishes a carbon tariff on carbon intensive products, such as steel, cement and some electricity, imported to the European Union
LULUCF, regulation on land use, land use change and Forestry	Enhances the role of natural carbon sinks and controls the management of land, the carbon pools and vegetation on land and the rules under which the land is considered as a sink or a source of CO_2
Energy taxation directive	Establishes the framework conditions of the European Union for the taxation of electricity, motor and aviation fuels and most heating fuels
Third energy package for gas	Includes rules designed to benefit European *energy* consumers and protect their rights among which the right to choose or change suppliers
Regulation of CO_2 emissions from new passenger cars and new light commercial vehicles	Compared to 2021, the regulation requires that the annual EU fleet-wide average CO_2 emissions from new cars and new vans be reduced: • By 15% for the 2025–2029 period • By 55% for new cars and 50% for new vans, for the 2030–2034 period • By 100% from 1 January 2035 onwards
Alternative fuels infrastructure regulation	Member States are obliged to ensure that, at the end of each year, starting from 2024, specific power output targets are provided through publicly accessible recharging stations

(continued)

Table 2.1 (continued)

Name of "fit for 55" element	Main contents and objectives
RefuelEU aviation regulation	Aviation fuel suppliers will have to blend increasing amounts of sustainable aviation fuels (SAF) with kerosene, starting with a 2% minimum blend in 2025, and rising to 70% in 2050
FuelEU maritime regulation	Aims to support the decarbonization of the shipping industry. Upon entering into force from 1 January 2025, it increased the share of renewable and low-carbon fuels in the fuel mix of international maritime transport in the European Union
Social climate fund	It was created alongside the ETS2 (*Emissions Trading System* 2) for emissions from fuel combustion in buildings, road transport and additional sectors. It provides Member States with dedicated funding so that the most affected vulnerable groups, such as households in energy or transport poverty, are directly supported, and not left behind during the green transition
Methane regulation	The development of legislation to reduce methane emissions from the energy sector. The proposed legislative acts include an impact assessment and will be supported by articles 192 and 194 of the Treaty on the Functioning of the European Union

an example of the rigorous and holistic type of legislation that would be needed for a net-zero situation by 2050.

2.1.1.2 Description of Individual Elements of the "Fit for 55" Package

The Renewable Energy Directive *(3rd Edition—RED III)*

This package aims to facilitate and incentivise the increased use of renewables in the EU's energy mix from 32 to 40% by 2030 (it was 19.7% in 2019). To achieve this overall goal, the RED Directive puts forward more detailed goals by category. Examples:

- A binding increase of 1.1% per year for renewables in heating and cooling at national level (Article 23).
- An indicative increase of 2.1% per year for renewables and waste heat and cold in district heating and cooling (Article 24 b).
- A new 13% reduction in the target for life cycle greenhouse gas emissions intensity (gCO_2/kWh) for transport (Article 25.1).
- An indicative increase of 1.1% per year in renewable energy use in industry (Article 22a).

- Setting a new benchmark to reach at least 49% renewable share in the energy used in buildings, to be measured in terms of final energy consumption (Article 15a).
- Removing certain multipliers associated with some renewable fuels and renewable electricity used in transport (Article 1. 16).
- Revise the calculation methods for GHG emissions savings from transport fuels including renewable electricity.
- New measures are included for boosting investment and deployment of renewables.
- Other measures such as:
  - Setting sub-targets and certification for renewable hydrogen (Articles 22a and 24).
  - Boosting electrification in transport.
  - Establishing a credit mechanism for transport (Article 1.14).
  - Obliging the Member States to put in place measures to deliver significant increases in Renewable Energy Systems (RES) in buildings, via self-consumption, renewable energy communities, local energy storage, etc. (Article 15a).
  - Stronger integration of electricity networks and markets with district heating and cooling systems (Article 24).
  - Introducing measures to facilitate renewable Power Purchase Agreements (PPAs) (Article 15).
  - Accelerating permissions for renewable energy projects (Article 3).
  - Promoting cross-border cooperation, including via a renewable energy financing mechanisms.

Of interest, are also the bioenergy related highlights of the RED III which are reinforcing sustainability criteria of the pre-existing EU biodiversity and other strategies, e.g.:

- New definition of "quality roundwood" (Article 1.1) i.e. wood in its natural state as felled, with or without bark (it may be round, split, roughly squared or in other forms).
- Prohibition of sourcing of biomass for energy production from primary forests, peatlands, and wetlands.
- Ending the support (subsidies) for using forest biomass in electricity-only installations (phase out from 2026 Article 1.2), unless it is in a "just transition region" or combined with Carbon Capture and Storage (CCS).
- Ending support (subsidies) for using saw or veneer logs, stumps and roots for energy generation (Art 3.3).
- Requiring all biomass-based heat and power installations to meet minimum GHG saving thresholds.
- Applying sustainability criteria to smaller heat and power installations (equal to or above 5 MW).

The Energy Efficiency Directive

With this directive, the EU "Fit for 55" package replaces the pre-existing non-binding 32.5% target for energy efficiency savings in primary and final energy consumption (relative to 2007 projections) by 2030 with binding targets of 36% and 39% savings in final and primary energy consumption by 2030 (savings were 17.0 and 17.4% in 2019). It also:

- Introduces indicative Member State contributions to the EU-level energy efficiency target.
- Requires each Member State to ensure 1.7% of energy savings per year in the public sector.
- Applies a policy and investment decisions with the "Energy Efficiency First Principle".
- Increases an annual energy savings obligation to 1.5% for all Member States.
- Obliges Member States to renovate at least 3% of total floor area of all public buildings annually.
- Encourages public bodies to use Energy Performance Contracts for renovation of large non-residential buildings.
- Obliges Member States to achieve energy savings among energy poor and vulnerable households, whilst alleviating energy poverty and boosting consumer empowerment.

The Emissions Trading System[1] (ETS) Directive

The EU Emissions Trading System (ETS) is strengthened under the "Fit for 55" package to deliver 61% GHG emissions reduction in the energy and manufacturing sectors by 2030 (up from 43% currently). In achieving this goal, it foresees (among other provisions), for a new—additional—emission trading system for buildings and road transport fuels that will focus on upstream fuel suppliers (rather than households and car drivers). It also aims to divert the additional funds that will result from this extended ETS, combined with other sources of finance (namely the Social Climate Fund and others), to alleviate energy poverty. It tries to create a level playing field for all heating options, by complementing the existing carbon price on electric heating and district heating with a carbon price on domestic heating. It furthermore uses the ETS revenues to double the size of the EU's *Innovation Fund* (IF), which supports first-time innovators to develop low-carbon technologies. Under this provision, the size of the IF is expected to increase to more than €50 billion over ten years. It also increases the *Market Stability Reserve* (MSR), which has been set up to address the surplus of emission allowances in the EU emission trading system since 2009, with

[1] This is the existing system in the EU that makes polluters pay for their greenhouse gas emissions. It covers today approximately 43% of the EU's emissions in the energy sector and manufacturing industry, as well as aircraft operators flying within the EU, i.e. emissions from around 10,000 installations. It is expected to also cover emissions from maritime transport.

an intake rate of 24% (from 12% before) and improves the system's resilience (by adjusting the supply of allowances to be auctioned). It finally, provides a gradual extension of the ETS for maritime transport, that started in 2023/24, following a 3-year phase-in period. This, maritime ETS, will focus on large ships (above 5000 gross tonnage) accounting for 90% of CO_2 emissions from intra-EU traffic and 50% of extra-EU voyages.

Carbon Border Adjustment Mechanism (CBAM) Regulation

This regulation applies a levy on EU imports from third countries in the sectors of electricity, iron/steel, aluminium, fertilisers, and cement. These sectors currently benefit from free allocations under the ETS, which will be phased out completely by 2036. The aim is to ensure that the EU climate ambition is not undermined by foreign firms that are subject to lower environmental requirements, while also encouraging foreign countries to decarbonise.

Energy Taxation Directive

The Energy Taxation Directive sets out rules and minimum excise duty rates for taxation of energy products and electricity used as motor fuel and heating fuel aiming to tax the phasing out of fossil fuels and induce introduction of renewable fuels. The taxes are to be linked to the energy content of fuels, and to their environmental performance rather than their volume, giving price signals that should encourage green innovation and investments. The way that energy products are categorised for taxation purposes is simplified to ensure that fuels most harmful to the environment are taxed the most.

The Land Use, Land Use Change and Forestry (LULUCF) Regulation

This part of the "Fit for 55" package addresses the land use and forestry sector and sets an overall EU target of net greenhouse gas removals in this sector at 310 million tonnes of CO_2 equivalent by 2030. The regulation reinforces the obligation of Member States to submit integrated mitigation plans for the land sector and aligns LULUCF policy objectives with related policies on biodiversity and bioenergy. Most importantly, the regulation is setting a new EU target to achieve climate neutrality for 2035 in the "land sector" (i.e., the LULUCF and the non-CO_2 agricultural sectors combined). Because of this very ambitious target and the actions that followed for its implementation, there were extensive demonstrations and strikes by the farmers and other agricultural sector professionals all over Europe at the end of 2023 and part of 2024 resulting in the postponement of the climate neutrality for 2035 target and the retraction of most of the first implementation measures.

The Climate Action Regulation (CAR)

This was previously called the Effort Sharing Regulation (ESR). It covers all sectors not covered by the ETS or the LULUCF regulations and aims at increasing the commitments for an average GHG emissions overall reduction in these sectors from 30 to 40% by 2030 (compared with 2005). The CAR addresses buildings and transport (including maritime).

CO_2 Emission Performance Standards for Passenger Cars and Light Commercial Vehicles

This part of the legislation provides for more ambitious CO_2 emission standards for new cars and vans to help grow the number of zero- and low-emission vehicles on European roads. Reducing CO_2 emissions in new cars and vans to zero by 2050, will inevitably mean banning sales of new Internal Combustion Engine (ICE) vehicles by 2035. This would lead to a zero emission EU car fleet by 2050, when all existing ICE vehicles will have reached their end of life. The new emission targets are to reduce CO_2 emissions for new cars by – 55% in 2030 (instead of – 37.5% as foreseen before) and to – 50% for vans (instead of – 31% as foreseen before). After 2035, both categories of vehicles would have to be banned if using internal combustion engines.

The Alternative Fuels Infrastructure Regulation (AFIR)

This regulation provides new and more ambitious targets for the development of infrastructures for clean fuels. The new regulation sets out mandatory national targets for the deployment of alternative fuels infrastructures for road vehicles, ships, and stationary aircraft, and lays down common technical specifications and requirements on user information, data provision and payment requirements. More specifically:

- For cars and vans—on the TEN-T core network (the EU's core highway network of an estimated length of 30,000 kms), there should be at least one 300 KW power output every 60 kms by 2025 and at least one 600 KW per 60 kms by 2030. On the so-called comprehensive network (much lengthier than the core one), at least one 300 KW power output every 60 kms by 2030 and at least one of 600 KW by 2035.
- Hydrogen refuelling stations are also foreseen under this regulation every 150 kms by 2030 in the TEN-T core network and in every urban node serving both light duty and heavy-duty vehicles by 2030.
- As regards the heavy-duty vehicles on the TEN-T core network at least one 1400 KW charging point every 60 kms by 2025 and at least one 3500 KW every 60 kms by 2030. On the TEN-T comprehensive network at least 1400 KW power

output every 100 kms by 2030 and at least one 3500 KW by 2035. Also, in every urban node and at every safe and secure parking by 2030.

The above provisions seem already unattainable but are kept here for reasons of completeness of the presentation and because they are included in the original "Fit for 55" package.

FuelEU Maritime Regulation

Under this regulation, there will be limits to the GHG intensity of energy used on ships, thereby encouraging the use of renewable and low-carbon fuels in maritime transport. It applies to ships arriving at, staying within, or departing from ports under the jurisdiction of an EU Member State. It also, introduces an obligation to use onshore power supply or zero-emission technology in ports under the jurisdiction of a Member State.

RefuelEU Aviation Regulation

This regulation refers to the aviation sector and defines or regulates the use of *sustainable aviation fuels* (SAF), with an obligation on fuel suppliers to distribute increasing levels of SAF at all EU airports. It also obliges airlines to fill up with SAF-blended fuel before each flight from an EU airport. The regulation focuses on the use of the most innovative and sustainable fuels, e.g. advanced biofuels and synthetic fuels (also known as electro-fuels) and requires a clean electricity supply for stationary commercial aircraft at all gates by 2025 and additionally at all outfield positions by 2030. As an indication, this regulation foresees by 2030 a 5% share of sustainable fuels in the fuels mix used for air transport, this percentage becomes 32% by 2040 with 8% electro-fuels (e-fuels) and 63% by 2050 of which 28% e-fuels.

The REPowerEU Legislation

Following the start of the war in Ukraine and the cut-off of several of its natural gas supplying pipelines, the EU reconsidered its energy policies. In May 2022, it issued this legislation which introduces big increases in the EU's renewable energy targets compared with those in its previous strategy. Several relevant questions were raised because of that, such as where the renewable power generators (mainly from sun and wind) should be placed onshore or offshore. Also, how much total electricity should the EU cater for and what will be the new economic model that will be required. For example, investors on renewable energy generators will need long term (e.g., minimum 30 years) Power Purchase Agreements—PPA contracts to deliver the required levels of confidence in renewables investments. Such long-term contracts will be accepted in the new EU electricity market rules, which will reduce the risks for investors. It must be stressed here, that the challenge in Europe today is not only investing in renewable electricity production but also investing in converting this electricity to heat or to industrial process fuels such as hydrogen (power2X).

Another challenge is energy storage. This has grown in importance owing to the increasing penetration of variable renewable electricity generation on the European grid. The potential for a more integrated approach to energy storage is becoming more widely recognised, including heat storage, power to liquid, and power to gas—notably hydrogen.

The EU Taxonomy Legislation

A final mention is due to the very important part of the EU climate change legislation that is known as the *Taxonomy* legislation. It was implemented in July 2020 and introduced a detailed classification system that sets out which activities can be defined as "environmentally sustainable", in the context of the European Green Deal. It also aims to provide certainty for investors, promote climate-friendly practices, mitigate market fragmentation, and facilitate sustainable investments to support the objectives of the European Green Deal. This legislation requires nonfinancial undertakings to disclose the proportion of their activities that are taxonomy-eligible and taxonomy-aligned in terms of their turnover, as well as capital expenditures (CapEx) and operating expenditures (OpEx). In other words, the taxonomy regulation is a framework to drive investments towards the achievement of the EU's 2050 climate goals. It sets out four overarching conditions that an economic activity must meet in order to qualify as environmentally sustainable (known as "Taxo4"): compatibility with the circular economy, protection of water and marine resources, pollution prevention and control, and respecting biodiversity and ecosystems. Pursuant to this legislation, the European Commission created an educational and user-friendly website offering a series of online tools to help users better understand the EU taxonomy in a simple and practical manner, ultimately facilitating its implementation and supporting companies in their reporting obligations.

To sum up, the European Union has put forward a massive and holistic program of legislative and other measures to achieve carbon neutrality by 2050 with very ambitious milestones to be achieved by 2030. Despite of that, the *Climate Action Tracker,* the independent scientific analysis organisation that tracks the progress of countries towards the Paris agreement goals (www.climateactiontracker.org) rated, in December 2024, EU's policies and targets for climate change as "*Insufficient*". This means that its policies are consistent with limiting global warming to below 2 °C, but not yet within the 1.5 °C limit of the Paris agreement. We return to the prospects of achieving net-zero GHG emissions in the EU, by 2050, in the last Chap. 10.

2.1.2 Plans and Proposed Actions in China

Like the EU, China's climate policy is of special interest and concern to the rest of the world because China has the potential to become a world leader in the transition to a low-carbon and resilient future and lead by example. Besides, it is one of the world's largest emitters of greenhouse gases (GHG) and with its *One Belt One Road* (OBOR) initiative, it can effectively influence its OBOR cooperation partners on climate actions. By supporting developing countries with finance, technology, and capacity building China can both help and motivate them towards taking climate change mitigation measures and reducing their carbon emissions but above all in taking adaptation measures. Chinese president, Xi Jinping, has recognized climate change to be one of his administration's top concerns and he has pledged to address it in a way that will make China a net-zero country by 2060 while maintaining economic growth and prosperity. This makes China's climate policies particularly interesting and important at world level.

Chinese climate mitigation policies are quite complex and evolving with a number of such policies being announced every year (Maizland, 2021; MEE, 2023; SCIO, 2022). Since 2008, China has released annual reports on its climate policies and actions. The 2023 report states several updated policies for both mitigation and adaptation and states notable achievements (MEE, 2023). According to this report, carbon emissions per unit of GDP were reduced by 2022, by over 51% compared to 2005 levels. By the end of 2022, non-fossil fuel energy consumption constituted 17.5% of total energy usage, and the total installed capacity of renewable energy reached 1213 TWh. In transportation, China has materialised an impressive increase in "new energy vehicles" (electric vehicles—EVs). The total number of EVs on Chinese roads are over 20 million currently which is almost 50% of the world's total.

The same report also describes China's "1 + N" carbon peaking and neutrality policy system and its impact on various sectors. The "1" means that this is the guiding idea and of top-level design, while the "N" denotes the various implementation schemes in key areas and sectors such as Energy, Industry, Transportation, Urban and Rural development, Agriculture, etc. As of 2022, twenty implementation policies were formulated (MEE, 2023). The top-level design of climate change mitigation policies is described in two documents issued in 2021, the *Working guidance for carbon dioxide peaking* and the *Carbon Neutrality in Full and Faithful implementation of the New Development Philosophy and Action Plan for carbon dioxide peaking before 2030* (SCIO, 2022). China's investments for mitigation and adaptation, aim at bringing payoffs not only in achieving the net-zero targets by 2060 but also in making China an indispensable global provider of these pivotal technologies, reaping both financial and geopolitical benefits. In parallel to its domestic goals for mitigation and adaptation, China has the unique opportunity of promoting, through its One Belt One Road—OBOR financing, many renewable energy projects in other countries. The Chinese government has also announced environmental standards for the OBOR infrastructures, but so far these standards have only been voluntary.

In the (MEE, 2023) report, the Ministry of Ecology and Environment is putting forward new targets and enhanced actions for both mitigation and adaptation. Among those the following key targets are mentioned:

(a) Peaking CO_2 emissions before 2030 and reach carbon neutrality before 2060.
(b) Lower CO_2 emissions per unit of GDP by 2030, by more than 65% from their 2005 level.
(c) Increasing the share of non-fossil fuels in primary energy consumption to more than 25%, by 2030. In 2022 this percentage had already risen to 28% (split as 14% hydro, 9% wind, and 5% solar).[2]
(d) Increasing the forest stock by 6 billion cubic meters from 2005 levels. A most notable example of successful reforestation in China, is the Saihanba National Forest Park north of Beijing near Mongolia. This is the world's largest planted forest also called the "Great Green Wall".
(e) Increasing the installed wind and solar power generation capacity to over 1.2 billion KW by 2030.
(f) Promoting nuclear power capacity, with 17 new reactors under construction until 2030.
(g) Incentivising further electric vehicle use (this includes battery electric vehicles, plug-in hybrids, and hydrogen fuel-cell vehicles), and
(h) Developing and implementing a new long-term low-greenhouse gas emissions development strategy for 10 key economic sectors including energy production, transport, and urban development.

China has also launched a national emissions trading scheme (ETS), which would force polluters to pay for environmental harm and incentivize them to reduce their emissions. This scheme was first announced in 2017 and came to operation in 2021. It has initially focussed on coal and gas-fired power plants for electricity and heat generation which emit almost 5 Gt of CO_2 annually (roughly 45% of China's and 15% of global CO_2 emissions). Then the government announced that the ETS will be expanded to include other sectors in the years ahead together with the national system for natural carbon sinks. The country's national carbon market has therefore been developing, with the cumulative trading volume of carbon emission allowances reaching 238 million tons, by mid-2023, amounting to a total value of RMB 10.9 billion (approximately 1.5 billion Euros).

China is paying particular attention to climate change adaptation measures. The strategy followed was announced, in June 2022, in the document *National Strategy for Climate Change Adaptation 2035*. According to this strategy China will establish a nationwide climate impact and risk assessment system, with improved early warning capabilities. At the same time, all major infrastructural projects in the country will have the obligation to perform climate and environmental impact assessments as part of their justification. Already, an early warning system has been established which is a "government-led, warning-triggered, cross-departmental coordination, and public

[2] According to Statista 2024: https://www.statista.com/statistics/1235176/China-distribution-of-electricity-production-by-source (accessed December 2024).

participation" system that is using mathematical models to give warning on major climate related events and assess the risk of meteorological disasters (see https://www.cma.gov.cn/en/service/highlight/EarlyWarning/—accessed December 2024).

China's climate change policies seem to have accorded greater priority to adaptation than to mitigation. This implies an assessment that, regardless of any mitigation efforts China will still have to cope with a climate crisis whose impacts can be devastating. China calls its adaptation policies with names such as "*ecological civilization*" or "*ecological security*" and the goals of its adaptation efforts as "enhancing sustainable harmony between humans and the natural environment". Among the main Chinese adaptation plans and policies there are some huge, in their scope and objectives, infrastructure projects. Examples include (Freymann, 2022):

- The largest water transfer system in human history, i.e. the South-to-North water transfer project. This is a mega project that ultimately aims to channel 44.8 billion cubic meters of fresh water each year from the Yangtze River in southern China to the more arid and industrialized north through three canal systems.
- Expanding and raising nearly 6000 miles of sea walls along its coasts.
- Building a strategic grain reserve (that is larger than the rest of the world's reserves combined).
- Carving wetland flood basins in the centres of its largest cities.
- Restoring coastal wetlands to act as buffers against storms, and
- Relocating hundreds of thousands of 'ecological migrants' in low-lying areas.

The above policies and measures taken by the Chinese government are encouraging, but are they enough? China is highly vulnerable to climate change. The Chinese government recognises this fact but while adopting the importance and urgency of climate mitigation and adaptation measures, it also has the ambition to sustain the high rates of economic growth and development that it has achieved in previous years. It will therefore not risk to severely impact the country's economic health, and energy security systems and so—for the time being—it still uses some climate polluting activities like for example the use of coal for its energy needs. According to reports by the Global Energy Monitor, China accounted for 95% of the world's new coal power construction activity in 2023 (GEM, 2024). Furthermore, there are earlier reports that coal power plants construction has been financed by China in several One Belt One Road (OBOR) countries in the past (Li & Shapiro, 2020).

It is difficult to foresee, at this stage, how Chinese efforts for climate mitigation as well as adaption will play out by 2030 and further. The *Climate Action Tracker*,[3] however, currently rates China's policies and targets for climate change as "*highly insufficient*" i.e., "not in line with its fair share of the global effort to reduce GHG emissions". We review the prospects of achieving net-zero GHG emissions in China by 2060 in more detail in Chap. 10.

[3] The independent scientific analysis organisation that tracks the progress of countries towards the Paris agreement goals (www.climateactiontracker.org). The ratings mentioned are for December 2024.

2.1.3 Plans and Proposed Actions in the USA

Plans to decarbonise the US economy and bring it to a net-zero state by 2050 were formed and legislated by the Biden administration of the US federal government during the period 2020–2024. This was a very ambitious program that had set as its goal the reduction of the net GHG emissions by 50–52% by 2030 and achieving net-zero emissions by 2050. All this would be achieved through investing in clean energy, transportation, and innovation as well as enhancing natural carbon sinks and reducing emissions from agriculture and forestry. In November 2021, the US government published its *long-term strategy for net-zero GHG emissions by 2050* (US Dept of State, 2021). In this document the US federal government declared its goal to reduce net GHG emissions in the country by 50–52% below 2005 levels in 2030 and set the decade 2020–2030 as the decisive decade to deliver on a set of new policies expanding rapidly the deployment of new clean technologies like electric vehicles and heat pumps and building the infrastructure for clean energy production and distribution. The strategy evolved around five key transformations, i.e. to:

1. *Decarbonise electricity* and transition to a clean electricity system, driven by solar and wind technologies. The goal was to have 100% clean electricity by 2035.
2. *Electrify end-uses and switch to other clean fuels* from cars to buildings and industrial processes. In areas such as aviation, shipping, and some industrial processes where electrification is difficult, priority would be given to the use of clean fuels like carbon-free hydrogen and sustainable biofuels.
3. *Cut energy waste* by using existing and new technologies that will enable use of less energy to provide the same or better service (e.g., through more efficient appliances and the integration of efficiency into new and existing buildings, and sustainable manufacturing processes).
4. *Reduce methane and other non-CO_2 emissions* by using low-cost options to reduce non-CO_2 sources, such as implementing methane leak detection and repair for oil and gas systems and shifting to climate-friendly working fluids in cooling equipment. Through the Global Methane Pledge, the U.S. would seek to reduce global methane emissions by at least 30% by 2030, which would eliminate over 0.2 °C of warming by 2050.
5. *Scale up CO_2 removal* from the atmosphere by removing carbon dioxide from the atmosphere, using processes and technologies that are rigorously evaluated and validated (land carbon sinks as well as engineered strategies).

To implement and materialise these transformations, the strategy relied on four “strategic pillars” to cooperate and act, namely:

A *The Federal government and national leadership,* i.e. the bodies at federal level that decide on investments and incentives to support the deployment of clean technologies in all sectors and other actions.

B *Innovation*, to drive federally supported research, development, demonstration, and deployment of competitive technologies as rapidly as possible and drive deeper GHG reductions in the late 2020s through 2050.

C *Non-federal leadership,* namely elected governments at subnational level such as the States, cities, counties, and tribal governments.

D *Society,* including nongovernmental organizations to support the overall transition of the U.S. economy. This pillar included universities, cultural institutions, investors, businesses, and other non-governmental organizations.

The most iconic legislation in implementing this strategy, came in 2022 with the *Inflation Reduction Act* (IRA) of 2022. This was a far reaching, rigorous legislation, and the largest program of state financing in the history of the US. It gave financial and other incentives worth many hundreds of billions of US dollars for climate related innovation and green energy production. The total output of federal state contributions that would be affected by 2033 would be more than 800 billion US dollars. The financial and other incentives provided in the IRA, included:

- Incentivising GHG emission reductions in various sectors by defining 70 different types of investment, production and excise credits designed to facilitate the transition to cleaner energy production, promote advanced manufacturing, encourage the adoption of clean vehicles and reduce greenhouse gas emissions using alternative fuels and energy efficient technologies.
- Financing of 270 projects for clean energy production.
- Subsidising the purchase of electric cars or the development of new types of batteries (a subsidy of up to 7500 US dollars is provided for new electric cars, 4000 US dollars for used ones).
- Development of renewable sources of energy (by installing 950 million sun panels, 120,000 wind generators, and 2300 battery parks by 2030).
- Other relevant items.

The IRA after its introduction in 2022, enacted a multitude of co-financing for clean energy and other climate friendly projects from private sources and so the total amount of resources devoted to the green transformation in the US is estimated to amount to several trillions of dollars. By 2025 it had unlocked construction of a multitude of new factories in the US—many of which diverted from outside the US, mainly the EU—to benefit from the financial incentives of the Act. The resulting competition was forcing the EU to announce new legislative and other actions to balance the competition and avoid the "distortions" that the IRA was creating in favour of the US and to the detriment of the EU economy.

Then, came the change of government following the elections of November 2024 and the new US federal administration under president D. Trump reversed all the previously mentioned policies of the previous administration. This was a major blow to the necessary international joint action for climate change mitigation. The new president in January 2025, hasten to declare in the very first days of his tenure that the US withdraws from the UN's Paris 2015 COP21 Agreement and he also declared his government's encouragement for more oil drilling and exploration and all this when California's Los Angeles was still burning under the unprecedent wildfires that erupted in early January 2025 due to the extreme dry weather of many months before and the strong winds. To make things worse the decisions and actions of the Trump administration in the US have set a dangerous negative precedent which may affect the climate policies of several other countries and also those of large private corporations in all economic sectors which had committed to adopting "green" policies. Already at the time of writing these lines all major banks in the US (JPMorgan, Citigroup, Goldman Sachs and others) have announced their withdrawal from their previous zero emission policies and the policies of these banks are likely to be followed by other competing banks in Canada and Europe who had committed to the "Net Zero Banking Alliance". Similar trends appear in the investment sector with withdrawals from the "Net Zero Asset Managers" alliance which may affect the investments in green technologies and installations to produce renewable energy.

This issue shows the difficulties and multiple impediments that exist on the road towards a common global approach to mitigating climate change and global warming. The repercussions of a negative US governance towards the need to take measures for mitigation and adaptation will be severe and far reaching at global level. Even under the measures taken during the Biden administration period, the *Climate Action Tracker*, had rated the policies and targets for climate change in the US as "*insufficient*". This means that they were consistent with limiting the global warming to 2 °C by the end of the century but not to the level of 1.5 °C as it was agreed in the Paris 2015 COP21 Agreement.

2.1.4 Plans and Proposed Actions in Other Countries

The other countries and world regions are smaller emitters of GHG than those examined separately above. Their plans and proposed actions for reducing their emissions are submitted and examined by the United Nations (UN) relevant database of the Framework Convention on Climate Change (UNFCCC). Over and above their overall UN responsibilities, the different countries demonstrate different levels of responsibility and capability to mitigate climate change, depending on their historical and current contributions to the problem, as well as on their economic and social development. According to the provisions of the Paris COP21 Agreement, all signatory countries have to formulate and maintain their own national plans for mitigating climate change, the so called *Nationally Determined Contribution* (NDC). Each NDC contains the national plan and targets for reducing emissions and the measures for adapting to the impacts of climate change. The NDCs are communicated to the UN which has established an *NDC registry* (https://unfccc.int/process-and-meetings/the-paris-agreement/) and monitors the progress made globally in terms of mitigation and adaptation.

An indicative summary presentation of the contents of the NDCs for three countries that are also major polluters, is as follows:

- *Russian Federation*: According to its Nationally Determined Contribution (NDC) document, which was submitted to the UN's registry on November 2020, the Russian government pledged several actions among which reduction of its GHG emissions by 30% (relative to 1990 levels) by 2030. It must be noted that the initial target was for a 70% reduction, but this was found as unrealistic and was reduced to 30%. However, even this 30% reduction is now judged as unrealistic after the latest developments in this country and the war economy that was developed after its war with Ukraine. The target year for zero emissions is set to 2060. According to the *Climate Action Tracker*, Russia's NDC of 2020 falls in the "Highly Insufficient" category (www.climateactiontracker.org/climate-target-update-tracker/russian-federation/).
- *India*: Plans to reduce the emissions intensity of its GDP by 33–35% by 2030, as compared to 2005 levels, have been submitted by India. In the same NDC, India pledges to achieve a level of 40% of its cumulative electric power installed capacity coming from non-fossil fuel sources by 2030. Like China, India also plans to create an additional carbon sink of 2.5–3 billion tonnes of CO_2 equivalent through additional forest and tree cover by 2030.
- *Brazil*: This country (home to the Amazon forest) pledged to reduce GHG emissions by 37% by 2025 and 43% by 2030 from 2005 levels. Also, to increase the share of renewables (over and above hydropower) in the energy mix to 23% by 2030. Also, to restore and reforest 12 million ha of forests by 2030 and promote low-carbon agriculture and biofuels.
- The rest of the 75 countries that have submitted NDCs, project similar plans and pledges for taking decarbonisation measures. The question is, how far all these countries (and even more the others that have not so far submitted NDCs), will move to concrete implementation actions to materialise their plans.

2.2 Despite Meaningful Momentum, the World Is not Doing Enough!

Based on the NDC registry, the UNFCCC secretariat publishes a synthesis report every five years to assess the aggregate effect of the NDCs and the progress towards the long-term goals of the Paris COP21 Agreement. The latest synthesis report was published in 2022 and showed that NDCs were submitted by only 75 countries (of the total 198 who have signed the Paris Agreement) representing about 30% of global GHG emissions. Even if this small number of NDCs are implemented, which is by no means certain, it is estimated that this would lead to a 21% reduction in emissions by 2030 (compared to 2010 levels) and this is far from the 43% reduction needed by 2030 to limit global warming to 1.5 °C by mid-century IPCC (2023). So, despite the almost unanimous (i.e. signed by all countries in the world) acceptance of the climate change as a problem to be tackled by way of priority and their participation in the UN's initiatives, so far little progress can be seen in terms of implementing actual measures and, even more, the major economies of the world do not move towards a coherent and concrete cooperation to face the problem through well-coordinated climate change actions. This section gives the necessary evidence that supports this statement.

In its regular *Global Stocktake* report which is used to monitor the Paris Agreement implementation and evaluate the collective progress made in achieving the agreed goals, the UN stresses that "even with increased efforts by some countries much more action is needed now to bend the world's emissions trajectory further downward and avoid the worst impacts of climate change" (UN, 2023). The UNFCCC (the UN's Framework Convention on Climate Change) on its Yearbook of Global Climate Action (UNFCCC, 2023), as well as the US Global Change (US Global Change, 2024), and several other organisations (Boehm et al., 2023), have raised warnings to the same effect: the world is NOT on track in implementing measures to mitigate climate change and reduce global warming. According to some calculations as in (McKinsey, 2022), or (IPCC, 2023), to limit global warming to 1.5 °C by 2050, as agreed in the COP21 Paris Conference, global GHG emissions should be falling by nearly 8% per year. This is by far not the case as currently, global GHG emissions show an increase rather than decrease compared to 1990 levels (McKinsey, 2022)!

At global level, as already mentioned, an overall picture of the progress in implementing climate change mitigation measures and policies is given by the data collected by the "*Climate Action Tracker—CAT*" (https://climateactiontracker.org/). This is an independent scientific project that tracks government climate action around the world and measures it against the globally agreed Paris Agreement target of "holding global warming below 2 °C and pursuing efforts to limit warming to 1.5 °C". Its assessments are renewed continuously, and the results can be seen in its web site. With data until December 2024, the rankings given by the *Tracker*, show that of the major economies that are world-level GHG emitters **none** qualifies for a rating of "*1.5 °C Paris Agreement compatible*". The EU, the US (under the Biden administration measures), Canada, Australia, S. Africa, Brazil, and a handful of others are rated as "*insufficient*" (the second-best rating), while China and India are rated as "*highly insufficient*", and Russia, Saudi Arabia, Turkey, Argentina and several other countries are rated as "*critically insufficient*". The CAT also assesses the targets for net-zero emissions that are set by governments around the world. This assessment is made in terms of scope, architecture, and transparency of the set targets and is shown in a diagrammatic form in Fig. 2.1. As shown there, only 4 countries (Chile, Colombia, Costa Rica, and the UK) and the EU (27 countries), which together represent 7% of global emissions, are assessed with "*acceptable*" net-zero target design. The net-zero targets of the largest group of countries, representing 68% of global emissions, are assessed with as "*Average*" or "*Poor*" net-zero targets (see Fig. 2.1).

Another major issue is the time necessary to develop and reach maturity of the new technologies and measures for decarbonisation. To avoid delays and long maturity periods there is a need for international cooperation and joint research and innovation development. Such collaboration—under the current global trade competition conditions—is difficult to materialise. It is reminded that the penetration of new low-carbon technologies into the global market, may take as long as 10 or even 15 years as such development follows a typical S-shaped curve that is characterised by three

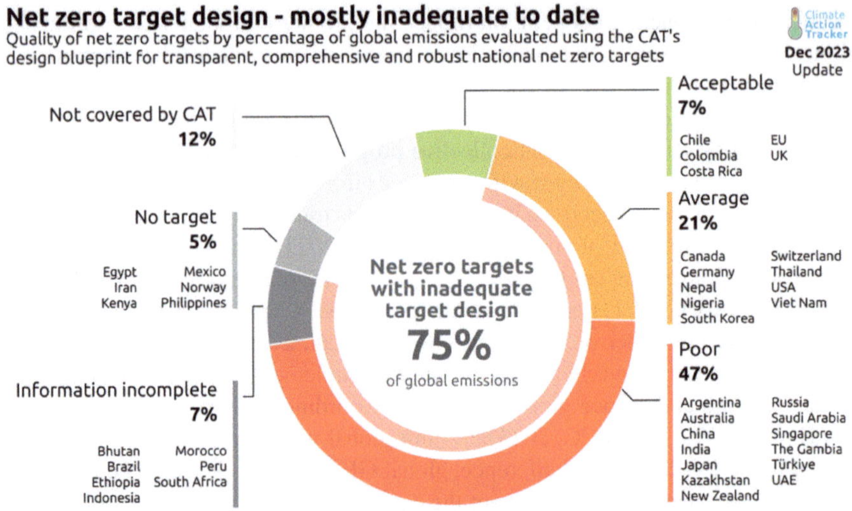

Fig. 2.1 Climate Action Tracker's evaluation of announced net zero targets (as of December 2023). *Source* Climate Action Tracker web platform (https://climateactiontracker.org/)

stages: (A) the emergence of the new technological system for low or no carbon transition (the new technology emerges and undergoes coordinated development and testing). (B) The diffusion stage, where coordinated diffusion and development of economies of scale takes place, and (C) the reconfiguration stage in which whole new markets are developed around the new technology or system with coordinated universal standards and mass production and diffusion into widespread use (Victor et al., 2019).

The prospect of not reaching the net-zero objectives by 2050, is also expressed most clearly in several studies and reports among which:

- The *Global Energy Perspective 2023 and 2024* reports by the McKinsey company (McKinsey, 2023b, 2024).
- The NGFS—*Network for Greening the Financial System* (a network of 114 central banks and financial supervisors that aims to accelerate the scaling up of green finance and develop recommendations for the role of central banks for climate change) (NGFS, 2024).
- The US Global Change Research Program, the Fourth National Climate Assessment report which concludes that even if the stated long-term strategies for decarbonisation are fully implemented on time the greenhouse gas emissions in 2050 would be roughly 63% lower than in 2019 and not 100% as desired (US Global Change, 2024).
- The Berkeley Earth observatory, a California-based non-profit research organization that has been preparing independent analyses of global mean temperature changes since 2013 (https://berkeleyearth.org/global-temperature-report-for-2024/).

- The International Energy Agency (2023), and
- Other reports and studies (IEA, 2023; IPCC, 2023; UN, 2023).

In all the above reports there are estimates made about the median level of global warming by the end of the century if current policies and (in)activities pursue. In the IPCC (Intergovernmental Panel on Climate Change) scenarios, the median level of warming by the end of the century is estimated at 2.9 °C, while in the McKinsey, NGFS, (Network for Greening the Financial System), and IEA scenarios it is estimated at 2.3 °C, 2.8 °C, and 2.4 °C, respectively. In further endorsement of the above findings, the *State of Climate Action 2023* report that was produced with funding from an independent group of international organisations, states that the progress made in closing the global gap in climate action remains "woefully inadequate" (Boehm et al., 2023). The report used 42 indicators to measure and demonstrate progress in achieving the goal of restricting global warming to 1.5 °C by 2050. Of these 42 indicators, 41 were found not to be on track to achieve their 2030 targets. Progress for more than half of these indicators remains well off track and they will necessitate a twofold increase in efforts for their correction during this decade. Another six indicators are heading in the wrong direction entirely, so overall it seems that global efforts to limit warming to 1.5 °C are failing across the board. This report also highlights actions that must urgently be taken in this decade to limit global warming and offers a roadmap that the world can follow to avoid increasingly dangerous and irreversible climate impacts. It translates the Paris Agreement's 1.5 °C temperature limit into 2030 and 2050 targets across sectors that account for roughly 85% of global GHG emissions (i.e., power, buildings, industry, transport, forests/land, food, and agriculture) as well as for carbon removal and climate finance actions.

There are several reasons why the global climate change action is lagging plans and expectations:

(a) There is still lack of clear and consistent policies and incentives for low-emission technologies in most countries of the world.
(b) User behaviour is very much unchanged attracted to high emission activities and disregard of the climate change consequences of such behaviour.
(c) There is much complexity and uncertainty about the necessary economic and societal transformations that are required for net-zero emissions.
(d) There are serious challenges in coordinating and collaborating across multiple actors, sectors, and regions that is necessary.
(e) There is strong inertia and resistance of the existing high-emissions systems and vested interests who invest and support politicians to deny climate change and not take measures to combat it. The result of the 2024 elections in the US are a most notable case of this situation.
(f) There are very high financial requirements involved in the implementation of the necessary climate change mitigation and adaptation measures. According to a 2023 McKinsey report, to successfully meet the global goals of the Paris Agreement, would require a vast increase in total capital spent each year, from $5.7 trillion spent on low- and high-emissions technologies in 2023, to as much as $9.2 trillion, on average, spent over the next three decades (McKinsey, 2023b).

The above financial assessment, points to a bleak future in which global warming will not be possible to be adequately restrained due to financial difficulties and so it will rise to higher levels.

(g) Finally, there are serious geopolitical events that erupt from time to time, in almost every corner of the globe. They start because of disputes between neighbouring countries and/or economic or commercial competitions but their repercussions are wide ranging and can be felt at global level. They diminish international cooperation and can even lead to local wars. Such events destabilise whole regions and cause diversion of funds to military or other expenditures that otherwise could have been devoted to climate mitigation and adaptation actions. They also distract or disrupt energy and other supply lines and divert the priorities and fund allocation from the implementation of decarbonisation projects to other causes.

In Europe, the disruptive force and painful implications of the geopolitical events mentioned in (f) above, were felt most vividly in the first half of the decade 2020–2030 when, starting in February 2022, the war erupted in Ukraine. Nearly two years later, the war in the Middle East started with the attack on Israel of October 7, 2023. The impacts of these two geopolitical events on slowing down the decarbonisation efforts in the EU are now clearly felt with the disruption of the energy lines and the reorientation of the EU energy policies. Furthermore, they are to a great extent responsible for delays in the implementation of some of the most prominent EU climate policies. One of the hardest hit sectors was agriculture. European farmers demonstrated for weeks and months in the period from the end of 2023 to the middle of 2024. They were demanding from the European Commission to retract or delay indefinitely, large parts of the climate change mitigation policies in its Common Agricultural Policy (CAP), something that at the end did happen.

Today, five years before the critical milestone year of 2030, the EU's target of a 55% reduction by 2030 of its GHG emissions (compared to 1990 levels), seems impossible to achieve not only because of the geopolitical events mentioned above but also of the social unrest that these policies raise. It seems that European citizens (like most citizens in other regions too) cannot easily accept the necessary changes in their daily lives that would occur from the implementation of the foreseen climate mitigation measures. As stated in the 2024 report by the *European Environment Agency* on the climate risks in the EU, "societal preparedness in the EU is still low, as policy implementation is lagging substantially behind quickly increasing risk levels. Most policies and actions to strengthen Europe's resilience to climate change are made for the long term, and some actions have long lead times. Urgent action is needed now to prevent rigid choices that are not fit for the future in a changing climate" (EEA, 2024).

In concluding, despite the ambitious plans and policies announced by national or international governments, the world seems to be at a practical inability to enforce the necessary decisions and policies for decarbonisation. The assessments made by various independent organisations show that at the middle of the critical decade 2021–2030 the world is not on track to achieve the goals of the Paris Agreement. The

UN's resolutions and decisions, remain largely unenforced and the UN does not have the jurisdiction or the power to enforce its decisions. In the absence of an effective and powerful international governance system, the successful implementation of the climate change policies and decisions that are taken during the UN's Conferences (COPs and other) remains at the full will of the governments of the participating countries. The result is reflected in the frustration expressed by the UN's Secretary-General Antonio Guterres remarks when releasing the 2022 report of the IPCC, "*we are firmly on track toward an unliveable world*"!

2.3 In the Wake of Failure Can Technology Help?

Given the delays and difficulties noted above, the timely and to the target implementation of the various mitigation actions raises the question if there are any innovative technologies (or "tools") that can enhance the effectiveness or facilitate the implementation of the various decarbonisation policies. This issue is examined in this section by distinguishing the technologies that can be used in two types, "hard" i.e., involving hardware systems and mechanisms as well as engineering structures and industrial installations, and "soft" i.e. involving computational or algorithmic techniques and software.

2.3.1 Hard Technologies

Of the "hard" category of technologies, two are the most prominently referred to in the bibliography in recent years. The first is the removal of CO_2 from the atmosphere using mechanical means, and the second is the obstruction to the sun's radiation that reaches the earth. Both are based on the premise that since global warming results from the combined effect of two elements i.e., the level of radiation-absorbing gases in the atmosphere (CO_2, methane, nitrous oxide, fluorocarbons, etc.) and the amount of solar radiation that reaches the earth, one can try to act on both these fronts and achieve a combined positive outcome. This means, that global warming can be reduced by reducing the radiation-absorbing gases in the atmosphere and/or by also reducing the solar radiation that reaches the earth.

For the first group of "hard" technologies, the systems developed are knows as NETs for *Negative Emission Technologies*. The most highly developed and promoted systems concentrate on CO_2 removal from the atmosphere which is known as CDR for *Carbon Dioxide Removal*. A range of NETs have been developed so far (Desport & Selosse, 2022; EASAC, 2018; Quang et al., 2023). These are industrial installations that absorb CO_2 from the atmosphere or from sources that emit it, such as energy intensive industries e.g. cement, steel, power generation, etc. The collected CO_2 is either stored in appropriate underground depositories (in which case we have *Carbon Capture and Storage*—CCS) or converted to other non-harmful gases. There are many alternative technologies that have been developed that go by the names *Bioenergy with Carbon Capture and Storage*—BECCS, or *Direct Air Capture with Storage*—DACS, or *Carbon Capture and Utilization*—CCU, etc., (Pires, 2019). In the "hard" group of technologies that aim at CO_2 capture from the atmosphere belong also those that are termed "natural" i.e., involving the development of carbon sinks through the planting of trees and forests (reforestation).

The effectiveness and the extent to which NET technologies can make a sizeable contribution to mitigating climate change is still unclear. The speed and the cost effectiveness of the various technical solutions for removing CO_2 from the atmosphere, or at the points where it is produced, will be crucial for their future acceptance and wider implementation. To this end, there are several simulation models that have been developed to investigate the impacts of NETs on meeting the Paris Agreement targets policies (Desport & Selosse, 2022; Höhne et al., 2021; Poggio et al., 2024). They invariably give positive results that show that the application of NETs can be a necessary additional measure, which together with the decarbonisation measures and policies can lead to the desired net-zero results and fulfilment of the COP21 Paris agreement targets. Whether the NET technologies will be cost-effective enough and economically viable is still to be proved, but the trends are that their costs are gradually reduced. Research on their development and implementation paths is going on, but their full development and implementation is envisaged in the medium-term future i.e., probably 2035–2040 and beyond. If the NETs prove cost-effective, they may be politically more appealing than engaging in the generally unpopular decarbonisation policies.

The other group of "hard" technologies that has been put forward is the "management" of solar radiation. The idea of these technologies is to reflect a small fraction of the sunlight back into space and thus cool the atmosphere. These techniques are also known as *solar geoengineering* and date back to the early 2010s. They include several individual techniques for effecting reflection of the sunlight back into space which go by the names of *marine cloud brightening*, *cirrus cloud thinning*, *space-based techniques*, and *stratospheric aerosol scattering*, amongst others (Ming et al., 2014). Besides the technical difficulties for the introduction of these technologies there is also the need for some sort of international governance for their implementation (Nicholson et al., 2018).

Besides all the above, a most important "hard" action for climate change mitigation that uses both "hard" and "soft" technologies, is the reduction of our energy consumption. We consume energy for a multitude of uses covering needs that some of them can be reduced or replaced by other not so energy thirsty activities. We have the technical potential to cut our energy demand considerably and by doing so reduce our polluting energy generation processes. Until we manage to decarbonise fully the power generation processes, using less energy is the easiest and cheapest "hard" type of "technology" to reduce carbon pollution of the atmosphere. Strangely, this option is often-neglected in current pathways to zero-carbon emissions.

2.3.2 *Soft Technologies and the Role of AI*

The role of "soft" technologies in addressing the climate change challenge is relatively less pronounced but more important and far reaching. The definition of "soft", denotes that these technologies are intangible and focusing on processes, strategies, and innovative techniques rather than physical systems and mechanisms. Soft technologies are primarily used to optimize operations and facilitate efficiency. They focus on streamlining processes and improving decision-making using advanced software systems, data analyses, and automation. "Soft" technologies are working in the background in almost every aspect of the climate change field from education and training programs (that train workforces with skills that green industries need), to the designing of new mitigation initiatives carefully and holistically with an eye toward how they interact with one another and with other related policies such as national security, industrial, innovation, and labour market policies.

In tackling the complexities of the net-zero transition, as it involves the radical transformation of most sectors of the national and global economy, "soft" technologies and software development are indispensable. Traditional software applications, based on less intelligent algorithms, suffered from slow speeds, bias amplification, increased need for human intervention, and the need to define predominately ex-ante problem solving processes. With the entrance of Artificial Intelligence (AI) these deficiencies are overcome as AI software development offers a great many more possibilities and options. AI software is now in the core of practically all "soft" technologies used for climate change mitigation and adaptation applications. Through machine-based learning, AI can make reliable predictions, recommendations, or decisions influencing real or virtual environments for any set of human-defined objectives. Compared to traditional software, climate change applications of AI—or "climate AI"—are more responsive to real needs, more big-data-friendly, better able to address large-scale hazards, and more focused on producing innovation (Dimock, 2022).

Potential areas of application of "climate AI" include:

1. *Climate risk assessment*. To assess the risk of occurrence of climate related extreme events for a specific country or region.
2. *Climate events prediction*. Aids the decision-making process by quickly predicting climate patterns and consequent extreme weather events and estimates their potential impact on the various ecosystems. It can also be used in formulating quickly evacuation plans or other prevention measures before the extreme events occur.
3. *Monitoring and mitigating air pollution*. Based on data from sensors and other sources that detect pollution levels, AI software can monitor air and water quality, detect real or potential sources of pollution, as well as the polluters, and enforce environmental regulations.
4. *Monitoring key environmental and climate events*. AI can be "trained" to measure and evaluate changes in key environmental indicators that are indicative of trends of climate change and can be used to evaluate the progress of mitigation efforts. For example, by continuously analysing satellite data and images, AI can present

data on the melting of icebergs and calculate their meltwater quantities. Similarly, it can monitor the net result of deforestation over large areas and calculate its impact on the climate.[4]

5. *Carbon footprint calculation.* Personalised AI, based on a persons' or item's movement information will estimate in real time the carbon footprint of an individual's or a freight item's movement. By doing so, it can help informed choices to be made for reducing the environmental impact. Similarly, it can help in producing low-carbon materials i.e., with lower total-cycle carbon footprints, for construction and other purposes.
6. *Optimising energy production and consumption.* By optimising the operations of renewable electricity generating assets such as wind turbines and solar panels, and at the same time optimising battery storage assets, AI can allow for a more balanced and secure renewable energy production. At the same time, it can offer optimised flexibility services to electricity grids by performing flexible management of millions of connected devices and electric cars. AI can also help optimise energy consumption by maximising the "energy efficiency" of buildings, factories, or transportation systems for reduced GHG emissions.
7. *Climate financing.* Intelligent management of financial data related to climate investments including tracking funding flows and optimal resource allocation for climate projects.
8. *Climate research.* Finally, AI will enhance climate research and modelling in many ways. Primarily, by processing vast amounts of data and discovering meaningful patterns and interrelations among them. AI may also accelerate the discovery of new energy-relevant materials, such as those used in photovoltaic cells, batteries, and electro fuels (for an example of the use of AI to increase the pace of battery experiments, see https://aionics.io).

All the above (indicative) areas of AI applications refer to what is known as "multimodal AI" that combines multiple types, or modes, of data to create insightful conclusions or make more precise predictions about a multitude of real-world problems. The multitude of ways and potential applications of multimodal AI in the transport and mobility field, are examined in detail in Chaps. 6 and 7. A special subset of multimodal AI that is known as "Generative", is the AI that generates content. Generative AI presents a particularly interesting set of potential climate change applications such as those listed below (McKinsey, 2023a):

(a) Summarising climate research information and knowledge for the purposes of dissemination across different audiences. Fact-based generative AI can provide a convincing "voice" for any environmental focus area providing accurate and timely information to millions more people than traditional communication means.

[4] "*Space Intelligence*" (https://www.space-intelligence.com/), a company dedicated to providing data and mapping information on forest coverage does exactly that by using software that calculates the carbon storage that is achieved against what is required to ensure transparency into the impact of nature-based mitigation solutions.

(b) Reviewing and synthesizing news and social media posts to fact-check or to understand general sentiments, to determine how to target future actions or campaigns on decarbonisation actions or other causes.
(c) Generating synthetic data (modelled after existing data sets, or imagery) as input for scientific research and laboratory testing.
(d) Creating foundation models that engage, with a deeper understanding of the laws of physics, to climate simulation and emulation.
(e) Creating many applications to serve with content generation in almost all areas of a company's operation, e.g.: customer operations, marketing and sales, software engineering, and R&D.
(f) Increasing labour productivity by supporting human activities across many economic sectors.

The era of "climate AI", as well as of "mobility AI", is just beginning. Excitement over these technologies is great and early pilots are almost everywhere. A full realization of the technology's benefits will take time and leaders in business and society still have considerable challenges to address (McKinsey, 2023a). These challenges include managing the risks inherent in generative and multimodal AI, determining what new skills and capabilities the workforce will need, and rethinking core business processes and retraining or developing new skills. AI is becoming increasingly useful in that it can help identify subtle patterns in very large amounts of data and allows to optimise and automate complex systems. There are also weaknesses associated with this ability in that the outputs can be strongly influenced by poor or biassed data, also that it is not always clear how AI arrives at its conclusions, and any answers it offers are only as good as the questions it is asked. We explore and investigate all these issues in the chapters that follow.

References

Boehm, S., Jeffery, L., Hecke, J., Schumer, C., Jaeger, J., Fyson, C., Levin, K., Nilsson, A., Naimoli, S., Daly, E., Thwaites, J., Lebling, K., Waite, R., Collis, J., Sims, M., Singh, N., Grier, E., Lamb, W., Castellanos, S., … Masterson M. (2023). State of climate action 2023. Berlin and Cologne, Germany, San Francisco CA, and Washington DC: *Bezos Earth Fund, Climate Action Tracker, Climate Analytics, ClimateWorks Foundation, NewClimate Institute, the United Nations Climate Change High-Level Champions, and World Resources Institute.* https://doi.org/10.46830/wrirpt.23.00010

Desport, L., & Selosse, S. (2022). Perspectives of CO_2 utilization as a negative emission technology. *Sustainable Energy Technologies and Assessments, 53*(Part C). https://doi.org/10.1016/j.seta.2022.102623

Dimock, W. C. (2022). What AI can do for climate change, and what climate change can do for AI. *Scientific American.* https://www.scientificamerican.com/article/what-ai-can-do-for-climate-change-and-what-climate-change-can-do-for-ai/. Accessed December 2024.

EASAC. (2018). Negative emission technologies: What role in meeting Paris agreement targets? *European Academies Science Advisory Council,* Policy Report 35. ISBN: 978-3-8047-3841-6. www.easac.eu/projects/details/

EEA. (2024). European climate risk assessment—Executive summary. *European Environment Agency*, report no. 01/2024. ISBN 978-92-9480-627-7. https://doi.org/10.2800/204249

Freymann, E. (2022). The adaptation advantage. *The Wire China*, cover story, July 17, 2022. https://www.thewirechina.com/2022/07/17/chinas-climate-adaptation-advantage/. Accessed December 2024.

GEM. (2024). Global coal plant tracker. *Global Energy Monitor*, October 2024. https://globalenergymonitor.org/projects/global-coal-plant-tracker/. Also in CarbonBrief blog 11 April 2024. https://www.carbonbrief.org/china-responsible-for-95-of-new-coal-power-construction-in-2023-report-says/. Accessed January 2025.

Höhne, N., Gidden, M. J., den Elzen, M., Hans, F., Fyson, C. l., Geiges, A., Louise Jeferry, M., Gonzales-Zuniga, S., Mooldijk, S., Hare, W., & Rogelj, J. (2021). Wave of net zero emission targets opens window to meeting the Paris Agreement. *Nature Climate Change, 11*, 820–822. https://doi.org/10.1038/s41558-021-01142-2

IEA. (2023). CO_2 emissions in 2022. *International Energy Agency* report, No.1 in the Global Energy Transitions Stocktake series, Paris, March 2023, Licence: CC BY 4.0. https://www.iea.org/reports/co2-emissions-in-2022

IPCC. (2023). AR6 synthesis report: Climate change 2023. *The Intergovernmental Panel on Climate Change.* Sixth Assessment Report of the Panel's 58th Session held in Interlaken, Switzerland, 13–19 March 2023. https://www.ipcc.ch/report/sixth-assessment-report-cycle/. Accessed December 2024.

Li, Y., & Shapiro, J. (2020). *China goes green: Coercive environmentalism for a troubled planet* (pp. 240). Wiley, September 2020. ISBN: 978-1-509-54311-3.

Maizland, L. (2021). China's fight against climate change and environmental degradation. *Council on Foreign Relations*, May 19, 2021. https://www.cfr.org/backgrounder/china-climate-change-policies-environmental-degradation. Accessed January 2025.

McKinsey. (2022). The net-zero transitions: What it would cost, what it could bring. *McKinsey Global Institute,* report authored by M. Krishnan, Hamid Samandari, J. Woetzel, S. Smit, D. Pacthod, D. Pinner, T. Nauclér, H. Tai, A. Farr, W. Wu, & D. Imperato. January 2022.

McKinsey. (2023a). The economic potential of generative AI: The next productivity frontier. *McKinsey & Company*, report, June 2023. https://www.mckinsey.com/capabilities/mckinsey-digital/our-insights/the-economic-potential-of-generative-ai-the-next-productivity-frontier#introduction. Accessed January 2025.

McKinsey. (2023b). Global energy perspective 2023. *McKinsey & Company* report, 18 October 2023. https://www.mckinsey.com/industries/oil-and-gas/our-insights/global-energy-perspective-2023. Accessed January 2025.

McKinsey. (2024). Global energy perspective 2024. *McKinsey & Company* report, 17 September 2024. https://www.mckinsey.com/industries/energy-and-materials/our-insights/global-energy-perspective. Accessed January 2025.

MEE. (2023). China's policies and actions to address climate change (2022). White paper report by the *Chinese Ministry of Ecology and Environment*, (unofficial translation), October 27, 2023. https://english.mee.gov.cn/Resources/Reports/reports/202211/P020221110605466439270.pdf. Accessed January 2025.

Ming, T., de Richter, R., Liu, W., & Caillo, S. (2014). Fighting global warming by climate engineering: Is the earth radiation management and the solar radiation management an option for fighting climate change? *Renewable and Sustainable Energy Reviews, 31*, 792–834. https://doi.org/10.1016/j.rser.2013.12.032

NGFS. (2024). NGFS scenarios: Purpose, use cases and guidance on where institutional adaptations are required, an explanatory note. *Network for Greening the Financial System—NGFS*, Technical document, January 2024. https://www.ngfs.net/sites/default/files/medias/documents/ngfs_guidance_note_on_the_scenarios.pdf

Nicholson, S., Jinnah, S., & Gillespie, A. (2018). Solar radiation management: A proposal for immediate polycentric governance. *Climate Policy, 18*(3), 322–334. https://doi.org/10.1080/14693062.2017.1400944

Pires, J. C. M. (2019). Negative emissions technologies: A complementary solution for climate change mitigation. *Science of the total environment, 672*, 502–514. https://doi.org/10.1016/j.scitotenv.2019.04.004

Poggio, M., Império, M., Bernardo Baptista L., Schaeffer, R., Lucena, F. P., Szklo, A., Rochedo, P. R. R., Hultman, N., McJeon, H., & Clarke, L. (2024). The role of bioenergy in Brazil's low-carbon future. *Energy and Climate Change, 5*. https://doi.org/10.1016/j.egycc.2023.100123

Quang, V., Milani, D., & Abu Zahra, M. (2023). A review of potential routes to zero and negative emission technologies via the integration of renewable energies with CO_2 capture processes. *International Journal of Greenhouse Gas Control, 124*. https://doi.org/10.1016/j.ijggc.2023.103862

SCIO. (2022). Responding to climate change: China's policies and actions. *State Council Information Office of the People's Republic of China*, white paper, October 27, 2021. http://english.scio.gov.cn/node_8026776.html. Accessed December 2024.

UN. (2023). New analysis of national climate plans: Insufficient progress made, COP28 must set stage for immediate action. *UN Climate Change News*, 14 November 2023. https://unfccc.int/news/. Accessed December 2024.

UNFCCC. (2023). Yearbook of global climate action 2023: Marrakech partnership for global climate action, 14 November 2023.

US Dept of State. (2021). *The long-term strategy of the United States: Pathways to net-zero greenhouse gas emissions by 2050*. United States Department of State and the United States Executive Office of the President.

US Global Change. (2024). Climate science—special report. *US Global Change Research Program*, Fourth National Climate Assessment report, Vol. 1. https://research.un.org/en/climate-change/reports

Victor, D. G., Geels, F. W., & Sharpe, S. (2019). Accelerating the low carbon transition: The case for stronger more targeted and coordinated international action. *Energy Transitions Commission, Brookings Institution*, November 2019. https://www.energy-transitions.org/publications/accelerating-the-low-carbon-transition/. Accessed December 2024.

Chapter 3
The Transport and Mobility Dimension

Abstract In terms of its CO_2 emissions, the transport and mobility sector is one of the top (and in some countries, *the* top) polluting sector globally. It accounts for almost 25% of all CO_2 emissions globally and higher than that in certain areas such as around 28% for the EU, 27% for China, 28% for the US etc. This chapter looks exclusively at the relation of the transport sector with climate change and examines the issues involved and the key elements of interaction of the sector vis-à-vis the unfolding climate change crisis. Primarily, it aims to consider the transport sector as an emitter of Green House Gasses—GHG, and to assess the main policies for its decarbonisation. It also considers the incoming innovations in the sector and their potential climate friendliness. The emphasis is on road transportation because this is the most polluting mode by far. The other transport modes are also considered but, in less detail, and only when necessary. The highlight of this chapter is that it presents a concise synthesis of the various proposals and suggestions for decarbonisation measures and policies in various studies by national or international agencies and highlights the types of transport decarbonisation measures and policies that are most promising and likely to be widely accepted. These analytical presentations follow the triptych: "*avoiding*", "*shifting*" and "*improving/deploying*" measures i.e., policies that aim to make the users of the transport system avoid making trips, shift to less emitting modes, and improve/deploy new mobility technologies that will enable low or no greenhouse gas emissions. Besides this analytical presentation of the necessary actions for transport decarbonisation, this chapter examines also the so called five pillars of future climate-friendly mobility, i.e. Electro-mobility (electric vehicle propulsion and battery storage systems), Automated mobility (driverless, intelligent vehicles communicating data and information with the infrastructure and other vehicles), Big data and mass information collecting and processing technologies, Artificial Intelligence (AI) and its many applications for mobility and transportation, and New cultural and socioeconomic mentalities that are developing in connection to the changing travel and mobility characteristics. The reader is then led through a first assessment of the "odds for success" i.e., an assessment of the measures and policies that are considered as enough to effect zero carbon emissions and their chances of been implemented in time to avoid global warming above the 2 °C by 2050 or'60. This is done by examining progress in implementing the various decarbonisation

G. A. Giannopoulos and Y. Li, *Human Mobility, Artificial Intelligence and Climate Change*, Lecture Notes in Mobility, https://doi.org/10.1007/978-3-032-08171-1_3

measures in the four categories: Managing the "demand" for travel; Improving the energy efficiency of conventional vehicles; Shifting passenger travel and freight to less carbon intensive modes; and, Transitioning to "clean" no-carbon emitting energy vectors. The chapter ends by stressing that all the different policies and measures for decarbonisation in the transport sector, must be implemented in a concerted and coordinated way as there is no "silver bullet" for the successful decarbonisation of the sector. In the uncertainty that reigns globally, the individual decarbonisation measures and initiatives that have been taken so far by governments around the world, are not considered to be enough.

Keywords Transport policy · Transport and climate change · Transport emissions · Climate-friendly mobility · Electric mobility · Automated mobility · Big data · Transportation demand management · Decarbonisation policies · Transport decarbonisation · Avoid actions · Shift actions · Clean electricity generation · Life-cycle analysis

3.1 The Transport Sector and Climate Change

Of all economic and industrial sectors that interact with climate change, the one that interacts more closely is the transport sector. In terms of its CO_2 emissions, the transport sector at global level accounts for almost 25% of all CO_2 emissions. At national or regional level this percentage may be higher, e.g. it is around 28% for the EU, 27% for China, 28% for the US (2023 data). This makes the transport sector one of the top polluting sectors globally. In some countries it is a close second to the power generation sector while for others it is indeed the top polluter. In any case, the transport and the power generation sectors are closely related as regards their CO_2 emissions because any effort to decarbonize the transport sector by use of clean fuels such as electricity or hydrogen, depends on how clean the electricity generation system is. If the carbon intensity of the electricity generation system (in gr of CO_2 eq/KWh) is high, the battery electric vehicles (BEVs) that will use this "not-clean" electricity to recharge their batteries will in effect be more polluting than internal combustion vehicles (ICEs) using fossil fuels! The critical value for the carbon intensity of the electricity generation system over which battery electric vehicles will be more polluting than ICEs, is around 700 gr of CO_2 eq/KWh (Cox, 2018; EASAC, 2019).

Besides been a high polluter, the transport sector is also the only economic sector where its Green House Gass (GHG) emissions have been increasing since 1990. Even in the EU, where GHG emissions have decreased in all economic sectors since 1990, the transport sector is the exception showing an increase instead of a decrease in emissions (Miles et al., 2004; EEA, 2024). This is so because, despite considerable increases in the overall energy efficiency of new passenger cars, vans and trucks, planes and ships, (with new engine types and cleaner fuels), the demand for transport services has also increased manyfold and so the level of transport related

GHG emissions increases overall. According to a 2023 report by the International Energy Agency (IEA), the whole of the transport sector (road, rail, aviation, maritime) continues to rely on oil products for approximately 90% of its final energy at global level (IEA, 2023a). This high percentage remains almost the same for the last 50 years and it is down only 3.5% from the early 1970s!

In addition to its diverging path in relation to climate-related goals and issues the current transport system in many countries, even at some developed ones, does not seem to follow a sustainable path. This means that the resources consumed, and the impacts caused by its operation, are not aligned with the environmental, economic, and social needs of the societies that support these systems and thus the benefits from their operation are outweighed by the environmental, economic, and social costs that are incurred from this operation. Achieving a sustainable transport system is an important goal to achieve besides the goal of making such system climate friendly through policies for its decarbonisation. Making our transport systems—especially the urban ones—both climate friendly and sustainable, will necessitate the formulation and implementation of integrated sustainability and climate change strategic plans for each area which would include and integrate sustainable traffic and transport management plans with mitigation and adaptation policies towards climate change. In the EU, the so called *Strategic Urban Mobility Plans* (SUMPS) which have been formed and are gradually being implemented in most urban areas, are examples of such strategic plans (see https://urban-mobility-observatory.transport.ec.europa.eu/sustainable-urban-mobility-plans_en).

This chapter aims to examine the relation of the transport and mobility sector with climate change. It examines the issues involved and the key elements of interaction of the sector vis-à-vis the unfolding climate change crisis. Primarily, it aims to consider the transport sector as an emitter of Green House Gasses—GHG, and to assess the main policies for its decarbonisation. It also considers the incoming innovations in the sector and their potential for sustainability and climate friendliness. The emphasis is on road transportation because this is the most polluting mode by far. The other transport modes are also considered but, in less detail, and only when necessary. So, unless otherwise specified, the word "transport" in the following, refers to road transport.

3.2 The Five Pillars of Future Climate-Friendly Mobility

In the field of transportation, we are currently faced with the advent of five "revolutionary" innovations both by way of the technologies used and by way of the transport system's operation. These innovations will define mobility in the twenty-first century and they will determine the success (or failure) of the decarbonisation effort of the sector. They have been defined as "revolutionary" or "disruptive" in (Giannopoulos & Munro, 2019) based on their potential for multiple and far-reaching consequences in the way the transport system operates and because each one of them can prove to be a "game-changer". These "pillars" of future mobility, are:

A *Electro-mobility* (electric vehicle propulsion and battery storage systems).
B *Automated mobility* (driverless "intelligent" vehicles communicating data and information with the infrastructure and other vehicles).
C *Big data* and mass information processing technologies.
D *Artificial Intelligence* (AI) technologies and their many applications for mobility and transportation.
E *New policies that will change cultural and socioeconomic mentalities* vis-a-vis the travel and mobility characteristics and will shape new types of demand for transport services in the future.

Each of the above innovations will be the result of the combined effect of several other smaller or "incremental" innovations focused on specific individual system elements or subsections. Together, they will define the future of the transportation sector for the bigger part of this century. At the same time, the success or failure of the plans and policies for decarbonisation will be determined by the success or failure in the way in which these five pillars will be imbedded and integrated into the future mobility systems especially in urban areas. Their full development and deployment will need continuous advances in related technologies and technical operations, appropriate technical and regulatory standards accepted at global scales, advances in sensors and telecommunications, and new traffic, transport, and travel management practices through the full use of AI.

Future developments will go well beyond new hardware and new software to include new ways of organizing mobility to ensure quality, reliability, and climate friendliness. We still have a long way to go to reach the ultimate level of development that is possible with these innovations. In 2025, these pillars are still at an initial stage of development just like the stage of development at which the internal combustion engine (ICE) vehicles were, 100 years ago. See the parallel by comparing the stage of development in which the ICE based mobility was in the 1925s and the stage of developed it achieved over the rest of the twentieth century until today. In the same way, one can imagine the level of development that is likely to be reached in all the above 5 pillars of climate friendly mobility by the end of the current century or earlier if the current pace of technological and socioeconomic progress (especially after the expected boost using AI), continues unabated. All this is discussed in detail in Chap. 10 but a brief look at each of the five pillars is also necessary at this point.

3.2.1 Electric Mobility

This is the mobility based on the movement of vehicles, on the ground or air, powered by electric motors that take electricity from batteries or fuel cells that convert the chemical energy of a fuel (e.g. hydrogen) and an oxidizing agent (e.g. oxygen) into electricity. The emissions output of these electric motors, in terms of what comes out of the tailpipe of the car, is zero. However, this does not mean that they are fully climate neutral in life cycle terms. The emissions produced for the construction or recycling of the batteries or fuel cells as well as for the production of the fuels used (electricity or hydrogen) as well as for the disposing of the vehicles at the end of their economic life, must also be considered. Electric mobility can be thought of as carbon-free only if the total emissions on a life cycle basis are zero (or close to it). Currently, for all the stages of an electric vehicle's full life cycle, we are still far from having an adequately "clean" record (Qiao et al., 2019). Narrow national economic interests and diverse policy priorities, as well as grave international geopolitical events and disruptions are creating a situation that points to a rather long period before we can have a truly decarbonised electric mobility.

The way the global situation evolves as regards clean electricity production is mixed. According to the International Energy Agency's (IEA) *World Energy Investment report* for 2023 (IEA, 2023a), wind and solar electricity are the fastest growing sources of clean electricity so far. They reached 10% of global electricity production in 2021 while at the same year all clean power sources (including nuclear) generated 38% of the world's electricity. The electricity generated by coal, globally, accounted for 36% (in 2021). The EU has set as its goal to achieve at least 80% of its electric power generation from "clean" sources by 2030. After the Covid-19 pandemic, the demand for electric power increased to a record high and as a consequence many countries postponed the planned closures of coal-based power generation units while others have decided to allow construction of new ones. Such is the case of the People's Republic of China where, in spite its impressive growth of wind and solar power generation capacity, unparalleled with any other country in the world, in first quarter of 2023 it also approved new coal electricity generating plants of more than 20 GW. This makes China the producer and consumer of more than half of the world's total coal generated power (Hawkins & Cheung, 2023; Li, 2023). According to the IEA, if the goal of keeping the global rise in temperatures to only 1.5–2 °C is to be achieved, the global coal powered electricity generation should be falling by 13% every year until 2030 (IEA, 2024b). Despite this, even in the absence of a fully clean electric power generation, the use of electric cars is beneficial for the environment as it is explained in later sections.

The global situation as regards the number of electric vehicles in circulation is improving but we are still far from the targets needed to achieve a sustainable and low-carbon mobility. To reach the Paris Agreement's goals, the global electric vehicle stock would need to reach 230 million by 2030 (IEA, 2025). We are still far from this target as in 2020 there were only 10 million new circulations of electric vehicles at global level (this was only 1% of the global car stock and 18% of the global new car registrations for that year). By 2024 there were 17 million new electric car registrations globally, bringing the total number of electric cars in the world to approximately 45 million (almost half of them in the P. R. China). So, we are still in need of a massive scale-up of policies, infrastructure, and innovation to support the adoption of electric mobility across all segments of the transport sector and regions of the world and at the same time clean-up electricity generation.

3.2.2 Automated Mobility

The second innovation pillar, *automated or autonomous mobility,* entails the introduction and wide circulation of automated (driverless) vehicles able to drive in all environments and under all conditions without the presence of a human driver. There are 6 levels of automated mobility according to the Society of Automotive Engineers'—SAE—international J3016 standard:

- Level 0: No automation.
- Level 1: Advanced Driver Assistance Systems (ADAS).
- Level 2: Partial Driving Automation.
- Level 3: Conditional Driving Automation (the vehicle drives by itself, but a human driver must actively monitor conditions and able to immediately take control of the vehicle if necessary).
- Level 4: High Driving Automation (Fully self-driving vehicles but within specific boundaries or Operational Design Domains).
- Level 5: Full Driving Automation (self-driving vehicles that require no driver assistance or monitoring and operate without boundaries or conditions).

The automated vehicle technology has been developed exponentially, and already automated vehicles of level 3 are commercially available while automated Level 4 vehicles operate as taxis in certain areas and countries (robotaxis). Full maturity of the technology, at automation level 5, is still quite far away due to many difficulties and obstacles that still have to be overcome. The most pronounced of these obstacles relate to the need to achieve a certified life-saving safety of operation to all involved, the need to settle all the legal, insurance, and cybersecurity issues involved, and the solving of the problems associated with the intermediate mixed phase of operation i.e., in which automated vehicles will be circulating together with vehicles driven by human drivers—a phase that may last several decades.

Automated vehicles are expected to be developed for air travel too. This will cover the needs for short distance urban or peri-urban travel to overcome congestion or cover missing links on the ground networks. For the flight of automated air vehicles, new air navigation and safety rules will have to be developed as this form of transport will be different from normal aviation and will circulate in a different air space than the one used by airplanes today.

3.2.3 Big Data and Mass Information Processing

Modern technology enables the collection of vast amounts of real-time traffic and transport-related data using sensors, surveillance cameras, radars, smartphones, social media apps, and other tools. They are collected in real-time using many types of sensors, surveillance cameras, radars, smartphones, social media apps and other. They can be processed and analysed using artificial intelligence (AI) techniques or more conventional methods (Willumsen, 2021). Real-time travel data can help understand travel behaviour and needs such as, for example, data collected for public transport use by tools collecting them from mobile apps and smart transportation cards and used for analysing user travel preferences, understanding travel demand during different time periods and locations, or optimizing the operation of public transportation systems and providing personalized travel services.

Data plays a critical role in supporting traffic management and decision-making in all modes. By collecting and analysing large amounts of data in the relevant networks and not only, one can gain comprehensive insights into the operation of the corresponding transportation systems, including traffic flows, bottlenecks, and travel patterns. This data helps transportation agencies develop effective traffic management strategies and make real-time decisions. Intelligent transportation systems, use AI for big data collection, processing, and communication technologies to enable many transport operations among them traffic signal timings, optimisation of traffic flows, provision of real-time traffic and navigation information, enhancing system efficiency and safety. In Chap. 7 (Data needs and utilisation) the reader will find a full account of the types of data used and the role of AI in creating and analysing large data bases for use in modern mobility operations.

3.2.4 Artificial Intelligence (AI)

The era of AI in the Transport sector is here and has already made significant contributions. Its relevance and applicability to transportation is well established and already many applications have been or are being developed in areas ranging from control and operation of automated vehicles to public transportation systems and operation, to traffic detection and management, and many other (Abduljabbar et al, 2019; Ivcevic & Grandinetti, 2024). Artificial intelligence is a branch of computer science that enables machines to perform tasks that normally require human intelligence. Typical human intelligence tasks undertaken by AI include, perception, reasoning, learning, and decision making. AI has the potential to play a crucial role in improving the quality of future transportation systems safety, efficiency, and sustainability and thus transform the transport sector. Many challenges and risks exist, and these are related to ethical, legal, and social issues, as well as cybersecurity threats, and human–machine interaction problems. Chaps. 5, 6, and 7 give more details on the AI systems and software being used in Transportation.

3.2.5 Policies to Affect Cultural and Socioeconomic Mentalities

This innovation pillar refers to all those aspects of human travel behaviour that affect the way an individual determines the elements or characteristics of his/her daily trip-making as well as the longer-term elements of his/her travel behaviour. The policies and the initiatives necessary to alter the human travel behaviour are known as *travel demand management* or TDM for short. TDM aims at changing the trip characteristics of travellers including their decision to travel or not, the hours of trip-making, the choice of destination, the choice of transport mode, etc., to achieve wider sustainability and climate related benefits. A particularly important area of TDM is to change the car ownership and car use perception especially for the urban traveller. This means convincing people to abandon owning a car and covering their mobility needs by public transport or sharing a car or other vehicle with others or using a car temporarily by purchasing the mobility service by specialised mobility broker companies. When a trip is made together with other travellers in the same car, we have *shared mobility* while when the trip is made by purchasing a mobility service by a mobility broker who arranges the whole transport chain, we have *car hailing* or *Mobility as a Service, MaaS.*

Taking measures and implementing policies for TDM is a very complex and sensitive issue because changing people's habits is difficult and needs time. Examples of TDM strategies:

(a) Provision of systematic, reliable, and ubiquitously available travel related information in real-time.
(b) Systematic information and training campaigns as well as incentives to convince car-owners to accept a more restricted use of their cars in favour of using shared mobility and/or MaaS.
(c) Changing the working conditions to align with the needs for less (or less intruding) travel. Typical example of such strategies is full or partial teleworking, the rescheduling of the working hours to shift some of the peak-hour trips to off-peak periods, and so on.
(d) Discouraging, by charging or even forbidding altogether, the use of polluting transport modes from environmentally sensitive areas.
(e) Imposing general restrictions on using polluting transport modes based on criteria such as the license plate number, or vehicle type, or emission levels, etc.

TDM is now profiting by strong technological backing and AI related software and techniques (EASAC, 2019; Li et al., 2022; Li & Lasenby, 2023; Oestreich et al., 2023; Shaik & Ahmed, 2022; Vagliasindi, 2023).

The benefits of shared mobility or use of MaaS, for the reduction of car ownership and use can be very significant. An earlier simulation study by the *International Transport Forum—ITF* that examined the impacts of shared mobility in several European cities by way of simulation under the relevant traffic conditions of each specific city, found that a wide-scale implementation of shared vehicles using cars, taxis and taxi-buses and satisfying all the travel demand in the area would have the following results (OECD-ITF, 2017):

- Congestion is practically eliminated.
- Traffic emissions can be eliminated by a third.

- The space required for public parking is greatly reduced (by 95%) leaving more road space available for re-allocation (e.g., broader sidewalks, bicycle lanes, etc.).
- The car fleet needed is only a very small percentage (e.g. 3%) of the size of today's car fleet needed to satisfy the same demand.
- The number of vehicle–miles travelled even during peak hours is reduced by as much as 37%.
- The trips are made less costly (to as much as – 50%) thanks to the highly efficient use of capacity.

Shared mobility and Mobility as Service are examined in more detail in Chap. 10 as they are expected to form the backbone of future mobility.

3.3 The Transport System as a Global GHG Emitter

Data on transport related CO_2 emissions at a global level are shown in Fig. 3.1. As shown there, the share of CO_2 emissions at global level (in Gt of CO_2 emitted per year) for the transport sector (which is depicted by the strip of lines second from the top), has been steadily increasing since 1970. In 2019 it reached the level of approximately 8 GT, which is nearly 22% of the total CO_2 emissions from all sectors in that period. After 2022 this percentage increased even further. Data analyzed by the *Partnership on Sustainable Low Carbon Transport* (SLOCAT) and presented in Table 3.1 shows the share of transport CO_2 emissions by region of the world and by two transport modes, international aviation and shipping. The data are given for two periods the 10-year period 2010–2019 and the 20-year period 2000–2019. As shown there, the share of Asia in the global transport related CO_2 emissions had the highest increase of + 41% during the 10-year period 2010–2019 with the second highest being Africa with a + 27% increase, the third one Oceania with a + 13% increase, and then North America with + 6%, Latin America + 3%. Europe (the EU) was the only region that showed a—2% decrease. At the same time, i.e. in the 10-year period 2010–2019, the share of the emissions from international aviation (air transport) increased by + 37% and from maritime transport by + 10%.

For the 20-year period 2000–2019 the above percentages are higher e.g., the growth of CO_2 emissions for Asia over this 20-year period was + 117% while those of Africa increased by + 105%, Latin America by + 39% and Oceania by + 32%. For the EU the overall increase was + 7% while for North America the increase remained practically the same (+ 5%). For international aviation the 20-year increase was + 77% and for international maritime + 47% (SLOCAT, 2021). Road transport (passenger and freight) contributed 77% of global transport CO_2 emissions in 2019 (Crippa et al., 2022).

So, before the COVID pandemic, the world experienced large increases in transport related CO_2 emissions with the highest increases being shown in Asia, Africa, and Latin America in that order. International Aviation and Shipping emissions also showed steep increases. The COVID pandemic abruptly reduced traffic flows and

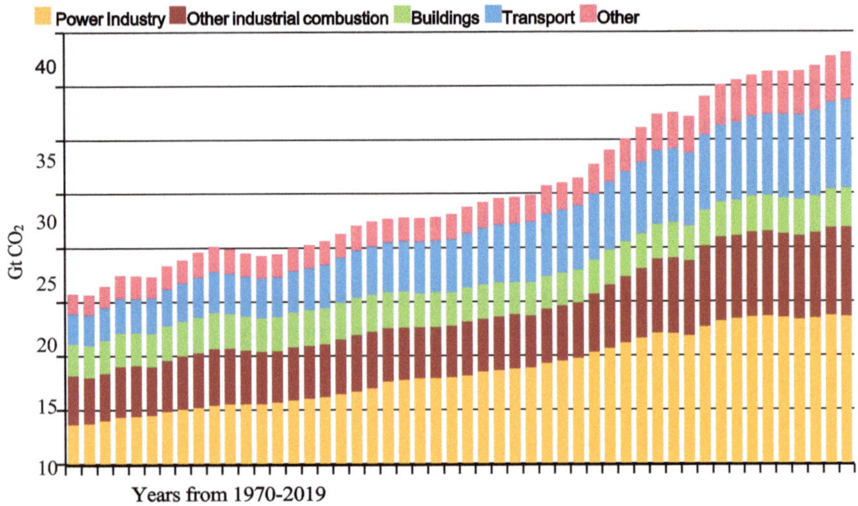

Fig. 3.1 Global annual emissions of fossil CO_2 in Gt CO_2/yr by sector from 1970 to 2019. *Source* European Commission's Joint Research Center (JRC, 2020; Crippa et al., 2022)

Table 3.1 Percent change in transport CO_2 emissions by region and in international aviation and shipping

Area or sector	10-year period 2010–2019 (%)	20-year period 2000–2019 (%)
Africa	+ 27	+ 105
Asia	+ 41	+ 117
Europe	− 2	+ 7
Latin America	+ 3	+ 39
North America	+ 6	+ 5
Oceania	+ 13	+ 32
International aviation	+ 37	+ 77
International shipping	+ 10	+ 47

Source SLOCAT (2021) based on Crippa et al., (2022)

mobility and by consequence caused large reductions in GHG emissions from the transport sector (of the order of − 50%) but only temporarily, i.e. for the years 2020 and (part of) 2021. After 2021 the transport sector had the strongest and faster rebound in emissions than any other sector (Cardama et al., 2023). As regards the expected effects of the proposed decarbonisation measures, the movement restrictions during the COVID-19 period and the resulting reductions in GHG emissions, can serve as a "window into a future" showing what could happen if the measures and policies for carbon-free mobility that are planned by the various governments are fully implemented.

Under business as usual, the global transport activity is projected to nearly double by 2050, rising 1.8 times for passenger transport and 2.0 times for freight transport compared to the 2019 levels (Boehm et al., 2023; SLOCAT, 2023). The current trends of car ownership, at global level, also point at the same direction i.e., global motorization rates will increase steadily year by year with the estimated global figure of car ownership to reach an estimated 450 vehicles per 1000 population if left unabated. Currently, in 2024, the global car ownership level was at 300 vehicles per 1000 people, from 243 vehicles per 1000 people in 2015 (Boehm et al., 2023). If shared mobility and mobility as a service are fully and globally introduced, global car ownership (and by consequence, use) is expected to decrease to 1990 levels i.e. approximately at a level of 100 vehicles per 1000 population by 2050 or 60.

According to the *Sustainable Low Carbon Transport* partnership, achieving carbon transport pathways that would be compatible with the limit of global warming to no more than + 1.5 °C by the end of the century, would require a 60% global reduction in transport-related CO_2 emissions by 2050 as compared to 2020 levels and at the same time the maximum "acceptable" increase in passenger transport activity, in the period 2020–2050, should be no more than + 50% and in freight activity no more than + 20%. Fossil fuel dependence in road transport would also need to decline drastically, from more than 90% in 2020 to 10% by 2050, with electricity becoming the dominant transport fuel by the early 2040s (SLOCAT, 2023). At the time of writing, none of these targets seems to be achievable and indications are that the + 1.5 °C limit will have been surpassed as early as by 2030 or 2035…

As already mentioned, there are great variations in the emissions patterns among the various regions of the world. As regards the emissions from the transport sector, high-income countries are responsible for most of the transport CO_2 emissions, while low-income countries contribute substantially less (IEA, 2024a). According to the IEA data, in 2021 high-income countries were responsible for 50.7% of transport CO_2 emissions world-wide, while low-income countries contributed less than 1%! The rest was contributed by middle-income countries where the per capita transport CO_2 emissions have doubled since 1980, while barely changing in low-income countries (IEA, 2024a).

Finally, as regards aviation and maritime transport emissions, in 2022 the CO_2 emissions from aviation (domestic and international) were 2.4% of the global CO_2 emissions (from all sectors). The non-CO_2 emissions from aviation were 3.5% of the total global ones. In aviation, it is worth noting that an estimated 1% of the world's population is responsible for more than 50% of the global CO_2 emissions from passenger air travel (SLOCAT, 2023). In the same year, 2022, maritime transport (domestic and international) accounted for 2.5% of the global CO_2 emissions. For both, aviation and maritime transport, the CO_2 emissions after 2022 were on the rise reaching the pre-Covid 2019 levels in 2023. The return to 2019 levels, by both air and maritime transport in 2023 shows the trend of a dynamic emissions growth in these two sectors. It is therefore no surprise that these two sectors are subject to increased scrutiny as regards their emissions which are not only CO_2 but contain several other pollutants such as methane, nitrous oxides, hydrofluorocarbons, perfluorocarbons, sulphur hexafluoride and nitrogen trifluoride.

Against this situation, the IEA (*International Energy Agency*), estimates that getting on track with the net-zero emissions scenario by 2050 would require that transport GHG emissions to fall by about 25% by 2030, to around 6 Gt/year, even with the anticipated growth in transport demand (IEA, 2024a). On a similar tone, the *Intergovernmental Panel on Climate Change* (IPCC) in its 6th Assessment Report estimates the necessary reduction, for transport CO_2 emissions, to be around − 20% by 2030 (IPCC, 2023). Achieving these reductions will require the formulation and implementation of an ambitious set of policies with specific measures and actions—primarily by the most advanced economies—as soon as possible and in the outmost urgency. It is to these potential measures and policies that we turn our attention now so that the scale and magnitude of the, urgently, needed actions become clear.

3.4 Measures and Policies for Net-Zero Transport Emissions

3.4.1 Introduction and Frame of Reference

Aligning transportation with the 2050 net zero scenarios i.e., for zero carbon emissions for the whole sector, would require implementing a broad set of policies to promote the use of non-carbon-intensive travel options and at the same time implement operational and technical energy-efficiency measures to reduce and eventually zero the carbon intensity of all transport modes. Repeated studies by national or international agencies have highlighted the types of transport decarbonisation measures and policies that are needed and most of these are now widely accepted. We have synthesised these measures and policies based on the relevant studies and reports by the following four organisations:

(a) The *European Academies Science Advisory Council* (EASAC). This is an organisation formed by the national academies of science of the EU Member States, plus Norway, Switzerland, and the United Kingdom with the purpose to enable them to collaborate in providing independent science advice to European policy-makers (www.easac.eu). The Energy Steering Group of EASAC commissioned a working group of experts to produce a report on the decarbonisation of the Transport sector in Europe. The report was submitted in 2019 and remains since then perhaps the most comprehensive scientific report of the potential measures and policies for the full decarbonisation in the transport sector by 2050 (EASAC, 2019).
(b) The *International Energy Agency* (IEA). Founded in response to the 1973–1974 oil crisis, this organisation has by now become the global authority on a broad range of energy issues including climate change and decarbonisation (https://www.iea.org/energy-system/transport). Its reports on Transport and Mobility (IEA, 2024a) as well as on the global energy transitions stocktake (IEA, 2024b) are well respected and relevant.

(c) The *Sustainable Low Carbon Transport partnership* (SLOCAT). This is a multi-stakeholder partnership of over 90 organizations, NGOs, Foundations, Academia, and the business sector (https://slocat.net/transport-sdgs/).
(d) The *McKinsey organisation.* A non-for-profit international organisation aiming to support policy makers as well as individuals and other non-profits in "unlocking the global potential for a more sustainable and inclusive world" (https://www.mckinsey.org/home). The McKinsey organisation regularly publishes reports on various subjects including transport and its potential pathway to net zero decarbonisation (e.g., McKinsey, 2023).

A useful frame under which one can consider the relevant transport decarbonisation measures and policies in a holistic way, is the one initially suggested in (Dalkmann & Brannigan, 2007) and utilised in IEA and EASAC reports (EASAC, 2019). It refers to the portfolio of measures and policies under the terms: ***"avoiding"***, ***"shifting"*** and ***"improving/deploying"*** (Fig. 3.2). As the meaning of these words implies, they are broad categories of measures and policies that, in the "*avoid*" case aim to make the users of the transport system avoid making trips or sending freight, so as to reduce the overall transport demand, in the "*shift*" case to induce shifting to less GHG emitting modes, and in the "*improve/deploy*" case to deploy new mobility technologies that will enable low or no GHG emissions.

We choose this frame of reference to present in the following our synthesis report of the most prominent and well discussed measures and policies for decarbonisation in the transport and mobility sector.

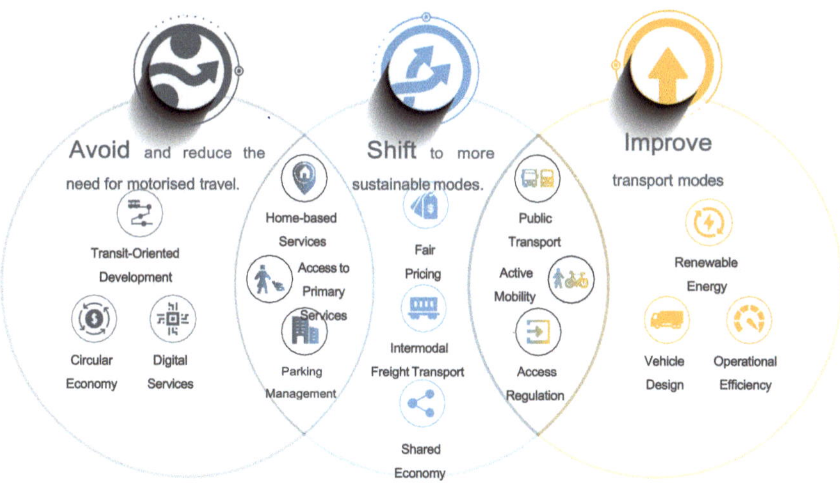

Fig. 3.2 The A-S-I frame of reference ("Avoid", "Shift", "Improve") for measures and policies for the decarbonisation in the transport sector (the measures and policies shown above are examples given for illustrative purposes). *Source* SLOCAT (2023), Fig. 15, based on Dalkmann and Brannigan (2007)

3.4.2 *Avoid Type of Actions*

3.4.2.1 Passenger Transport

Convincing people to abandon their trip making habits and reduce the number of their daily trips, is perhaps the most difficult and long-term set of options. Most of these actions form part of the *demand management* policies that were mentioned earlier. A set of valid incentives that aim to induce people in travelling less, or at off-peak hours, must be formulated to make people reconsider their options. The same incentives or their expansion could be used to make people choose less-polluting modes for their trips or even reallocate their residence closer to their work to minimise travel distances/times, and so on. The incentives must be well researched and documented followed by a wide and consistent information dissemination campaign aimed at informing and convincing people that the changes are indeed necessary. When the incentives do not work or if the pollution situation is beyond bearable limits (probably at the confluence of both), the incentives can turn to direct restrictions in the corresponding harmful trip making practices.

We can refer here to three of the most relevant types of "avoid" actions which, subject to taking special attention to provide exceptions—or special provisions—for vulnerable citizens such as low-income groups, the sick, the handicapped or the elderly, are the following:

A1: *Teleworking/tele meeting*. The COVID pandemic gave a big push to the practice of teleworking which is now here to stay since more and more companies as well as governmental organisations or individuals choose to tele-work or tele-meet as a substitute to a physical presence at work or at a meeting. Teleworking has been proved to be beneficial for both the employing organisation as well as for the employee (Haan & Main, 2023). In China, a recent study by researchers at Tsinghua university found that in Beijing 84% of the employees are positive about choosing teleworking, or some other form of flexible work mode, and that teleworking can lead to an average of 1.32 million metric tons of CO_2 reduction per year in Beijing alone. This is 7.05% of the total carbon emissions by road transport in that area (Li et al., 2023). Studies in the US reveal similar results. A study by the *Forbes Advisor* showed that in 2023 in the US, 12.7% of full-time employees work from home while 28.2% work a hybrid model. A 16% of companies are operating without a physical office while the percentage of workers that want to work remotely, at least for some of the time, is as high as 98% (Haan & Main, 2023). In Europe, the overall percentage of workers teleworking varies by country but generally it is quite high e.g., 34% in Italy, 25% in Germany (35% in the services sector) and so on (Eurofound, 2024). The sector with the highest percentage of teleworking is, understandably, the Information Technology (IT) sector.

A2: *Relocate the home residence closer to work*. This can be quite an effective action to reduce the distances travelled every day (and thus reduce fuel consumption and emissions) but it is also the most difficult to implement. It entails the organised and incentivised relocation of workers close to their workplace to avoid long commuting hours to and from work. So far, only some very big companies have attempted relocation of employees (e.g., Amazon) but so far, the measure has not been attempted by any local or central (federal) authority to any significant scale.

A3: *Restricting, wholly or partly, or charging for certain types of trips.* Several countries have introduced in the past, prohibitions or restrictions on daily trips to and from certain well designated areas to reduce air pollution levels (mainly in congested city centres). For example, in the Beijing urban area the circulation of cars is permitted based on the last digit of their number plate numbers. Every weekday only two such numbers are permitted to circulate (electric vehicles are excluded). Similar types of prohibitions exist in Rome, Athens, and other cities in Europe as well as in Asia. In other cases, all car trips within a certain geographic area, usually the urban centre, are either prohibited or charged for entry (the practice is known as *congestion charging* or *congestion relief zones*). As the various milestone years for climate change mitigation measures approach and the expected results of mitigation measures are not satisfactory, such prohibitions may have to be applied more often and more rigorously.

3.4.2.2 Freight Transport

Effecting "avoid" measures in freight transport involves, primarily, behavioural challenges for those who own the goods to be transported. They will need to learn to avoid using transport modes that emit heavily or avoid making the transport altogether and replace it with alternative actions. The freight "avoid" options can be grouped into the following three broad categories with some indicative measures mentioned for each one of them (McKinnon, 2018):

A4: *Reduce the demand for vehicle-km on the roads,* by:

- Change the location of the production of goods and the procurement of materials (suppliers) to bring them closer to consumption.
- Create more inventories (stock of finished goods or materials used in production) and place them closer to the production/consumption sites.
- Increase the carbon emission mitigation costs assigned to supply chains.
- Increase the use of digital technologies such as smart tags, internet of things, big data analytics and AI software tools for supply chain management, asset tracking and delivery scheduling.
- Greater and wider use of additive manufacturing (3D printing) to avoid the physical movement (transport) of certain goods.

A5: *Prioritise goods with increased "material efficiency"*, by:

- Construct products using modules from which, at the end of their lives, it is easy to recover materials and components that can be re-used in the frame of the future circular economy.

- Replace physical consignments with electrons by using digital technologies. For example, distribute information, stories, music, and films digitally instead of in books, magazines, newspapers, compact discs and DVDs, which require transport and supply chains for timber, pulp, paper, chemicals, plastics and dyes.
- Promote the use of products with less material content (miniaturization, light-weighting).
- Use 3D printing technologies to eliminate physical transportation of goods.

A6: *Reduce the growth in demand for transported goods*, by:

- Promote the use of goods with long lives, which can be readily repaired or renovated.
- Require product suppliers to pay for end-of-life disposal costs.
- Encourage consumers to shift their non-essential expenditure from the consumption of material goods towards recreation and other experiences, which do not require significant use of freight transport. This is also called "*substituting experiences*", but it must be applied with extreme care not to increase passenger transport flows—especially air transport—and in this way increase instead of lower GHG emissions.

The demand for freight transport services after the COVID pandemic has increased rapidly and by 2023 the volumes transported were back to pre-COVID levels. For the EU, the earlier forecasts for a more than 30% average overall increase in freight volumes by 2050 as compared to 2019, are still valid and if there is no other major disruption in economic activities these forecasts may prove even modest (EC, 2016). For 2023 the modal split figures for European freight were (approximately), 80% of all inland freight transported by road (compared to 76.7% pre-COVID), 17% used rail and the remainder used inland waterways. Correspondingly, the share of freight Transport in the CO_2 emissions in the EU of 27 member states was of the order of 7–8% of the total (passenger and freight) transport sector emissions (Rodrigue, 2024).

3.4.3 Shift Type of Actions

3.4.3.1 Passenger Transport

These types of measures aim to induce passenger transport users to shift from carbon polluting transport modes to non-polluting ones. There is a wide range of such measures and policies the most widely used and accepted of which is given in the following. Their impact, in terms of percent reductions in emissions, cannot be given as it depends on the prevailing local weather and application conditions as well as on the specific configuration of the measures to be applied.

S1: *Incentives for increased use of (low carbon) public transport* (PT) modes and *shared transport services*. PT related incentives include properly located and easy to access and use intermodal transfers, real time travel information systems, easy-to-use multi-operator ticketing systems, subsidised low fares for PT services or even introducing no-fare policies.[1] The current experience shows that merely reducing or even eliminating tickets may not be sufficient as a stand-alone policy for inducing substantial numbers to use public transport services instead of their car. Complimentary measures such as, more stringent restrictions or increased charges for parking, provisions for the last-mile section of the PT trips, and even outright restrictions in the use of ICE cars will also be necessary. As regards *shared mobility services* there is, as already mentioned, a large body of research work as well as applications that demonstrate the value of shared mobility for reducing urban vehicle trips and congestion. There are also already in operation many shared mobility apps and shared mobility service applications (Giannopoulos & Li, 2024; Heineke et al., 2021; Jia-Wei & Creutzig, 2021; Liu et al., 2022).

S2: Development of safe cycle lanes, pedestrian zones, and walkways to promote micromobility. Micromobility ("slow traffic system" in Chinese), has become a modern way of traveling in congested urban streets used mainly by young or relatively young persons. It includes several types of "light" travel modes such as walking or jogging, bicycle, tricycle, electric bicycle or scooter riding, roller skating, and similar. This is a slow traffic system with speeds below 15–20 km/h which nevertheless requires use of a well-planned and implemented network of cycling lanes and walkways. Many cities around the world have implemented such infrastructures and are rigorously promoting the use of micromobility modes with Copenhagen being the best example in Europe and probably one of the best in the world. In Chinese cities, too, micromobility is used widely (Liao & Correia, 2021). There are, however, increasing doubts about this type of transport mode as it tends to generate many accidents involving the young persons who use this type of mode. For these reasons, micromobility is becoming the subject of strict safety regulation in many countries.

S3: Implementation of new mobility services and initiatives. These include *Mobility as a Service* (MaaS) packages offered by appropriate businesses operating under old or new business models, offering tailor-made mobility services to users and employing "clean" (electric) or automated vehicles. MaaS have already started to be implemented and are expanding in several parts of the world (China, the US, the EU, Japan). They are expected to expand further in the medium-term future. In China, the Beijing's MaaS scheme is operating under the *Beijing Municipal Commission of Transport* (BMCT) through a public–private collaboration model, leveraging companies such as *Gaode Maps* (AutoNavi) and *Baidu*. Through the *Beijing MaaS platform*, operators and mobility service integrators provide integrated mobility services

[1] The abolition of fares for public transport and the replacement of the necessary funding by other sources is an increasingly researched practice and has already been applied in several cities in Europe e.g. Luxemburg, Dunkirk (France), Tallinn (Estonia), Cluj-Napoca (Romania), Aubagne (France), Avesta (Sweden) and recently for the whole of Germany (introduction of the "9 Euro ticket"). The first of the current authors has suggested this measure for the Athens public transport system since the early '80s.

across an impressive array of transport modes and services (bus, metro, commuter train, cycling, ride-hailing, ridesharing). Even aviation, railway, and long-distance bus services are included in this platform. Their aim is to provide citizens with a comprehensive and convenient "door-to-door" smart mobility encompassing pre-trip smart decision-making support, in-trip guidance, and intercity travel planning services (WRI-Ch, 2024). In the EU, the *European MaaS Alliance* has been formed several years ago by prominent European businesses with the aim of promoting the implementation of MaaS in European urban areas (https://maas-alliance.eu/). The potential benefits from the full introduction of MaaS may be many and varied ranging from the reduction of peak hour travel to the reduction of space needed for parking etc. (see also previous section). Rebound effects may occur in the sense that if MaaS would result in making transport faster, cheaper, and more convenient it may generate new demands for trips that were not present before and consequently reduce the initial benefits.

S4: Develop and implement advanced traveller information systems. Smart, real time, dynamic, and personalised travel information is the main outcome of these systems whose purpose is to inform, influence, and ultimately manage a more environmental and climate sensitive and sustainable travel behaviour. The information is formed in real-time by analysing the collected big data by several sources as it was discussed earlier. AI can play a primary role here and be a game-changer especially within an overall intelligent and real time traffic and travel management environment.

S5: Develop the smart—AI based—urban environment. This is known as a *"smart city"* i.e. a technologically state-of-the-art urban area that uses different types of AI based electronic services and sensors to collect and analyse data so as to produce real time information to feed-in user information channels or data to manage assets, resources, and services efficiently (https://smart-cities-marketplace.ec.europa.eu/). A principal characteristic of a smart city is the *positive energy district* (PED) concept i.e., a district that produces at least as much (clean) energy as it consumes on an annual basis.

S6: Develop holistic, well-integrated strategic plans for sustainable clean mobility. The use and effectiveness of any of the previously mentioned measures and policies for decarbonisation and sustainability will be maximised if they are all embedded within a well-designed and integrated plan developed after a thorough study of the local conditions. By local conditions is meant not only the physical and functional characteristics of the available infrastructures but also the administrative and legal environment for implementation of such measures and the estimated economic, social, and environmental implications of the plans for decarbonised mobility. An example of such type of plans from the EU, is the *Sustainable Urban Mobility Plans* or SUMPs that all urban areas in the EU, even smaller towns, are recommended to develop and implement. The idea of creating an overall plan, aims at adopting and implementing not one but a comprehensive set of sustainable urban mobility measures and implement them simultaneously, so that their carbon emission impacts are maximised. AI has an important role to play here, both in the formulation and documentation of the specific measures to be included in the plan and in their evaluation as well as in monitoring and coordinating their implementation.

3.4.3.2 Freight Transport

Freight transport emissions data show that road freight modes have the second highest emission levels per t-km. The highest emissions per t-km come from air freight transport, while trains and large container ships produce the lowest (EASAC, 2019, Table 2.2). For road freight, GHG emissions depend primarily on the type of trucks and fuel used but also on factors such as, the size of the vehicle and of the shipment, the degree of utilisation of the loading capacity of the vehicles, the energy vector used by the trucks (diesel, LNG, H_2 or electric) and the overall driving conditions e.g. whether the driving is made in urban or inter-urban roads and whether the driver follows eco-driving rules or not. Therefore, the *Shift* actions in the freight transport sector take a broader meaning i.e. not only shifting modes from road to other cleaner modes but also shifting within the road sector to vehicles with higher load carrying capacities or to increasing the load factors of existing vehicles through sharing or pooling, and so on. Summarising the multitude of potential actions, the following measures for freight transport "shifting" actions can be mentioned:

S7: Promote and pursue a modal shift to "cleaner" modes. This is the traditional—so to speak—policy of promoting the transfer of freight from road to rail or inland waterway transport. This is one of the first policies adopted by the European Commission (since its first Transport White Paper of 2001) but so far without great success. Several studies have tried to explain why these policies failed to achieve substantial modal shifts (Kemp, 2016; Nelldal & Andersson, 2012; Nordenholz et al., 2016).

S8: Incentivise and promote the use of "clean" vehicles for road freight. Here, the aim is to cause a substantial shift from polluting internal combustion trucks for road transport to electric or hydrogen ones. In the EU, the Euro 7 Regulation is the legislative act that regulates the permitted emissions from internal combustion engine trucks, vans, and cars pushing for minimisation of their pollutant emissions. The same Regulation introduces a direct ban, from the EU's internal market, of all internal combustion engine heavy-duty buses and trucks from 2035 onwards. Incentives for the use of cleaner road vehicles may include imposing a pollution tax on road freight performed with non-clean vehicles or monitoring and restricting the use of internal combustion trucks in certain pollution vulnerable areas, imposing legal restrictions on the transport of certain goods by road, and so on.

S9: Promote the systematic improvement of road vehicle load factors and use of additive manufacturing. The aim here is to reduce the number of freight vehicles on the road by better packaging arrangements, reduction in size (miniaturization) and light-weighting of final products, or by a more efficient grouping of delivery destinations and/or recipients. Reducing the number of trucks on the roads can also be made by promoting additive manufacturing practices also known as 3D printing.

If additive manufacturing (3D printing) becomes a widespread practice globally, decreases in road freight transport energy demand from 5% to as much as 27% have been estimated (Verhoef et al., 2018).

S10: Improvement of road vehicle routings. By implementing "environmentally sensitive" routings for the delivery trucks, in urban or interurban areas, rather than choosing to satisfy other routing criteria some reduction in GHG emissions by these vehicles can be achieved. This applies to all types of road freight movements either long-haul, or short haul.

S11: Train the truck drivers to eco-driving and platooning. By observing eco-driving rules, road freight vehicles can reduce emissions by as much as 15% (EASAC, 2019). These rules entail using steady speed with the minimum of acceleration or deceleration, avoiding roads with steep gradients, avoiding breaking with the engine, and so on. With the advent of commercially available Level 3 and up automated vehicles, truck platooning on motorways is now possible and it takes place in many countries. Truck drivers must be trained for the platooning form of driving so that they choose to use it as it is another (and better) form of eco-driving. For truck platooning, the European Automobile Manufacturers Association estimated back in 2017, when this type of movement first started, reductions of up to 10% in truck related emissions (ACEA, 2017).

S12: Develop S*ustainable Urban Logistics Plans* (SULPs) that like their passenger transport counterparts (the Sustainable Urban Mobility Plans—SUMPs) will be the strategic plans containing a complete set of measures and policies to address the decarbonisation of urban logistics. The measures to be involved are the ones already mentioned earlier (or similar). For Europe the practice of SULPs as well as their main contents and recommended procedures can be found in (NOVELOG, 2019).

The difficulties of achieving the kind of GHG emissions reductions in the freight transport sector, that would be compatible with the overall goal of limiting global warming to + 2 °C by the end of the century, have been demonstrated in a book by Prof. M. McKinnon as early as 2018 (McKinnon, 2018). In this book, it is claimed that for achieving the target of net-zero transport related emissions by 2050, a sixfold reduction in road freight transport emissions will be needed between 2018 and 2050. To achieve such level of reductions, at EU level, a plausible scenario would require the following sub-sector targets to be achieved, at minimum, in the period until 2050 (McKinnon, 2018):

- 30% modal shift from road to rail,
- 20% increase in routing efficiency,
- 30% higher load factors,
- 50% greater energy efficiency of the vehicles used, and
- 50% less carbon content in the energy vectors.

Achieving any of these sub-targets individually by 2050 would be a very challenging task, but to achieve all of them simultaneously would be a tremendously difficult task—a scarcely realistic expectation.

3.4.4 Improve/Deploy Type of Actions

The most pronounced measures and policies in this category, involve options that are common for both passenger and freight transport and involve the improvement of the existing types of ICE vehicles or deployment of totally new ones with carbon-free emissions (electricity, hydrogen, or other) and construction of their relevant infrastructures. A large section of these measures are primarily transitional i.e., aimed at improving the existing vehicles for better aerodynamics, light-weighting, energy efficient tyres, etc., or improving the energy efficiency of their powertrains including hybridisation. For most of the measures that refer to road vehicles, existing research can give us estimates of their potential impacts and these are mentioned together with the relevant options. The main relevant measures in this category are listed in summary form in the following.

I1: Improve vehicle body design. There are three types of actions that are included here for reducing energy consumption and emissions:

(a) *Use of light materials to reduce the weight of vehicles.* Reducing the weight of a vehicle can typically reduce the energy consumed by up to about 10% (SCEER, 2017). Reducing vehicle weight has also been shown to reduce the emissions of particulates (PM_{10} and $PM_{2.5}$) that are typically emitted from tyres and brakes, and this is an additional benefit in terms of improved air quality (Timmers & Achten, 2016).
(b) *Improving the aerodynamic performance.* This can reduce the energy consumed by up to 10% depending on the type of vehicle. For example, using devices to reduce "air drag" for trucks on long-haul routes can produce reductions in the fuel costs of a large truck of between 1 and 9% for a vehicle doing an annual mileage of 80,000 miles (Zulfaa & Filippone, 2010).
(c) *Improving the rolling resistance,* of the vehicle by using energy-efficient tyres. This can reduce energy consumption and emissions especially of small particulates by up to 10% for light duty vehicles (LDVs) as well as passenger cars (SCEER, 2017).

For heavy duty vehicles (HDVs) of the Euro 7 category, most of the above options is typically included in the design. Generally, the above improvements are giving better results, the larger the vehicle. So, they are more pronounced for HDVs rather than LDVs.

I2: Improve conventional powertrains. Improvements to conventional powertrains aimed at reducing carbon emissions are already a priority for vehicle manufacturers. Despite their growing focus on producing electric vehicles, car manufacturers are also focusing on improving conventional powertrains as these are expected to continue serving the conventional vehicle fleet in the short-to-medium term. The level of CO_2 emissions of conventional powertrains is being reduced and going below the 100 g/ km limit. The ultimate in terms of further reductions in emissions is estimated to be a further 25% improvements in reductions i.e. somewhere around the 75 g/km limit. This will be reached primarily through improved powertrain design and hybridisation (Cox, 2018; Elgowainy et al., 2018; Onder et al., 2011).

I3: Reduce the carbon content of vehicle fuels. The carbon content of vehicle fuels can be reduced by mixing or substituting them with less polluting alternatives such as *methane*, *natural gas* or renewable fuels such as *biofuels*. In European road transport, only approximately 5% of existing internal combustion engines run on methane. In China, internal combustion engines (ICEs) that run on methane are part of a broader strategy to reduce greenhouse gas emissions and transition towards cleaner energy sources. Methane-powered ICEs are being used across various sectors such as urban public transport and shipping. The methane comes from fossil sources but is a relatively clean fuel offering a cleaner alternative to traditional fossil fuels for the transitional period. It can be stored on the vehicle in tanks either as compressed natural gas (CNG) or as liquefied petroleum gas (LPG). Spark ignition internal combustion engines (ICEs) can be optimised to burn methane with a high compression ratio and then perform with efficiencies which are up to 25% higher than similar engines burning gasoline (van Basshuysen, 2015). The use of biofuels in the transport sector will need to be prioritised for applications in which alternative low carbon solutions with competitive costs are difficult to find, such as in aviation and long-haul shipping.

I4: Use of alternative energy vectors: Battery Electric Vehicles (BEV)

The market for BEVs is growing but—for the time being—it is negatively impacted in many countries by their limited driving range and a lack of battery charging facilities. Passenger cars and vans that run entirely on electricity from batteries typically have today a driving range of 300–400 kms under real-world conditions. This makes many car owners to hesitate to buy BEVs because they also want to use their vehicles to make much longer journeys and worry about finding a battery charging station. Many BEV owners can charge their vehicles overnight at home using “slow” chargers (usually 11 kW) but when batteries must be charged very quickly, for example at motorway service stations, then “superchargers” with powers of up to 120 kW need to be used.

As the numbers of electric vehicles grow, smart charging systems with time dependent tariffs will be needed to manage the supply of power to BEVs because otherwise the peaks of total power demand for BEV charging can overwhelm the power distribution network and can reach as much as twice the current peak electricity demand. For the batteries, continued research and innovation are expected to bring the development of advanced cathode and anode materials for batteries and new battery chemistries such as for example the sodium-ion (Na-ion) batteries which are built on lower value materials than the current Lithium ion (Li-ion) ones. The first electric car to use Na-ion batteries was on the market in 2023 (the BYD Seagull). A special, most notable, category of BEVs is electric buses used for public transport services in urban areas as well as electric trucks. The market for battery electric buses is growing (most of them are now manufactured and operating in China). Electric trucks accounted for just over 1% of global truck sales in 2022, a sales share that should reach around 30% by 2030 in the net-zero scenario.

I6: Use of alternative energy vectors: Plug-in Hybrid Electric Vehicles (PHEV)

These vehicles are fitted with both a conventional internal combustion engine and a battery, which can be charged either by the engine or by being connected to the electricity grid. They can therefore run on electricity for short journeys but can also be used on long journeys without having to be recharged at public charging stations. Most PHEVs have short electric driving ranges of 20–85 kms, so they can be used mainly as urban electric vehicles because car journeys in urban areas are normally of the range of 25–40 kms (JRC, 2013). The batteries in PHEVs are typically much smaller than those in BEVs, so the CO_2 emissions embedded in these batteries (and consequently their life cycle emissions) are lower (Cox, 2018).

I7: Use of alternative energy vectors: Hydrogen and Fuel Cell Electric Vehicles (FCEV)

Hydrogen can be used as fuel in internal combustion engines, where its performance has been studied for many years, but it is more efficient to use hydrogen in fuel cells to produce electricity that will then move the electric motor very much as in a battery electric car. The gravimetric energy density of hydrogen is higher than that of batteries, so the driving range of hydrogen FCEVs is typically longer than that of BEVs. Also, filling a fuel tank with hydrogen can be done much more quickly than the re-charging of a battery. However, the efficiency of FCEVs across the whole energy conversion chain from electricity generation to wheel is 2–2.5 times lower than that of BEVs (EASAC, 2019). The costs of fuel cells are also still relatively high. *Hydrogen* is largely produced today from natural gas by steam methane reforming. This involves large GHG emissions of approximately 10 kg of CO_2-eq per kg of hydrogen produced (EASAC, 2019). In the future, "green" hydrogen will be increasingly produced by electrolysis of water using low carbon electricity (e.g. from hydro, nuclear, solar or wind power). Some major industry sectors, such as steel making, are already investing in research on reducing the costs of green hydrogen production and storage to meet their own decarbonisation obligations. The full commercial application of hydrogen cars or hydrogen FCEVs is likely to occur at a later stage (most probably beyond 2035) when the current problems regarding the manufacture of "green" hydrogen and its transportation to fuel stations and the vehicle, have been solved. Some international automotive vehicle manufacturers are developing hydrogen vehicles today mainly heavy-duty road vehicles and inter-city buses. For the time being, the use of hydrogen in transportation is likely to be limited to those vehicles for which battery power is not well suited, such as heavy-duty long-distance trucks, long distance buses, trains on non-electrified routes, ferry boats and medium distance ships.

I8: Construct the refuelling infrastructures for items I4–I7

The relevant refuelling infrastructures for the fuels mentioned in I4–I7 above are of critical importance for the successful implementation and proliferation of all "new energy" sources such as battery electric vehicles as well as hydrogen ones. In the EU, new energy refuelling infrastructures are regulated by the *Alternative Fuels Infrastructure Regulation* (AFIR) which was issued in July 2023 (Regulation no. PE-CONS 25/23/Brussels 13 July 2023). According to this Regulation all member countries of the EU should ensure:

- Fast recharging stations of at least 150 kW for cars and vans to be installed every 60 kms along the EU's main road transport corridors from 2025 onwards.
- Recharging stations for heavy-duty vehicles with a minimum output of 350 kW every 60 kms along the core EU motorway network and every 100 kms on the larger (so called comprehensive) road network from 2025 onwards, with complete network coverage by 2030.
- Hydrogen refuelling stations serving both cars and lorries in all urban nodes and every 200 kms along the EU's core motorway network from 2030 onwards.
- Maritime ports servicing a minimum number of large passenger vessels, or container vessels, by 2030.
- Airports, for stationary aircraft at all gates by 2025, and at all remote stands by 2030.

Similar regulations exist for China where the relevant regulatory authority, the *China Certification Centre for Automotive Products* have issued two new regulations the GB/T 18487.1-2023 on Conductive charging system for electric vehicles—Part 1: General Requirements and GB/T 27930-2023 Digital communication protocols between on-board conductive charger and electric vehicle. These have now come into force on 1 April 2024.

I9: Electric road systems (ERS): According to this technology, vehicles powered by electric motors are continuously supplied with (green) electricity as they drive along the road through overhead wires or through power cables impeded in the pavement below. This can be done by using overhead lines and pantographs, or conductor rails buried in the road. Inductive connection options are developed to provide a continuous supply of electricity to moving vehicles. The overall efficiency of *electric road systems* depends on the effectiveness of their connection to the electricity grid and on the extent to which they employ on-board batteries. In general, they can be significantly more efficient than BEVs.

I10: Technologies to reduce the emissions from ships. For ships, the use of alternative fuels like LNG (Liquified Natural Gas), or the various biofuels including bio-methanol and biodiesel options are now being introduced as well as other more innovative technologies like wind propulsion or assistance (re: International Windship Association—https://www.wind-ship.org/). For existing powertrains, retrofit of ship engines to reduce carbon emissions is also an option and promising demonstrations have been made as well for alternative fuels such as hydrogen, ammonia, or methanol. It must be mentioned here that battery electric ships (BES) are being tested for short to medium distances. Already the Chinese state-owned company *COSCO*

Shipping has launched what it calls the world's largest river-to-sea electric container ship. It is the *Green Water 01*, a 10,000 + -ton fully electric ship that sets a new benchmark in sustainability in the marine logistics industry. Its battery capacity is 50,000 kWh and this can be increased by adding battery modules depending on the length of the voyage at sea. The Green Water 01 has commenced weekly service between Shanghai and Nanjing (as published in www.electrek.co, May 2, 2024).

I11: Technologies to reduce the emissions of airplanes. In air transport, entirely new airframe and engine designs for aircraft such as blended wing-body aircraft or ultra-high bypass ratio jet engines are claimed to reduce fuel burned by more than 20% (Barbosa, 2021). Also, hybrid electric aircraft (Xie et al., 2021) and aircraft running on hydrogen (AIRBUS Zero-e concept aircraft) including via direct combustion, via onboard fuel cells, or a hybrid of the two, are currently at various stages of conceptual design and prototyping.

I12: Direct bans for air polluting vehicles. The ultimate measures in this *improve/deploy* category will inevitably be—at some stage—direct bans on polluting internal combustion engine (ICE) vehicles running on fossil fuels. The movement to ban ICE vehicles in urban areas is gaining momentum. The EU has approved a law to ban all new sales of ICE vehicles from 2035 onwards and several US states are planning to do the same with the lead of California. In China there are no such policies announced yet, but the sale of new ICE vehicles is strongly regulated with strict quotas being placed on new ICE vehicle licences. A version of the total ban would be the banning of ICE cars from specific areas such as city centres. A milder version is the exclusion of ICE vehicles from specific lanes on busy roads unless they have more than a given minimal number of passengers on board.

3.4.5 The Importance of Clean Power Generation

The availability of clean electricity generation is a sine qua non condition for all actions involving the introduction and use of electric vehicles. This is revealed when using *Life Cycle Analysis* (LCA) to detect and trace back the totality of GHG emissions of battery or other types of electric vehicles (Qiao et al., 2019). Figure 3.3 shows the results of a specific LCA analysis reported in (Cox, 2018) in which the overall life cycle emissions (in gr CO_2-eq/km) of various types of road transport vehicles are presented in relation to the carbon intensity of the electricity grid (in gr CO_2-eq/kWh). As shown in that Figure, the higher the carbon intensity of the grid electricity system, the higher the total life cycle emissions of the various types of electric vehicles. According to these data, when the GHG footprint of the electricity grid is approximately 700 gr CO_2-eq/kWh or greater, BEVs can be more polluting in terms of their total LCA emissions than ICE vehicles using fossil fuels.

A similar crossing point between BEVs and hybrid electric gasoline vehicles (HEV-gasoline) occurs when the GHG footprint of the electricity grid is 370 gr CO_2-eq/kWh. Another interesting point illustrated in Fig. 3.3, is that when the GHG

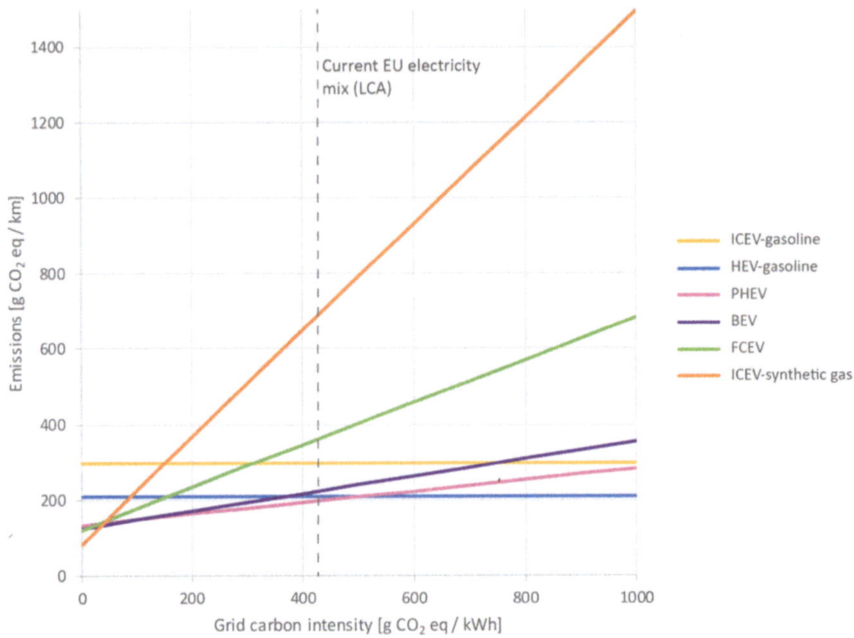

Fig. 3.3 Sensitivity of emissions of different passenger car technologies to carbon intensity of the electricity generation mix. *Source* Adapted from Cox (2018)

footprint of the electricity grid is ~ 600 gr CO_2-eq/kWh then the BEVs have higher life-cycle emissions than both gasoline hybrid electric (HEV-gasoline) and Plug-in-hybrid electric vehicles (PHEV). Figure 3.3 is given primarily for illustrative purposes to explain the importance of clean electricity generation. Extrapolations or other use should not be made without independent checking. The encouraging news is that several governments around the world—especially in the developed countries—are now committing billions in clean energy production, a policy that seems to get priority over other decarbonisation actions (IEA, 2024b).

3.5 Assessing the Odds for Success

Each of the different policies and measures for decarbonisation in the transport sector that are mentioned above, can reduce CO_2 and other GHG emissions by a certain amount but so far, we have no reliable way of forecasting this amount. Special difficulties arise for measures whose success depends on the reaction of individual travellers. Also, it is difficult to estimate impacts in cases where there is a combination of measures to be applied, and we are seeking their combined effect. Artificial Intelligence can be of particular help here, for example by:

1. Analysing and integrating vast amounts of data from various sources, such as satellite imagery, sensor networks, and historical emissions data.
2. Simulating the effects of various policies and technologies by use of machine learning algorithms that can be trained on historical data to predict future emissions under different scenarios.
3. Analysing patterns of human behaviour and consumption to identify the capacity for reducing emissions under a specific stimulus (incentive or other policy).

When policy makers can have a more accurate estimate of the potential results of their policies, there will be greater capability for planning and implementing decarbonisation measures in a more efficient way.

The time that will be required to fully decarbonise the transport sector will depend on the speed with which investment decisions are made and projects implemented to build infrastructures for low carbon energy vectors, including renewable power generation, production, storage and distribution of new energy vectors like hydrogen and synthetic fuels, also on the penetration of low or no carbon vehicles. It will also depend on the success of the CO_2 emissions abatement policies such as pricing of CO_2 emissions in the transport sector to account for their external costs on climate change and the successful distribution of the costs for the decarbonisation of the power generation industry to all sectors that will benefit from it, primarily the three sectors transport, buildings, and industrial. The costs and benefits for the decarbonisation of the (electric) power industry sector should be distributed between all the relevant beneficial sectors, and primarily among the previously mentioned three so that the final decarbonisation objective is shared in a fair and affordable way. This is called in Europe the "*coupling*" of the sectors and it is a field of research and legislative actions that is particularly suited to AI applications.

Two main questions are pertinent here. Will the measures and policies outlined in the previous sections be enough? and will they be implemented in time to avoid global warming above the 2 °C limit? The answer to the first of the above questions is already derived by the presentations and the discussions in all previous sections and is a negative one. Achieving zero carbon emissions in the transport sector by mid-century by any combination of the available measures, will be a very challenging and largely unachievable task. A definitive answer to the second question cannot, also, be given but all indications—at global level—are that the critical period for implementation that will determine success or failure, is the next 5–10 years. So, the progress to be made by the intermediate milestone year of 2030, seems particularly important. Unfortunately, as it has been made clear in previous sections, the indications at the time of publication of this book (2025) are that the trends are diverging from these desired targets.

We will further substantiate the above, by examining progress in the various decarbonisation measures mentioned earlier. This can be done more effectively, by classifying the measures and policies mentioned earlier in the following four categories:

A Managing the "demand" for travel.

B Improving the energy efficiency of conventional vehicles.

C Shifting passenger travel and freight to less carbon intensive modes, and
D Transitioning to "clean" fuels i.e., no-carbon emitting energy vectors.

Managing the "demand" for travel especially by use of motorised transport modes, involves making changes to the way people behave, act and live. Current progress in this area is minimal and the chances are that such changes will be difficult to plan and implement anytime soon in any of the major polluting countries. So far, any demand management measures that have been applied, have had only limited impacts on emissions. An estimate of the potential reductions in emissions that can be achieved due to this type of actions is of the order of 4–5% reduction per individual demand management measure (EASAC, 2019). When a combination of several such measures is implemented, the impacts will be higher than 4–5% but there is no way to estimate in any reliable way the result of such combined measures with current data and technologies. So, the degree to which passenger transport demand can be incentivised (and the potential travellers convinced) to make substantial changes in their trip making behaviour, is contested and the efforts so far have had rather weak results. Digital technologies such as AI will provide the necessary leverage for creating better awareness and planning of travel demand management measures and will help decision makers decide which will be the most effective incentives to use. AI will definitely help estimate the impact of such measures.

Improving the energy efficiency of existing ICE vehicles. This group of measures and policies can produce considerable improvements and in the short term. Such improvements will come mainly from the hybridisation of ICE vehicles and by lowering engine performance characteristics such as, for example, increasing the time needed for passenger cars to accelerate from 0 to 100 km/h. by increasing this time by 2.5 s, a maximum of 8% savings in fuel consumption and corresponding emissions can be achieved (EASAC, 2019). Also, the gradual introduction of Euro-7 types of freight vehicles as well as improvements in the body design of all vehicles to reduce their weight and the air drag, as well as their rolling resistance, could produce emission reductions of the order of 5–15% depending on the vehicle type and the types of improvement made (SCEER, 2017). Improving the emission performance of conventional powertrains or using "drop-in" fuels such as natural gas, or advanced biofuels is another option. These types of improvements are being brought about rather quickly by the vehicle manufacturers and can be helpful in the transition period (until full decarbonisation).

Shifting passenger travel and freight to less carbon intensive modes. This option has proved to be the most difficult to implement in the past and it can be costly too. So far, the results from more than three decades of relevant EU policies[2] have failed to achieve a significant shift of passengers or freight from road to rail. In the

[2] The European Commission set the shifting of the balance between the different modes of transport as one of the main objectives in its first 10-year Transport Policy white paper, as early as 1992. In 2001, it confirmed the importance of revitalising the rail sector in order to attract more traffic from the road while, in its 2011 white paper, it has set a target of shifting as much as 30% of road freight being transported further than 300 km, to other modes of transport such as rail or waterborne transport by 2030, and more than 50% by 2050.

last decade, 2011–2020, the modal split of freight in the EU has remained the same at approximately 75% road, 18% rail and 7% inland waterways, despite a series of relevant EU legislative measures, and despite some substantial financial contributions to improve rail services in Europe. Several explanations have been given for why these policies have failed to deliver substantial results (EP, 2015). However, the issue merits more in-depth analysis and study.

Transitioning to less or non-carbon intensive energy vectors. This is the area where the most substantial reductions in GHG emissions can be expected to be made. The use of *Battery Electric Vehicles* (BEVs) is increasing in markets across the world, but their use will be truly beneficial for decarbonising the transport sector if the carbon intensity of electricity production is low. "*Green*" electricity production is a sine qua non condition for achieving the full positive effect of electric mobility. *Plug-in Hybrid Electric Vehicles* (PHEVs) can exhibit significantly lower overall CO_2 footprints to those of BEVs due to the smaller size of their batteries, provided they are driven regularly on their battery power, something that is not always the case. Also, PHEVs produce less emissions from their internal combustion engines than conventional ICE vehicles because they employ energy recovery and powertrain optimisation systems.

The use of *Hydrogen* as a new energy vector directly or in Fuel Cell Electric Vehicles (FCEVs) can provide a potential future breakthrough should its production process improve to allow hydrogen production from green electricity and should efficient methods be developed for its transport, distribution, and storage to hydrogen filling stations and vehicle tanks. The rate of deployment of green hydrogen for transport will depend largely on the rate of development and installation of its distribution infrastructure. However, the use of green hydrogen for transport is always likely to be more expensive than using green electricity from batteries because of the limited efficiencies of the hydrogen processes involved. The primary market for green hydrogen technology are inter-city buses, long-distance heavy-duty trucks, trains and ferry boats (i.e., forms of transport that cannot easily use electricity for battery charging).

Future roles are also expected for advanced biofuels, *synthetic hydrocarbon fuels and (green or blue)*[3] *ammonia* as transport energy vectors (IEA, 2023b). The use of advanced biofuels is constrained mainly by the limited availability of sustainable biomass resources for their production, whilst the use of synthetic hydrocarbon fuels and low carbon ammonia is limited mainly by their high costs, primarily due to inefficiencies in their production processes. As a rule of thumb, it requires about five times as much green electricity to drive a vehicle using synthetic fuels as compared to driving the same vehicle as a BEV. So, the future role of synthetic fuels will be, mainly, in the transport sector niches where the vehicles with battery electric motors are not feasible due to their type of operation. This is, for example, the case of aviation (airplanes), open sea maritime vessels, and long-haul road trucks. For these applications, synthetic fuels can form an intermediate alternative to fossil fuels but

[3] "Green" ammonia is produced from nitrogen and hydrogen using low carbon electricity, whilst "blue" ammonia is produced from fossil fuels together with carbon capture and storage.

for that, steps will have to be taken to make them affordable and competitive (e.g., through strengthening of the Emissions Trading System levies on carbon pollution in the transport sector, or some form of specific carbon pricing for these types of transport).

As regards the option of delivering (green) electricity directly to vehicles using *Electric Road Systems* (ERS), this option can be energy efficient and climate change compatible but requires major investments in new grid infrastructure before even one vehicle can operate. The ERS using inductive energy by transfer from wires imbedded in the asphalt, is the most promising case of such systems for the future as the transfer of electric energy is done without overhead wires which cause visual pollution especially in environmentally sensitive areas. Also, the impeded in the pavement charging infrastructure can also provide the necessary energy for contactless charging of the batteries in BEVs. Once the necessary infrastructures are in place, the operational and environmental (in terms of pollution) costs of ERS mobility can be very competitive to those of BEVs or ICE vehicles.

Lastly, there are two outlier options which can only be seen as a "plan B" for the case when the climate crisis is hitting hard with extreme weather events and widespread damage and decision makers realise that the measures and policies already implemented will not deliver the urgently required net-zero situation in time. The two "Plan B" measures are the following:

(a) *Containing the mobility of people and goods* at all income levels. This is the most unpopular and thus undesirable course of action because it entails direct restrictions in the movement of persons and goods within well specified areas or (even worse) horizontally. The restrictions that were enforced during the COVID pandemic is a good example of these types of measures and how difficult it will be to reintroduce them, this time in the face of an intense and wide-ranging climate crisis.

(b) Trying to remove the CO_2 and other GHG from the atmosphere using the *Negative Emission Technologies* (NETs) mentioned earlier. It is unlikely that any commercially viable NETs will be deployed at a sufficiently large scale to make an impact on climate warming unless if seen as urgently required due to extreme weather events and a climate crisis. For this reason, a relevant report by the *European Academies Science Advisory Council* concluded that NETs have "limited realistic potential" to halt increases in the concentration of greenhouse gases in the atmosphere at the scale envisioned in the Intergovernmental Panel on Climate Change (IPCC) scenarios (EASAC, 2018). Another analysis, published in 2020, suggested that implementing NETs at the necessary scale will require very high expenditure as well as new R&D to improve the understanding of mineralization processes used in Direct Air Capture (DAC) and improving other land-based approaches (agriculture and forestry) for CO_2 capture especially as regards their impacts on food production and ecosystems (Kramer, 2020). The difficulties of implementing NET technologies are shown by the relative failure of US legislation to provide effective incentives for advancing NET technologies through the so called, *45Q Carbon Capture Tax Credit*. This credit was first enacted in

2008 to incentivize the capture of CO_2 in the oil industry, but it was expanded in 2019 to make a tax credit eligible both for CO_2 captured and stored ($50/ton) and CO_2 captured for other uses ($35/ton).

The inevitable conclusion from all the above and the assessment of the different policies and measures for decarbonisation in the transport sector, is that there is no "silver bullet" for a successful decarbonisation of the sector. Already, many initiatives and relevant legislation have been and are being introduced for climate change mitigation actions in all major economies, but the results of these legislation are still modest and largely unknown. What is more unsettling is the uncertainty about the continuing implementation of such measures and policies by future governments. In the uncertainty that reigns, what appears to be certain is that—unfortunately—the decarbonisation measures and initiatives announced so far and even more the concrete actions that are (not) taken by governments around the world, will not bring the desired results by 2050 and perhaps later…

References

Abduljabbar, R., Dia, H., Liyanage, S., & Bagloee, S. A. (2019). Applications of artificial intelligence in transport: An overview. *Sustainability, 11*(1). https://doi.org/10.3390/su11010189

ACEA. (2017). *EU roadmap for truck platooning*. European Automobile Manufacturers Association—ACEA, 9 May 2017 (based on a report of 10 March 2016). https://www.acea.auto/publication/eu-roadmap-for-truck-platooning/. Accessed January 2025.

Barbosa, F. (2021). *Ultra-high bypass ratio engine technology review—The efficiency frontier for turbo fan propulsion*. SAE Technical Paper 2021-36-0032, 2022. https://doi.org/10.4271/2021-36-0032

Boehm, S., Jeffery, L., Hecke, J., Schumer, C., Jaeger, J., Fyson, C., Levin, K., Nilsson, A., Naimoli, S., Daly, E., Thwaites, J., Lebling, K., Waite, R., Collis, J., Sims, M., Singh, N., Grier, E., Lamb, W., Castellanos, S., …, Masterson, M. (2023). *State of climate action 2023*. Bezos Earth Fund, Climate Action Tracker, Climate Analytics, Climate Works Foundation, New Climate Institute, the United Nations Climate Change High-Level Champions, and World Resources Institute. https://doi.org/10.46830/wrirpt.23.00010

Cardama, M., Krapp, A., Medimorec, N., & You, A. (2023). *Transforming transport and mobility to achieve the targets of the Paris agreement and the sustainable development goals*. SLOCAT Transport, Climate and Sustainability Global Status Report, 3rd ed. Brussels. https://tcc-gsr.com/

Cox, B. L. (2018). *Mobility and the energy transition: A life cycle assessment of Swiss passenger transport technologies including developments until 2050*. PhD thesis, ETH Zurich, Swiss Competence Center for Energy Research (SCCER).

Crippa, M., Guizzardi, D., Banja, M., Solazzo, E., Muntean, M., Schaaf, E., Pagani, F., Monforti-Ferrario, F., Olivier, J. G. J., Quadrelli, R., Risquez Martin, A., Taghavi-Moharamli, P., Grassi, G., Rossi, S., Oom, D., Branco, A., San-Miguel, J., & Vignati, E. (2022). *CO_2 emissions of all world countries*. JRC Science for Policy Report no. JRC/IEA/PBL 2022, Publications Office of the European Union, Luxembourg.

Dalkmann, H., & Brannigan, V. (2007). *Transport and climate change: Module 5e, sustainable transportation sourcebook: A sourcebook for policymakers in developing countries*.

German Federal Ministry for Economic Cooperation and Development, Transport Policy Advisory Servicers (GTZ), Eschborn. http://changing-transport.org/wp-content/uploads/2007_dalkmann_brannigan_transportandclimatechange.pdf. Accessed January 2025.

EASAC. (2018). *Negative emission technologies: What role in meeting Paris agreement targets?* European Science Advisory Council, policy report 35, February 2018. ISBN: 978-3-8047-3841-6.

EASAC. (2019). *Decarbonisation of transport: Options and challenges*. European Science Advisory Council, policy report 37, March 2019. ISBN: 978-3-8047-3977-2.

EC. (2016). *EU reference scenario 2016: Energy, transport and GHG emissions trends to 2050*. EU-Directorate-General for Energy, based on study by E3M-Lab (Prof. P. Capros) using the PRIMES and GEM-E3 as well as the IIASA–GAINS and IIASA–GLOBIOM/G4M models.

EEA. (2024). *Transport and mobility*. European Environment Agency report, by Stefan Bakker, 19 February 2024. https://www.eea.europa.eu/en/topics/in-depth/transport-and-mobility

Elgowainy, A., Han, J., Ward, J., Joseck, F., Gohlke, D., Lindauer, A., Ramsden, T., Biddy, M., Alexander, M., Barnhart, S., Sutherland, I., Verduzco, L., & Wallington, T. (2018). Current and future United States light-duty vehicle pathways: Cradle-to-grave lifecycle greenhouse gas emissions and economic assessment. *Environmental Science Technology, 52*(4), 2392–2399. https://doi.org/10.1021/acs.est.7b06006. Accessed January 2025.

EP. (2015). *Freight on road: Why EU shippers prefer truck to train*. European Parliament, Directorate General for internal policies, a study by Steer Davies & Gleave, March 2015. https://www.europarl.europa.eu/RegData/etudes/STUD/2015/540338/IPOL_STU(2015)540338_EN.pdf. Accessed January 2025.

Eurofound. (2024). *The changing structure of employment in the EU: Annual review 2023*. Eurofound research paper, by Hurley J., & Litardi Ch. Publications Office of the European Union, Luxembourg, 28 March 2024.

Giannopoulos, G. A., & Li, Y. (2024). Transport research implementation: Current issues and lessons learned from Europe and China. *Frontiers in Future Transportation, 5*, 1339893. https://doi.org/10.3389/ffutr.2024.1339893

Giannopoulos, G. A., & Munro, J. F. (2019). *The accelerating transport innovation revolution: A global, case study-based assessment of current experience, cross-sectorial effects, and socioeconomic transformations* (pp. 390). Elsevier, April 17, 2019. ISBN: 9780128138045.

Haan, K., & Main, K. (2023). Remote work statistics and trends in 2024. *Forbes Advisor*, 12 June 2023. https://www.forbes.com/advisor/business/remote-work-statistics/. Accessed January 2025.

Hawkins, A., & Cheung, R. (2023). China on course to hit wind and solar power target five years ahead of time. *The Guardian*, Thursday 29 June 2023. https://www.theguardian.com/world/2023/jun/29/china-wind-solar-power-global-renewable-energy-leader. Accessed January 2025.

Heineke, K., Kloss, B., Möller, T., & Wiemuth, C. (2021). *Shared mobility: Where it stands and where it's going*. McKinsey Centre for Future Mobility. Available at: shared-mobility-where-it-stands-where-its-headed.pdf (mckinsey.com). Accessed January 2025.

IEA. (2023a). *World energy investment 2023*. International Energy Agency, report (revised version), May 2023. https://www.iea.org/energy-system/transport. Accessed January 2025.

IEA. (2023b). *Net-zero roadmap: A global pathway to keep the 1.5 °C goal in reach*. IEA. https://www.iea.org/reports/net-zero-roadmap-a-global-pathway-to-keep-the-15-0c-goal-in-reach. Accessed February 2025.

IEA. (2024a). *Transport related statistics*. International Energy Agency, background data and information for its 50th Anniversary Ministerial Meeting, February 2024. https://www.iea.org/energy-system/transport. Accessed January 2025.

IEA. (2024b). *Global energy transitions stocktake*. International Energy Agency report. https://www.iea.org/topics/global-energy-transitions-stocktake. Accessed January 2025.

IEA. (2025). Global EV Outlook 2025 Expanding sales in diverse markets. International Energy Agency report, July 2025.

IPCC. (2023). *AR6 synthesis report: Climate change 2023*. The Intergovernmental Panel on Climate Change. Sixth Assessment Report of the Panel's 58th Session held in Interlaken, Switzerland, 13–19 March 2023. https://www.ipcc.ch/report/sixth-assessment-report-cycle/. Accessed January 2025.

Ivcevic, Z., & Grandinetti, M. (2024). Artificial intelligence as a tool for creativity. *Journal of Creativity, 34*(2). https://doi.org/10.1016/j.yjoc.2024.100079

Jia-Wei, H., & Creutzig, F. (2021). A systematic review on shared mobility in China. *International Journal of Sustainable Transportation, 16*(1), 374–389. https://doi.org/10.1080/15568318.2021.1879974

JRC. (2013). *Analysis of national travel statistics in Europe*. European Commission, Joint Research Centre, Institute for Prospective Technological Studies, report no. JRC83304/EUR 26054 EN.

JRC. (2020). *Fossil CO_2 emissions of all world countries 2020*. Joint Research Centre—JRC, Science for policy report no. JRC 121460 (EUR 30358 EN) by Crippa M., Guizzardi D., Muntean M., Schaaf E., Solazzo E., Monforti-Ferrario F., Olivier J.G.J., Vignati E. ISBN 978-92-76-21514-1

Kemp, R. (2016). *Scope for reduction in transport CO_2 emissions by modal shift*. Lancaster University, Energy technologies Institute. https://s3-eu-west-1.amazonaws.com/assets.eti.co.uk/legacyUploads/2016/02/Scope-for-reduction-in-transport-CO2-emissions-by-modal-shift-ETI-Comms.pdf. Accessed February 2025.

Kramer, G. (2020). Negative carbon dioxide emissions. *Physics Today, 73*(1). https://doi.org/10.1063/PT.3.4389

Li, D., & Lasenby, J. (2023). Investigating impacts of COVID-19 on urban mobility and emissions. *Cities, 135*. https://doi.org/10.1016/j.cities.2023.104246

Li, Q., Bai, Q., Hu, A., Yu, Z., & Yan, S. (2022). How does COVID-19 affect traffic on highway network: Evidence from Yunnan Province, China. *Journal of Advanced Transportation, 2022*, Article ID 7379334, 23. https://doi.org/10.1155/2022/7379334

Li, S. (2023). Is China really leading the clean energy revolution? Not exactly. Article in *The Guardian*, Thursday 6 July 2023. https://www.theguardian.com/commentisfree/2023/jul/06/china-clean-energy-revolution-coal-power. Accessed February 2025.

Li, W., Liu, N., & Long, Y. (2023). Assessing carbon reduction benefits of teleworking: A case study of Beijing. *Science of the Total Environment, 889*(2023), Article 164262. https://doi.org/10.1016/j.scitotenv.2023.164262

Liao, F., & Correia, G. (2021). Electric carsharing and micromobility: A literature review on their usage pattern, demand, and potential impacts. *International Journal of Sustainable Transportation, 16*(3), 269–286. ISSN 1556-8318. https://doi.org/10.1080/15568318.2020.1861394

Liu, Z., Li, R., & Dai, J. (2022). Effects and feasibility of shared mobility with shared autonomous vehicles: An investigation based on data-driven modelling approach. *Transportation Research Part a: Policy and Practice, 156*, 206–226. https://doi.org/10.1016/j.tra.2022.01.001

McKinnon, A. (2018). *Decarbonising logistics: Distributing goods in a low carbon world* (328 pages). Kogan Page, 3 June 2018. ISBN: 978 0 7494 8047 9.

McKinsey. (2023). *An affordable, reliable, competitive path to net zero*. A McKinsey & company report by M. Krishnan, H. Tai, D. Pacthod, S. Smit, T. Nauclér, B. Houghton, J. Noffsinger, & D. Simon, November 2023. https://www.mckinsey.com/capabilities/sustainability/our-insights/an-affordable-reliable-competitive-path-to-net-zero. Accessed January 2025.

Miles, T., Delle Site, P., & Meyer-Rühle, O. (2004). Decoupling transport from economic growth: Towards transport sustainability in Europe. *European Journal of Transport and Infrastructure Research—EJTIR, 4*(4), 381–404.

Nelldal, B., & Andersson, E. (2012). Mode shift as a measure to reduce GHG emissions. *Procedia—Social and Behavioural Sciences, 48*(2012), 3187–3197.

Nordenholz, F., Winkler, C., & Knorr, W. (2016). Analysing the modal shift to rail potential within the long-distance passenger travel market in Germany passenger travel market in Germany. Paper in 44th *European Transport Conference*, Barcelona, 5–7 Oct 2016.

NOVELOG. (2019). Sustainable urban logistics planning. *European Platform on Sustainable urban mobility* plans, *NOVELOG project*, Grant Agreement no 636626, June 2019. *ELTIS the urban mobility observatory.* https://civitas.eu/projects/novelog. Accessed January 2025.

OECD-ITF. (2017). *Transition to shared mobility: How large cities can deliver inclusive transport services.* International Transport Forum—Corporate Partnership Board, Policy Insights, 30 May 2017. https://www.itf-oecd.org/search/site/shared%20mobility%20report. Accessed February 2025.

Oestreich, L., Sandri, R. P., da Silva Vieira, J., & Ruiz-Padillo, A. (2023). Impacts of the COVID-19 pandemic on the profile and preferences of urban mobility in Brazil: Challenges and opportunities. *Travel Behaviour and Society Journal, 31*, 312–322. https://doi.org/10.1016/j.tbs.2023.01.002

Onder, C., Ott, H., & Guzzella, T. L. (2011). *How much fuel can hybrid electric vehicles Save?* International Advanced Mobility Forum (IAMF), ETH, Department of Mechanical and Process Engineering, 8–9 March 2011. https://doi.org/10.3929/ethz-a-006886477

Qiao, Q., Zhao, F., Liu, Z., He, X., & Hao, H. (2019). Life cycle greenhouse gas emissions of Electric Vehicles in China: Combining the vehicle cycle and fuel cycle. *Energy, 177*, 222–233. https://doi.org/10.1016/j.energy.2019.04.080

Rodrigue, J.-P. (2024). *The geography of transport systems*—(6th ed, 402p). Routledge. ISBN 9781032380407. https://doi.org/10.4324/9781003343196

SCEER. (2017). *Towards an energy efficient and climate compatible future Swiss transportation system.* Swiss Competence Center for Energy Research—SCEER, Working Paper, 9 May 2017.

Shaik, M. E., & Ahmed, S. (2022). An overview of the impact of COVID-19 on road traffic safety and travel behaviour. *Transportation Engineering.* https://doi.org/10.1016/j.treng.2022.100119

SLOCAT. (2021). Tracking trends in a time of change: The need for radical action towards sustainable transport decarbonisation. Transport and Climate Change Global Status Report—2nd edition. www.tcc-gsr.com

SLOCAT. (2023). *Transport, climate and sustainability global status report* (3rd ed). Sustainable Low Carbon Transport Partnership (SLOCAT). www.tcc-gsr.com

Timmers, V., & Achten, P. (2016). Non-exhaust PM Emissions from electric vehicles. *Atmospheric Environment, 134*, 2016.

Vagliasindi, M. (2023). *The Impact of COVID-19 on mobility and congestion.* World Bank, Policy Research Working Papers; 10351. Washington DC. http://hdl.handle.net/10986/39533. Accessed January 2025.

Van Basshuysen, R. (2015). *Erdgas und erneuerbares Methan für den Fahrzeugantrieb.* Springer Vieweg, March 2015. https://doi.org/10.1007/978-3-658-07159-2

Verhoef, L. A., Budde, B. W., Chockalingam, C., Nodar, B. G., & van Wijk, J. M. (2018). The effect of additive manufacturing on global energy demand: An assessment using a bottom-up approach. *Energy Policy, 112*, 349–360.

Willumsen L. (2021), *Use of big data in transport modelling.* International Transport Forum Discussion Paper, No. 2021/05, OECD Publishing, Paris.

WRI-Ch. (2024). *Introduction to mobility-as-a-service (maas) practice guide and case book.* World Resources Institute China. https://doi.org/10.46830/wrirpt.21.00063

Xie, Y., Savvarisal, A., Tsourdos, A., Zhang, D., & Gu, J. (2021). Review of hybrid electric powered aircraft, its conceptual design and energy management methodologies. *Chinese Journal of Aeronautics, 34*(4), 432–450.

Zulfaa, M.-K., & Filippone, A. (2010). Fuel savings on a heavy vehicle via aerodynamic drag reduction. *Transportation Research, Part D: Transport and Environment, 15*(5), 275–284.

Chapter 4
Transport System Resilience in Climate Crises

Abstract This chapter begins with a review of existing research work on transport resilience in general and against climate change in particular. Understanding how transportation systems respond to and recover from the various disruptions, including the climate related ones, is crucial for developing effective strategies to enhance system resilience. This review reveals the various approaches and methodologies that researchers have employed to evaluate transport system vulnerability and adaptive capacity including in the face of climate challenges. Building upon these assessment findings, the adaptation policies that have been proposed or implemented by governments in the three major geopolitical blocks of our time (namely China, the European Union, and the United States under Biden administration) are then examined. Due to the distinctive economic, social, and political contexts of the decision-making structures in each of the corresponding administrations, many of the policies followed for resilience differ from each other. Nevertheless, by analysing these adaptation policies this chapter has identified common themes, reliable practices, and potential areas for improvement in developing resilient and sustainable transport systems globally. In this way it is able to give, in the concise form of a Table, typical policy instruments of each type that have been implemented in all three regions. Monitoring the success of the various adaptation policies is of crucial importance and this, heavily relies on the ability to quantitatively measure and monitor system resilience through well-defined indicators. Following this recognition, the chapter provides a brief overview of the various indicators that have been proposed and adopted to evaluate transportation system resilience, mainly focusing on topology-based, trip-based, and macroscopic fundamental diagram (MFD)-based indicators. These can reflect the real system resilience from a specific aspect or multiple aspects. At the end, an analysis is given of the applicability of the indicators of resilience in various types of disruptions and their integration with artificial intelligence (AI)-powered approaches. The material in this chapter, offers guidance in selecting appropriate resilience indicators and prioritising development investments, ensuring that resources are allocated effectively to enhance system performance and resilience. Lastly, the various AI applications for transport resilience are considered. The focus is on the current state and future advancements of AI-powered adaptation policies aiming to inspire stakeholders to re-evaluate the practical value and

G. A. Giannopoulos and Y. Li, *Human Mobility, Artificial Intelligence and Climate Change*, Lecture Notes in Mobility, https://doi.org/10.1007/978-3-032-08171-1_4

transformative potential of AI technologies in enhancing climate change adaptation strategies and redefining short-term response mechanisms.

Keywords Resilience · Transport resilience · Climate change resilience · Climate change adaptation · Adaptation policies · Adaptation policy instruments · Transport resilience indicators · AI for transport resilience

4.1 Introduction

As it was demonstrated and documented in the previous three chapters, human activities, especially those associated with substantial greenhouse gas (GHG) emissions, are causing climate change and increase global warming. These activities interfere with the planet's climate system causing the greenhouse effect and leading to changes in seasonal temperatures, alterations in precipitation patterns, and sea level rise, to name a few (Markolf et al., 2019). Climate change is expected to increase the frequency of various extreme weather events such as typhoons, floods, wildfires, and droughts (Wang et al., 2020). A typical example is the mudslides along the Big Sur section of Highway 1 in California in 2017, which resulted from a combination of an extended drought and heavy rainfall or the January 2025 wildfires in the Pacific Palisades region California, and many others. These disasters can exert severe challenges to the planning, operation, and management of transportation systems. Due to the propagation nature of traffic flows, their impacts can propagate from a region to other regions across the network. This will result in greater disruptions in terms of severity and area of influence, especially in the cases of multi-modal transport systems (Wang et al., 2020).

In addition to "direct physical impacts"—such as asphalt cracking due to extreme heat events and road flooding due to intense precipitation and sea level rise—the impact of extreme weather events can also cascade from the transportation sector to other sectors (e.g., public health, logistics) and even the entire urban system given the crucial role transportation plays. These additional impacts, often referred to as "indirect impacts", include both physical and non-physical effects. In particular, disruptions caused by climate change and extreme weather events to the transportation system can be categorised as follows based on the pathways they affect in transportation (Markolf et al., 2019):

A *Direct physical disruptions.* Influence on the physical infrastructure (e.g., road pavement) belongs to this category, such as the washout of a bridge due to flooding.

B *Indirect physical disruptions.* Disruptions resulting from interconnected or co-located infrastructure of other interdependent systems. For example, water pipeline break may disrupt road traffic.

C *Direct non-physical disruptions.* Disruptions affecting public health, human behaviour, and decision-making.

D *Indirect non-physical disruptions.* Disruptions resulting from loss of information, social, and financial resources, such as the interruptions of routing/mapping services.

Transportation system disruption is usually defined as the state in which the system is not operating at optimal efficiency or the cost to maintain desired levels of mobility is significant (Lu et al., 2024a; Markolf et al., 2019). Disruption induced by extreme weather can cause extensive economic and societal losses (Chalkiadakis et al., 2022), making the resilience of transportation systems an urgent global concern (Cinderby et al., 2024). In consequence, developing transport system resilience against climate crises is increasingly recognised as a critical component of adaptation policies essential for ensuring the sustainability of future transport infrastructure and services and for achieving the United Nations (UN) Sustainable Development Goals (SDGs) (Cinderby et al., 2024). The U.S. Federal Highway Administration (FHWA) defines resilience in transportation systems as "*the ability to prepare for changing conditions and withstand, respond to, and recover rapidly from disruptions*". Resilience is fundamental not only to maintaining the functionality of transport networks during and after adverse events but also to ensuring the long-term viability of economic activities and social welfare that depend on reliable transportation. As Glock et al. (2021) pointed out, there cannot be a climate-resilient society without a resilient transportation system.

This chapter begins with a review of existing research work on transport resilience against climate change. Understanding how transportation systems respond to and recover from climate-related disruptions is crucial for developing effective strategies to enhance system resilience. This review reveals the various approaches and methodologies that researchers have employed to evaluate transport system vulnerability and adaptive capacity in the face of climate challenges. Building upon these assessment findings, the adaptation policies that have been proposed or implemented by governments in major geopolitical blocks of our time are then examined (namely the United States, the European Union, and China). Each of these regions has developed pragmatic strategies and frameworks to strengthen the resilience of their transportation system (Ministry of Ecology & Environment of the People's Republic of China, 2022; US DoT, 2021). Due to their distinctive geographical, economic, social, and political contexts, many of their policies differ from each other. However, given the fact that climate change is a global challenge and due to the nature of climate change, their major strategies are more or less the same, i.e., reducing GHG emissions and enhancing transportation infrastructure resilience. By analysing these adaptation policies, we have identified common themes, reliable practices, and potential areas for improvement in developing resilient and sustainable transport systems globally, so we give in the concise form of a Table, the typical policy instruments of each type that have been implemented in all three regions.

Monitoring the success of these adaptation policies is of crucial importance and this, heavily relies on our ability to quantitatively measure and monitor system resilience through well-defined indicators. These indicators can serve as essential tools for policymakers to assess policy effectiveness and guide necessary adjustments in their adaptation strategies. Therefore, following this recognition, we provide a brief overview of the various indicators that have been proposed and adopted to evaluate transportation system resilience, mainly focusing on topology-based, trip-based, and macroscopic fundamental diagram (MFD)-based indicators. These can reflect the real system resilience from a specific aspect or multiple aspects. Their viability and reliability usually depend on the disruption faced. By examining these indicators, we can understand how to enhance the resilience of transport systems regarding the corresponding aspects they represent.

At the end, an analysis is given of the applicability of these indicators in various types of disruptions and their integration with artificial intelligence (AI)-powered approaches. Given the significant potential of AI techniques in transport-related applications, including real-time traffic control and infrastructure monitoring, this discussion is crucial for decision-makers. It offers guidance in selecting appropriate indicators and prioritising development investments, ensuring that resources are allocated effectively to enhance system performance and resilience. Lastly, we discuss AI applications in transport resilience, focusing on the current state and future advancements of AI-powered adaptation policies aiming to inspire stakeholders to re-evaluate the practical value and transformative potential of AI technologies in enhancing climate change adaptation strategies and redefining short-term response mechanisms. The exploration of the potential of AI in this chapter comes in addition to a more detailed analysis and documentation given exclusively on this same subject in Chap. 6.

4.2 Transport Resilience Research and Climate Change Adaptation

Studies on transportation system resilience vary significantly in their definitions, indicators, and assessment methods, largely due to diverse nature, scale, and impact of disruptions (Serdar et al., 2022). This section provides an overview of the main categories of disruptions and approaches in the field of transport resilience against climate change. Global climate change exerts profound effects on transport infrastructure through phenomena such as sea level rise and the increased frequency of extreme weather events. Flood resilience emerges as a critical focus in the literature due to its severe consequences. Researchers have investigated various flood scenarios, including those induced by hurricanes (Dong et al., 2023) and precipitation (Wassmer et al., 2024). These studies typically employ empirical analyses using historical data to evaluate a floods' impact on network topology, reliability, and stability (Dong et al., 2023; Wassmer et al., 2024). Notably, Zhang and Alipour (2023) proposed a two-stage stochastic programming approach that integrates data-driven simulation to optimise pre-disruption mitigation and post-disruption recovery measures simultaneously, enhancing resilience to flood hazards.

In addition to flood resilience, attention has grown towards resilience against snowstorms and typhoons in recent years, as evidenced by studies such as Mirjalili et al. (2023), Santiago-Iglesias et al. (2023), and Fang et al. (2022).

Inequality in resilience has also been a subject of exploration. Transportation systems in low-income African countries are particularly vulnerable to extreme weather events resulting from climate change (Cinderby et al., 2024). Byun and D'Ayala (2022) conducted a probabilistic analysis of disruption inequality across urban areas, focusing on seismic resilience. The disparity in disruptions has also been examined in terms of socioeconomic groups and access to amenities (Anderson et al., 2022). Moreover, disparities in transportation development investments, influenced by political, social, and economic factors, have exacerbated regional disparities.

The evaluation of resilience in multi-modal transportation systems has emerged as a significant research area due to the interdependencies among various urban transport modes such as subway, bus, and taxi services. For instance, Wang et al. (2023) developed a dynamic resilience index integrating passenger demand and infrastructure supply to assess multi-modal transportation systems. Chen et al. (2023) focused on the resilience of bus-metro interdependent networks, considering factors like network topology damages and cascading failures in node recovery prioritisation. Meanwhile, Zhang et al. (2024) investigated node reliability in multi-modal transportation networks, while Aparicio et al. (2022) conducted a comparative analysis of network metrics to evaluate the topological robustness of multi-modal systems. Fang et al. (2022) proposed a multi-phase model to assess the resilience of complementary transportation systems across multiple modes in response to typhoons. These studies collectively contribute to a deeper understanding of resilience dynamics inherent in multi-modal transportation systems, offering insights to enhance overall resilience against disruptions.

Various models have been employed for resilience-based evaluation and optimisation, including probabilistic models, mathematical optimisation models, and data-driven methods. Probabilistic models typically integrate hazard components with probability distributions of disturbance intensity, resilience evaluation components, and simulation components for mapping resilience strategies (e.g., Amini and Padgett (2023) for hurricane-induced debris impact assessment). Mathematical optimisation models aid in planning pre-disruption mitigation and post-disruption recovery strategies under practical constraints such as budget and time (Serdar et al., 2022). Representative models include two-stage stochastic programming and bi-level optimisation. However, these models may face computational inefficiencies in large-scale networks. Data-driven methods utilise sensor data like loop detectors and GPS data over extended periods, offering real-time insights into resilience dynamics. For instance, Diab and Shalaby (2020) used metro system data to analyse the impact of outdoor tracks and weather conditions on system resilience. For a comprehensive review and comparison of these model categories, including their merits and limitations, readers are referred to Serdar et al. (2022).

4.3 Adaptation Policies to Improve Transport Resilience

The United States, the European Union, and China are key contributors to global GHG emissions and collectively generate about half of the world's GDP. Consequently, their climate strategies and policies are critical not only for reducing future GHG emissions but also for influencing climate policy developments worldwide. The climate adaptation policies in the transport sector of these three regions highlight both shared strategies and distinct approaches driven by their unique political, social, and economic contexts. These three regions are taken as an example to analyse the special characteristics of their adaptation policies as well as the limitations of such adaptation strategies.

4.3.1 United States: Prioritising Social Equity in Transport Adaptation

The U.S. Department of Transportation (DoT)'s Climate Action Plan, which was composed under the Biden administration, emphasised integrating resilience into infrastructure planning while specifically addressing social equity concerns. The plan recognised that climate impacts disproportionately affect vulnerable communities and mandates adaptation strategies that prioritise disadvantaged populations. This approach, while commendable for its social consciousness, faced implementation challenges even before the advent of the Trump administration. The federal structure of the U.S. implied that state-level buy-in and coordination are crucial for effective implementation. The focus on equity, while necessary, sometimes comes at the cost of delayed infrastructure improvements in areas that might be technically more vulnerable to climate impacts but serve fewer disadvantaged populations.

The "2024–2027 Climate Adaptation Plan" formulated by the U.S. Environmental Protection Agency (EPA) in 2023 built on this framework by emphasising operational resilience against extreme weather events. The effectiveness of these plans is potentially limited by:

- Fragmented jurisdiction between federal, state, and local authorities.
- Inconsistent funding mechanisms across different regions.
- Complex environmental review processes that can delay crucial adaptation projects.

4.3.2 European Union: Cross-Border Coordination Challenges

The EU's adaptation strategy, exemplified by the revised Trans-European Transport Network (TEN-T) regulation of 2024, showcases an ambitious commitment to building climate-resilient infrastructure. A cornerstone of this approach is the establishment of climate-resilient standards for transportation infrastructure, marking a significant step toward enhancing regional preparedness. However, the strategy faces several critical challenges. Coordinating adaptation efforts across member states is complex, given their differing economic capacities and climate vulnerabilities. Additionally, standardising adaptation measures across diverse geographic regions with distinct climate risks proves difficult. Balancing the urgency of adaptation needs with the EU's often lengthy bureaucratic decision-making processes further complicates the implementation of this ambitious approach.

The effectiveness of the EU's adaptation policies is particularly tested in cross-border infrastructure projects, where different national standards and priorities must be reconciled. While the TEN-T framework provides a common standard, its implementation varies significantly across member states, potentially creating vulnerabilities in the network's weakest links.

4.3.3 China: Technology-Driven Pragmatic Adaptation

China's National Climate Change Adaptation Strategy 2035 adopts a pragmatic approach that balances regional flexibility with central coordination. It emphasises technological advancements in infrastructure resilience, aligning with the country's broader modernisation objectives. Key aspects of the strategy include the development of regional adaptation plans tailored to specific local climate risks, a strong focus on establishing robust technology standards for infrastructure, and the integration of adaptation measures with ongoing urbanisation efforts.

However, this approach has certain limitations. The heavy reliance on technological solutions may neglect softer adaptation measures such as fostering behavioural change and building system redundancy. Additionally, the rapid pace of infrastructure development carries the risk of locking in vulnerable designs if future climate projections turn out to be more severe than anticipated. Furthermore, the strategy's emphasis on economic growth can sometimes prioritise short-term gains over the long-term resilience needed to address climate challenges effectively.

The effectiveness of China's approach is enhanced by its centralised decision-making structure, allowing for rapid implementation of adaptation measures. However, this same structure might limit the flexibility needed to address unexpected climate challenges at the local level.

4.4 Adaptation Policy Instruments in Transport

Efforts to develop sustainable transport systems involve using various planning, regulatory, economic, information, and technology instruments. The development of climate-resilient transport systems requires the setting of a comprehensive set of policy instruments that focus on enhancing the transport system's ability to withstand, respond to, and recover from climate-related disruptions. The following categorisation of adaptation policy instruments is proposed with each category of instruments aimed to serve a specific role in building transport resilience:

A *Planning Instruments.* They focus on incorporating climate resilience into infrastructure design and network configuration from the earliest stages. These instruments ensure that transport systems are designed to withstand projected climate impacts while maintaining essential connectivity during extreme events.

B *Regulatory Instruments.* They establish mandatory requirements for climate adaptation, ensuring that resilience considerations are integrated into all aspects of transport infrastructure development and operation. These requirements create a framework for consistent application of adaptation measures across different jurisdictions.

C *Economic Instruments.* They provide the financial mechanisms necessary to implement and maintain resilient infrastructure. They ensure adequate funding for both preventive measures and post-disaster recovery.

D Information Instruments. They enable proactive responses to climate threats through improved monitoring, communication, and preparedness. These tools are crucial for both operational resilience and long-term adaptation planning.

E *Technology Instruments.* They leverage advances in materials science, monitoring systems, and predictive analytics to enhance transport resilience. These tools support both immediate operational responses and long-term adaptation strategies.

The implementation of these instruments varies across regions based on their institutional frameworks, climate vulnerabilities, and economic resources. While some measures, such as climate-resilient design standards, are widely adopted across the world, others show significant regional variation in their application and effectiveness. Table 4.1 summarises some typical policy instruments of each type that have been implemented in the three regions discussed in the previous section.

4.5 Indicators of Transport Resilience

4.5.1 General

Mathematically, system resilience can be assessed by integrating the deviation of system functionality from its optimal state over the disruption period. For ease of understanding, Fig. 4.1 provides a graphical illustration of the calculation of resilience loss, commonly referred to as the "resilience triangle" in the literature. After a disruption happens, the system experiences a decline in functionality. Then, owing to the system's inherent resilience and the implementation of recovery measures, this decline eventually stabilises and initiates a process of restoration

Table 4.1 Policy instruments for climate-resilient transport systems

Category	Adaptation policy instruments	Implementation examples
Planning instruments	• Climate-resilient infrastructure design standards • Redundant transport network planning • Flood-resistant road and rail construction • Emergency evacuation route planning • Heat-resistant material specifications	• US: Climate resilient design guidelines • EU: Technical guidance on the climate proofing of infrastructure in the period 2021–2027 • China: Enhanced infrastructure standards for extreme weather
Regulatory instruments	• Mandatory climate risk assessments for new infrastructure • Emergency response protocols • Infrastructure maintenance regulations • Resilience requirements in procurement • Weather-related operational protocols	• US: NEPA guidance on consideration of GHG emissions and climate change • EU: Climate adaptation requirements in major infrastructure projects • China: Regional climate risk assessment mandates
Economic instruments	• Resilience infrastructure funding mechanisms • Climate insurance requirements • Emergency response funds • Maintenance reserve requirements • Disaster recovery funding	• US: PROTECT grant program • EU: Climate-resilient infrastructure funding • China: Special funds for adaptation projects
Information instruments	• Early warning systems • Real-time weather monitoring • Emergency communication systems • Public awareness of climate risks • Staff training for extreme events	• US: DOT emergency response guidebook • EU: Copernicus climate change service • China: National climate monitoring system
Technology instruments	• Smart infrastructure monitoring systems • Weather-resilient materials research • Predictive maintenance systems • Climate modelling integration • Infrastructure vulnerability assessment tools	• US: FHWA VAST tool • EU: Horizon research programs • China: Smart infrastructure monitoring

Note NEPA = National Environmental Policy Act; PROTECT = Promoting Resilient Operations for Transformative, Efficient, and Cost-saving Transportation; TEN-T = Trans-European Transport Network; VAST = Vulnerability Assessment Scoring Tool

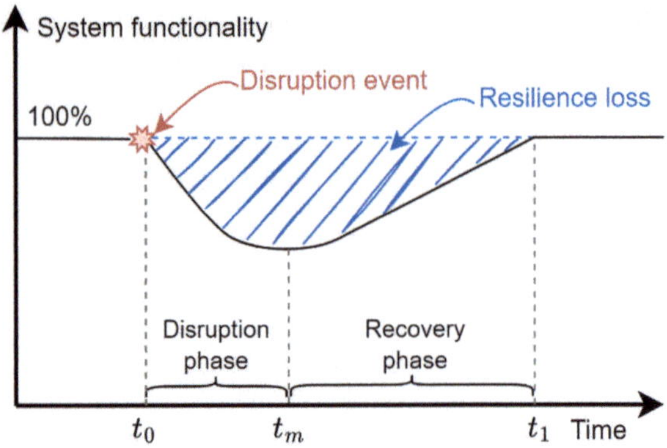

Fig. 4.1 Graphical definition of system resilience

towards a new equilibrium state. Hence, the entire disruption period can often be divided into a disruption phase and a recovery phase.

A reliable indicator of system functionality is essential for evaluating system resilience. Although the concept of resilience has been applied to transportation systems for over two decades, there remains no universally accepted definition of system functionality within the context of urban road transportation. Conventionally, topological measures derived from complex network theory, which capture structural properties, such as connectivity and accessibility, have been employed. However, these applications typically do not consider traffic dynamics. To address the limitations of topology-based indicators, trip-based indicators have been adopted, including network average travel time, average speed, and demand served, etc. However, these indicators, which rely on direct trip information, are often sensitive to variations in travel demand levels and patterns.

The MFD, defined by Geroliminis and Daganzo (2008) and its successful applications to perimeter control (e.g., Chen et al., 2022, 2024), offer a robust framework for constructing traffic resilience indicators that relate MFD dynamics to system functionality. The MFD describes the relationship among the space-mean production, the space-mean state, and spatial homogeneity in traffic flow, providing a method to evaluate network performance without requiring detailed traffic physics. Given that the MFD is an intrinsic property of a homogeneously congested transportation network (Huang et al., 2020; Su et al., 2020), indicators based on the MFD have the advantage of integrating information on transportation network characteristics, traffic dynamics, and travel demand patterns.

In the following the focus is on these three types of resilience indicators, i.e. topology-based, trip-based, and MFD-based, as these have been commonly discussed and applied in relevant studies. Moreover, considering their distinctive characteristics, we also specifically discuss their applicability in different types of disruptions caused by climate change and their potential for integrating into AI-powered systems.

4.5.2 Topology-Based Indicators

Transportation networks, as fundamental infrastructure systems, exhibit performance characteristics that are closely tied to their topological design. This relationship becomes particularly crucial when considering climate change impacts, as extreme weather events can significantly disrupt network connectivity. Network layouts vary from grid structures common in new planned urban centres to irregular patterns in historically developed areas, each responding differently to climate-related disruptions.

Recent research by Pan et al. (2021) demonstrates how network topology significantly influences traffic flow patterns and system vulnerability to climate-related disruptions. For instance, grid networks, while lacking clear centres, often provide multiple alternative routes that can maintain connectivity during localised flooding. In contrast, hub-and-spoke networks concentrate connections at central points, potentially creating critical vulnerabilities. This understanding of topological characteristics has become increasingly important for climate adaptation planning, particularly in developing resilient urban transportation systems. Table 4.2 lists some representative topological attributes that have been extensively utilised as measures of transport resilience in the literature.

When evaluating transport resilience through topology, traditional measures derived from complex network analysis such as connectivity and accessibility metrics provide initial insights. However, recent developments have shifted toward more sophisticated indicators that better align with climate adaptation needs and AI applications.

The limitations of static topological indicators have become increasingly apparent in recent years, particularly in the context of climate change adaptation. These measures fail to capture three critical aspects of modern transportation challenges: the dynamic nature of climate-related disruptions, variations in traffic demand patterns, and the cascading effects of network failures. Additionally, they inadequately evaluate partial capacity losses, such as reduced vehicle speeds during rainfall events or heat-induced speed restrictions. To address these limitations, recent studies have begun integrating dynamic indicators that combine topological structure with operational performance. For example, Ma et al. (2020) integrate measures of the shortest path, congestion degree, and passenger flow intensity to evaluate the structural and functional vulnerability of the network.

These indicators alone are insufficient for comprehensive resilience assessment in the context of climate change. Their primary limitation lies in their inability to capture the temporal dynamics of network congestion and the complex interactions between infrastructure damage and traffic flow patterns. This limitation becomes particularly significant when evaluating adaptation strategies for extreme weather events, where both immediate impacts and recovery patterns need consideration.

The future development of topology-based indicators is centered on their integration with AI systems and dynamic data sources. Advanced machine learning algorithms can significantly enhance these indicators by processing real-time sensor

Table 4.2 Typical topological attributes and indicators of transport resilience

Attribute	Indicator	Description
Attributes from basic statistics		
Network statistics	Number of nodes	
	Number of links	
	Average node degree	
	Total length of all edges	
	Average length of all edges	
Attributes from directed graph		
Assortativity	Degree assortativity	The assortative coefficient is the Pearson correlation coefficient of degree between pairs of linked nodes. It indicates the similarity of connections with respect to node degree
	Average neighbour degree	The average neighbour degree of a node measures the average degree of all nodes within the neighbourhood of a specific node
Centrality	Average degree centrality	The average of the fraction of nodes that a node is connected to
	Load centrality	The average of the fraction of all shortest paths passing through a node
	Edge load centrality	The average of the fraction of all shortest paths passing through an edge
	Harmonic centrality	The average of the sum of the reciprocal of the shortest path distances from all other nodes to a node
Connectivity	Alpha index	Ratio of existing circuits to the maximum possible circuits
	Beta index	Ratio between the number of links and number of nodes
	Gamma index	Ratio of the number of links to the maximum possible number of links
Reciprocity	Reciprocity	Ratio of the number of edges pointing in both directions to the total number of edges
Attributes from undirected graph		
Clustering	Average clustering coefficient	The clustering of a node is the fraction of possible triangles through that node that exists. For a given node, a triangle is formed if it is connected to its two neighbours. The average clustering coefficient of a graph is then the mean of local clustering
Efficiency	Global efficiency	The efficiency of a pair of nodes in a graph is the multiplicative inverse of the shortest path distance between the nodes. The average global efficiency of a graph is the average efficiency of all pairs of nodes

Source Lu et al. (2024a)

data to continuously update network vulnerability assessments. They also have the potential to predict possible failure patterns under various climate scenarios, thereby enabling proactive measures. Furthermore, these algorithms can optimise network design to improve climate resilience and identify critical links within the network that require priority adaptation measures.

For transportation planners and decision-makers, topology-based indicators serve as foundational tools for understanding network vulnerability to climate change. While they should not be used in isolation, they provide essential insights for initial vulnerability assessments and long-term adaptation planning. Their greatest value emerges when integrated with AI systems and complemented by other dynamic indicators, creating comprehensive frameworks for climate resilience assessment and adaptation strategy development.

4.5.3 Trip-Based Indicators

Trip-based indicators provide direct measurements of transportation system performance through metrics that capture actual travel experiences. In a comparative analysis by Balal et al. (2019), several traffic dynamics-oriented resilience measures were investigated, including queue length, link speed, link travel time, frontage road delay, and detour route delay. The study revealed a relatively low correlation among these measures, indicating that the significance of links to network resilience varies depending on the specific measure used. This highlights the necessity of selecting appropriate indicators for different types of disruptions. It also suggests that evaluating network resilience at the link level might result in contradictory conclusions and offer limited insights. As a result, the need for reliable network performance measures in both spatial and temporal dimensions has shifted the focus from local traffic performance towards developing network-wide measures.

Bucar and Hayeri (2020) proposed a framework for assessing the impact of precipitation-induced flood events in urban areas using a two-layer simulation model. Metrics such as vehicle miles travelled, vehicle hours travelled, and trips completed were utilised to evaluate network resilience. Similarly, Hoogendoorn et al. (2015) assessed network resilience by examining changes in the level of service due to density deviations caused by incidents. In this context, average speed served as a proxy for the level of service, making network resilience a function of density and density deviation. However, these indicators fail to compare disruption scenarios with normal operation conditions and neglect the time required for recovery, which are crucial elements in defining resilience. Additionally, they complicate the calculation of optimal system functionality (i.e., the equilibrium state before disruption) as they are highly dependent on demand levels and patterns.

Zhang et al. (2015) employed the ratio of the maximum post-disaster throughput to the pre-disaster throughput as a resilience indicator. This indicator was further applied to optimise post-event recovery actions and evaluate the significance of network topology in resilience. Yao and Chen (2023) utilised the percolation theory, using the giant component size to evaluate network resilience to random failures. Nonetheless, these indicators overlook traffic dynamics and their temporal variability, thus failing to provide an accurate depiction of resilience over time.

Trip-based indicators can be improved along three key aspects for climate adaptation. First, the integration of real-time data with historical performance patterns allows for more accurate assessment of weather-related disruptions. This approach enables transportation managers to distinguish between normal performance variations and climate-induced disruptions, supporting more targeted adaptation strategies. Second, the incorporation of recovery time metrics has gained prominence. Recovery patterns after weather-related disruptions can vary significantly based on network characteristics and implemented adaptation measures. Third, the development of composite indicators that combine multiple trip-based metrics has emerged as a promising approach. These indicators provide a more comprehensive view of system performance during climate-related disruptions while remaining interpretable for decision-makers.

Trip-based indicators encounter several challenges when applied to climate adaptation. One major issue is their difficulty in establishing consistent baseline conditions for comparison, as travel demand patterns constantly evolve. Additionally, their strong dependence on demand levels and patterns complicates the evaluation of system functionality under varying climate scenarios. Traditional trip-based metrics also tend to fall short in capturing the cascading effects of climate-related disruptions across the transportation network.

Integrating AI technologies with trip-based indicators offers significant potential to overcome these limitations. AI applications can process large volumes of trip data to identify patterns and anomalies related to weather conditions, predict system performance under various climate scenarios, and optimise response strategies during weather-related disruptions. They also enable real-time decision-making during extreme weather events. By leveraging advanced machine learning algorithms, trip-based indicators can be further enhanced to identify complex patterns in travel behaviour during weather events, predict potential disruptions before they occur, and optimise recovery strategies in real-time. This integration supports adaptive management approaches and strengthens the effectiveness of trip-based indicators when paired with complementary metrics.

For transportation planners and decision-makers, trip-based indicators provide valuable insights and opportunities for climate adaptation planning, particularly when enhanced by AI capabilities. Their direct connection to user experience makes them especially useful for:

- Quantifying the immediate impacts of weather-related disruptions.
- Evaluating the effectiveness of implemented adaptation measures.
- Communicating system performance to stakeholders.
- Justifying investment in climate resilience measures.

4.5.4 Macroscopic Fundamental Diagram-MFD Indicators

In recent years, there has been substantial scholarly interest in performance indicators based on the MFD theory. One of the pioneering efforts in this area, particularly concerning transportation resilience, was undertaken by Kim and Yeo (2017). They proposed a network performance loss indicator that leverages MFD to evaluate the criticality of network links. This indicator acknowledges that disruptive events exacerbate spatial heterogeneity in traffic flow, thereby modifying the shape of the MFD. The performance loss FR is quantified as the proportion of the aggregated flow reduction caused by disruptions and is given by

$$FR = \frac{\int_{k_0}^{k_c} q(k) - q^d(k)\, \mathrm{d}k}{\int_{k_0}^{k_c} q(k)\, \mathrm{d}k}$$

where $q(k)$ and $q^d(k)$ are the weighted space-mean flow across the network under normal and disruption operation conditions, respectively; k_0 and k_c represent the lower and upper bounds for comparison. However, in this definition, both $q(k)$ and $q^d(k)$ are weighted based on the total length of the original network without accounting for link removal during disruptions. This weighting scheme can lead to inaccurate comparisons (Amini et al., 2018). Furthermore, the comparison solely focuses on the shape of the MFD, neglecting the temporal dynamics of the network.

To address these limitations, Amini et al. (2018) introduced an enhanced MFD indicator by incorporating changes in trip length into the evaluation of network resilience. They defined the service rate of a network as the number of completed trips per unit of time. Building upon the linear relationship discovered by Geroliminis and Daganzo (2008) between trip production (P) and trip completion rate (D), expressed as $P = \mathcal{L}D$, where $\mathcal{L}$ represents the average trip length, and leveraging the relationship between trip production and weighted flow, expressed as $P = \sum_{i \in L} q_i l_i = qL$, where L denotes the length of all links equipped with detectors, they derived a resilience index that reflects the network's service rate using weighted flow as a surrogate. The index is calculated as follows:

$$RI(t) = q(t) - q^d(t) \times \frac{\mathcal{L}}{\mathcal{L}_d}$$

where $\mathcal{L}$ and $\mathcal{L}_d$ are the average trip length under normal and disruption conditions, respectively. It is important to note that this index has been normalised by $L/\mathcal{L}$ without differentiating between the network's state before and after disruptions. This assumption implies that the length of the network remains unchanged during disruptions, which contradicts the reality of infrastructure disruptions.

Furthermore, Gao et al. (2022) proposed a novel definition of traffic resilience for a system comprising two reservoirs. This definition is based on the total congestion deviation, which is quantified as the difference between the actual vehicle accumulation during operation and the optimal value. By assuming parabola-shaped MFDs, the traffic resilience TR is computed using the following equation:

$$TR = -\int_{t_0}^{t_c} \left| n_1(t) - \frac{n_1^{\max}}{2} \right| + \left| n_2(t) - \frac{n_2^{\max}}{2} \right| \mathrm{d}t$$

where $n_i(t)$ is the vehicle accumulation at reservoir i at time t, and $n_i^{\max}$ is the corresponding maximum accumulation. The time interval from t_0 to t_c represents the congestion period. A larger congestion deviation (the term inside the integral at a specific t) represents more severe congestion or more spare capacity. However, using accumulation as a metric for transportation system performance, particularly during disruptions, has limitations. Firstly, the number of vehicles within the network, represented by accumulation, does not provide a comprehensive reflection of transportation network functionality. The performance/functionality of a system/facility is generally represented by its service rate (i.e., the trip completion rate in MFD) rather than the number of customers (i.e., the accumulation in MFD). Secondly, the optimal accumulation during disruptions can exceed or equal that of normal operations, implying that the same accumulation value may result in lower resilience loss during disruptions, leading to inaccurate estimations.

Inspired by the MFD-based indicators discussed above, Lu et al. (2024a) proposed a traffic resilience indicator for urban transportation systems based on MFD dynamics, treating trip completion rate as the system's service rate. This indicator fits with various supply disruptions that impact transportation networks, whether locally or globally. For instance, local disruptions may include regional link closures due to road maintenance, such as repairing the asphalt cracking issue caused by drought, while network-wide disruptions can result from large-scale flood disasters.

During supply disruptions, the shape of the MFD typically undergoes a "shrinkage" phenomenon, as observed in previous studies (Gao et al., 2022; Kim & Yeo, 2017). The traffic resilience indicator proposed by Lu et al. (2024a) measures resilience loss by comparing MFD dynamics before and after the disruption. Specifically, at any given moment, resilience loss is quantified by the reduction in trip completion rate compared to its optimal value.

The traffic resilience loss due to supply disruptions can be quantified by integrating the reductions in trip completion rates over the disruption period. Mathematically, it is expressed as:

$$R^s = \int_{t_0}^{t_s} \max\{D(t) - D^s(t), 0\} \, \mathrm{d}t$$

where $D(t)$ and $D^s(t)$ are the trip completion rates under normal and disruption conditions, respectively, and the time from t_0 to t_s indicates the time period of interest.

This indicator have been applied to optimise post-disruption recovery measure optimisation in Lu et al. (2025).

For decision-makers considering MFD-based indicators for climate resilience assessment, several key considerations should be taken into account. The system requires comprehensive traffic flow data from strategic network locations, weather monitoring stations, and infrastructure condition assessments. Technical infrastructure needs include real-time data processing capabilities and integration with existing traffic management systems. Resource planning must address staff training needs, system maintenance requirements, and regular calibration and updating procedures.

From an AI implementation perspective, MFD-based indicators offer significant advantages over traditional metrics. They can process large-scale network data efficiently, work with aggregated measurements (reducing computational complexity), and provide clear mathematical frameworks for machine learning models. These characteristics make them particularly suitable for developing AI-driven tools for:

- Real-time resilience monitoring and early warning systems.
- Predictive analytics for climate impact assessment.
- Optimisation of adaptation strategies.
- Automated response system development.

4.5.5 Selecting the Appropriate Indicators: A Decision-Making Framework

The selection of appropriate resilience indicators is crucial for effective climate adaptation planning in transportation systems. This section provides a structured framework to help choose suitable indicators based on their specific context, available resources, and AI capabilities.

4.5.5.1 Context-Based Selection

The choice of resilience indicators should primarily be driven by the specific climate threats facing the specific transportation system. For acute climate events such as flash floods or extreme storms, trip-based indicators offer direct and immediate insights into system performance degradation and recovery patterns. These indicators can effectively capture the real-time impact of sudden disruptions on travel times, speeds, and completed trips. However, for chronic stressors like sea-level rise or increasing heat waves, topology-based indicators become more valuable as they help identify structural vulnerabilities that may compromise long-term system functionality.

The scale of analysis also significantly influences indicator selection. At the neighbourhood or district level, trip-based indicators provide detailed operational insights useful for local adaptation strategies. For citywide or regional assessments, MFD-based indicators offer a more comprehensive understanding of system-wide performance and recovery patterns, particularly valuable for coordinating large-scale adaptation efforts.

4.5.5.2 Resource Considerations

Data availability and quality fundamentally shape the feasibility of different indicators. Topology-based indicators require primarily network configuration data, making them accessible for systems with limited operational data collection capabilities. Trip-based indicators demand continuous monitoring of traffic conditions through sensors, cameras, or floating car data (FCD). MFD-based indicators necessitate comprehensive traffic monitoring systems capable of capturing network-wide traffic states, though recent advances in AI and machine learning have made their implementation more achievable through data fusion and estimation techniques (e.g., Jin et al., 2024).

Computational resources and technical expertise requirements also vary significantly. Topology-based analyses can often be conducted with standard GIS tools (e.g., ArcGIS, QGIS) and basic network analysis software. Trip-based indicators typically require more sophisticated traffic modelling capabilities and real-time data processing systems. MFD-based approaches demand advanced analytical capabilities and specialised expertise, though they offer the most comprehensive understanding of system dynamics.

4.5.5.3 AI Implementation Potential

The integration of AI significantly enhances the utility of different indicator types. For topology-based indicators, AI can automate the identification of critical infrastructure components and predict potential cascade failures during climate events. Machine learning algorithms can process vast amounts of historical data to identify patterns in network vulnerability and recommend structural improvements.

Trip-based indicators benefit from AI through improved real-time data processing and prediction capabilities. AI algorithms can forecast travel time impacts during different types of climate disruptions and suggest optimal routing strategies for system recovery. These capabilities are particularly valuable for emergency response planning and adaptive traffic management during extreme weather events.

MFD-based indicators show the highest potential for AI integration. Machine learning techniques can enhance the estimation of network states, predict MFD evolution under different climate scenarios, and optimise recovery strategies. AI can also help address the challenge of spatial heterogeneity in traffic states, improving the accuracy of MFD-based resilience assessments.

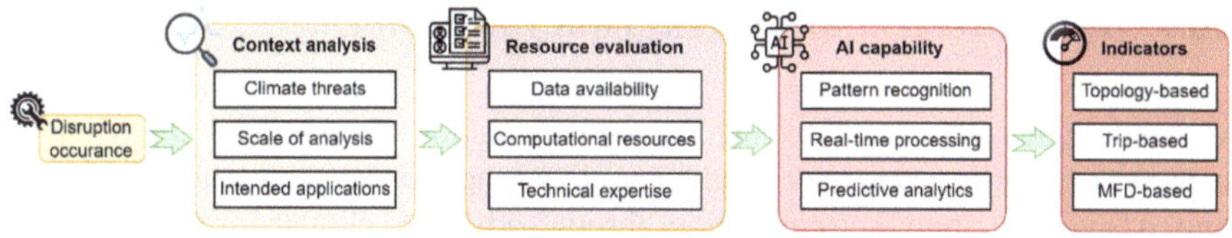

Fig. 4.2 The decision-making framework for resilience indicator selection

4.5.5.4 Implementation Strategy

For organisations beginning to implement resilience monitoring, a staged approach is recommended. Starting with basic topology-based indicators provides a foundation for understanding system vulnerabilities. As data collection capabilities expand, trip-based indicators can be incrementally incorporated to capture operational dynamics. MFD-based indicators can be implemented in phases, beginning with pilot areas and expanding based on results and resource availability.

The selection of indicators should also consider the intended applications. For infrastructure investment planning, topology-based indicators combined with AI-driven predictive analytics can identify priority areas for adaptation measures. For operational resilience, trip-based indicators supported by AI-enabled real-time monitoring systems provide actionable insights for immediate response. For strategic planning and policy development, MFD-based indicators offer comprehensive understanding of system dynamics and potential adaptation strategies.

Figure 4.2 depicts the decision-making framework for selecting appropriate resilience indicators when facing climate-induced disruptions. This framework emphasises that indicator selection is not a one-size-fits-all decision. Instead, it should be viewed as a dynamic process that evolves with changing climate threats, technological capabilities, and organisational needs. The integration of AI technologies can significantly enhance the implementation and utility of all indicator types, though the specific applications should be carefully matched to local contexts and capabilities.

4.6 Artificial Intelligence Applications for Transport Resilience

The integration of AI into transport resilience strategies represents both a significant advancement in current adaptation practices and an unprecedented opportunity for future innovation. This section examines how AI enhances existing policy instruments while enabling transformative approaches to climate resilience in transport systems.

4.6.1 Enhancement of Current Policy Instruments

AI technologies are revolutionising the implementation of traditional policy instruments across the transport sector and for resilience applications in particular. In planning, AI systems strengthen capabilities through advanced climate impact modelling and network optimisation, enabling planners to anticipate vulnerabilities and design more resilient systems. These systems can rapidly evaluate multiple adaptation strategies under different climate scenarios, providing insights that would be impossible to obtain through conventional methods.

The regulatory landscape benefits from AI through automated monitoring systems and real-time assessment capabilities. These systems not only track adherence to resilience standards but also adapt to changing conditions, ensuring that regulatory frameworks remain effective as climate challenges evolve. The integration of predictive analytics helps identify potential regulatory violations before they occur, enabling proactive enforcement measures.

Economic instruments have gained new sophistication through AI-driven analysis. Predictive maintenance systems optimise resource allocation, while advanced cost–benefit analysis tools evaluate adaptation investments across different time horizons. Risk assessment models inform insurance and funding mechanisms, creating more nuanced and responsive financial instruments for resilience projects.

Information gathering and dissemination have been transformed by AI capabilities. Real-time monitoring systems track infrastructure performance and environmental conditions, while early warning systems predict and communicate potential disruptions. These systems ensure that critical information reaches stakeholders efficiently, enabling rapid response to emerging threats.

In the technological realm, AI drives advancement through smart materials and sensors that adapt to changing environmental conditions. Autonomous maintenance systems respond to infrastructure stress, while intelligent traffic management systems optimise network utilisation during disruptions. These technological applications represent the current state of the art in resilience enhancement.

4.6.2 Regional Implementation and Opportunities

The implementation of AI-driven resilience solutions varies significantly across regions, reflecting different technological capabilities and institutional frameworks. In the US, the fragmented jurisdictional structure presents opportunities for AI-powered coordination platforms that harmonise adaptation efforts across state lines. The emphasis on social equity in adaptation planning demands innovative applications that ensure fair distribution of resilience benefits.

The EU's complex multi-national framework has spurred the development of sophisticated cross-border AI systems. These platforms enable coordination while preserving member state autonomy, creating a model for international cooperation in transport resilience. The EU's strong emphasis on ethical AI guidelines and data protection will shape how these systems are implemented.

China's centralised approach enables rapid deployment of experimental AI technologies in transport resilience. The integration of these systems with existing smart city initiatives creates opportunities for comprehensive urban resilience strategies. China's focus on scaling solutions across diverse geographic regions provides valuable insights for other large-scale implementations.

4.6.3 Future Directions and Recommendations

The future of AI in transport resilience points toward increased integration with Internet of Things (IoT) devices, more sophisticated predictive modelling, and enhanced automation in adaptation response systems. Success in these areas requires addressing several critical challenges, including data quality, technical expertise requirements, and system integration issues.

It is recommended to establish international standards for AI in transport resilience while maintaining flexibility for regional customisation. Investment in capacity building and the fostering of public–private partnerships will be crucial for driving innovation. Additionally, the development of open-source AI frameworks for adaptation could accelerate the adoption of these technologies globally. The potential of AI in transport resilience extends far beyond current applications. By embracing these innovative approaches, transport systems can become not just more resilient to climate impacts, but actively adaptive to changing conditions. The key lies in viewing AI not merely as a tool for implementing existing adaptation strategies, but as a catalyst for transforming how we approach transport resilience entirely.

References

Amini, K., & Padgett, J. E. (2023). Probabilistic risk assessment of hurricane-induced debris impacts on coastal transportation infrastructure. *Reliability Engineering & System Safety, 240*, Article 109579.

Amini, S., Tilg, G., & Busch, F. (2018). Evaluating the impact of real-time traffic control measures on the resilience of urban road networks. In *2018 21st International Conference on Intelligent Transportation Systems (ITSC)* (pp. 519–524).

Anderson, M. J., Kiddle, D. A. F., & Logan, T. M. (2022). The underestimated role of the transportation network: Improving disaster & community resilience. *Transportation Research Part d: Transport and Environment, 106*, Article 103218.

Aparicio, J. T., Arsenio, E., & Henriques, R. (2022). Assessing robustness in multimodal transportation systems: A case study in Lisbon. *European Transport Research Review, 14*(1), 28.

Balal, E., Valdez, G., Miramontes, J., & Cheu, R. L. (2019). Comparative evaluation of measures for urban highway network resilience due to traffic incidents. *International Journal of Transportation Science and Technology, 8*(3), 304–317.

Bucar, R. C. B., & Hayeri, Y. M. (2020). Quantitative assessment of the impacts of disruptive precipitation on surface transportation. *Reliability Engineering & System Safety, 203*, Article 107105.

Byun, J.-E., & D'Ayala, D. (2022). Urban seismic resilience mapping: A transportation network in Istanbul, Turkey. *Scientific Reports, 12*(1), 8188.

Chalkiadakis, C., Perdikouris, A., & Vlahogianni, E. I. (2022). Urban road network resilience metrics and their relationship: Some experimental findings. *Case Studies on Transport Policy, 10*(4), 2377–2392.

Chen, C., Geroliminis, N., & Zhong, R. (2024). An iterative adaptive dynamic programming approach for macroscopic fundamental diagram-based perimeter control and route guidance. *Transportation Science, 58*(4), 896–918.

Chen, C., Huang, Y. P., Lam, W. H. K., Pan, T. L., Hsu, S. C., Sumalee, A., & Zhong, R. X. (2022). Data efficient reinforcement learning and adaptive optimal perimeter control of network traffic dynamics. *Transportation Research Part c: Emerging Technologies, 142*, Article 103759.

Chen, C., Wang, S., Zhang, J., & Gu, X. (2023). Modelling the vulnerability and resilience of interdependent transportation networks under multiple disruptions. *Journal of Infrastructure Systems, 29*(1), 04022043.

Cinderby, S., Haq, G., Opiyo, R., Muhoza, C., Ngabirano, A., Wasike, Y., Mwamba, D., & Cambridge, H. (2024). Inclusive climate resilient transport challenges in Africa. *Cities, 146*, Article 104740.

Diab, E., & Shalaby, A. (2020). Metro transit system resilience: Understanding the impacts of outdoor tracks and weather conditions on metro system interruptions. *International Journal of Sustainable Transportation, 14*(9), 657–670.

Dong, S., Gao, X., Mostafavi, A., Gao, J., & Gangwal, U. (2023). Characterizing resilience of flood-disrupted dynamic transportation network through the lens of link reliability and stability. *Reliability Engineering & System Safety, 232*, Article 109071.

Fang, C., Chu, Y., Fu, H., & Fang, Y. (2022). On the resilience assessment of complementary transportation networks under natural hazards. *Transportation Research Part d: Transport and Environment, 109*, Article 103331.

Gao, S., Li, D., Zheng, N., Hu, R., & She, Z. (2022). Resilient perimeter control for hyper-congested two-region networks with MFD dynamics. *Transportation Research Part b: Methodological, 156*, 50–75.

Geroliminis, N., & Daganzo, C. F. (2008). Existence of urban-scale macroscopic fundamental diagrams: Some experimental findings. *Transportation Research Part b: Methodological, 42*(9), 759–770.

Glock, J. P., Hartl, R., Krieg, M., & Becker, U. J. (2021). No climate-resilient society without a resilient transport system. In R. C. Brears (Ed.), *The Palgrave handbook of climate resilient societies* (pp. 1147–1175). Springer International Publishing.

Hoogendoorn, S. P., Knoop, V. L., Van Lint, H., & Vu, H. L. (2015). Applications of the generalized macroscopic fundamental diagram. In M. Chraibi, M. Boltes, A. Schadschneider, & A. Seyfried (Eds.), *Traffic and granular flow '13* (pp. 577–583). Springer International Publishing.

Huang, Y. P., Xiong, J. H., Sumalee, A., Zheng, N., Lam, W. H. K., He, Z. B., & Zhong, R. X. (2020). A dynamic user equilibrium model for multi-region macroscopic fundamental diagram systems with time-varying delays. *Transportation Research Part b: Methodological, 131*, 1–25.

Jin, L., Xu, X., Wang, Y., Lazar, A., Sadabadi, K. F., Spurlock, C. A., Needell, Z., Don, D. R., Amirgholy, M., & Asudegi, M. (2024). Macroscopic traffic modeling using probe vehicle data: A machine learning approach. *Data Science for Transportation, 6*(3), 17.

Kim, S., & Yeo, H. (2017). Evaluating link criticality of road network based on the concept of macroscopic fundamental diagram. *Transportmetrica a: Transport Science, 13*(2), 162–193.

Lu, Q.-L., Sun, W., Dai, J., Schmöcker, J.-D., & Antoniou, C. (2024a). Traffic resilience quantification based on macroscopic fundamental diagrams and analysis using topological attributes. *Reliability Engineering & System Safety, 247*, Article 110095.

Lu, Q.-L., Sun, W., Lyu, C., Schmöcker, J.-D., & Antoniou, C. (2024b). Post-disruption lane reversal optimization with surrogate model to improve urban traffic resilience. *(Under Review)*.

Lu, Q.-L., Sun, W., Lyu, C., Schmöcker, J.-D., & Antoniou, C. (2025). Post-disruption lane reversal optimization with surrogate model to improve urban traffic resilience. *Transportation Research Part B: Methodological 197*, 103237.

Ma, F., Liang, Y., Yuen, K. F., Sun, Q., Zhu, Y., Wang, Y., & Shi, W. (2020). Assessing the vulnerability of urban rail transit network under heavy air pollution: A dynamic vehicle restriction perspective. *Sustainable Cities and Society, 52*, Article 101851.

Markolf, S. A., Hoehne, C., Fraser, A., Chester, M. V., & Underwood, B. S. (2019). Transportation resilience to climate change and extreme weather events—Beyond risk and robustness. *Transport Policy, 74*(October 2018), 174–186.

Ministry of Ecology and Environment of the People's Republic of China. (2022). *National climate change adaptation strategy 2035, May 2022.*

Mirjalili, R., Barati, H., & Yazici, A. (2023). Resilience analysis of New York city transportation network after snow storms. *Transportation Research Record: Journal of the Transportation Research Board, 2677*(1), 694–707.

Pan, S., Yan, H., He, J., & He, Z. (2021). Vulnerability and resilience of transportation systems: A recent literature review. *Physica a: Statistical Mechanics and Its Applications, 581*, Article 126235.

Santiago-Iglesias, E., Carpio-Pinedo, J., Sun, W., & García-Palomares, J. C. (2023). Frozen city: Analysing the disruption and resilience of urban activities during a heavy snowfall event using Google Popular Times. *Urban Climate, 51*, Article 101644.

Serdar, M. Z., Koç, M., & Al-Ghamdi, S. G. (2022). Urban transportation networks resilience: Indicators, disturbances, and assessment methods. *Sustainable Cities and Society, 76*, Article 103452.

Su, Z. C., Chow, A. H. F., Zheng, N., Huang, Y. P., Liang, E. M., & Zhong, R. X. (2020). Neuro-dynamic programming for optimal control of macroscopic fundamental diagram systems. *Transportation Research Part c: Emerging Technologies, 116*, Article 102628.

US DoT. (2021). *Climate action plan: Revitalizing efforts to bolster adaptation and increase resilience.* US Department of Transportation, Office of the Secretary of Transportation, Office of the Deputy Assistant Secretary for Climate Policy, August 2021.

Wang, N., Wu, M., & Yuen, K. F. (2023). A novel method to assess urban multimodal transportation system resilience considering passenger demand and infrastructure supply. *Reliability Engineering & System Safety, 238*, Article 109478.

Wang, T., Qu, Z., Yang, Z., Nichol, T., Clarke, G., & Ge, Y.-E. (2020). Climate change research on transportation systems: Climate risks, adaptation and planning. *Transportation Research Part d: Transport and Environment, 88*, Article 102553.

Wassmer, J., Merz, B., & Marwan, N. (2024). Resilience of transportation infrastructure networks to road failures. *Chaos: An Interdisciplinary Journal of Nonlinear Science, 34*(1), 013124.

Yao, K., & Chen, S. (2023). Percolation-based resilience modeling and active intervention of disrupted urban traffic network during a snowstorm. *Journal of Transportation Engineering, Part a: Systems, 149*(5), 04023027.

Zhang, N., & Alipour, A. (2023). A stochastic programming approach to enhance the resilience of infrastructure under weather-related risk. *Computer-Aided Civil and Infrastructure Engineering, 38*(4), 411–432.

Zhang, X., Miller-Hooks, E., & Denny, K. (2015). Assessing the role of network topology in transportation network resilience. *Journal of Transport Geography, 46*, 35–45.

Zhang, Y., Zheng, S., & Chen, Y. (2024). Identification of key nodes in comprehensive transportation network: A case study in Beijing-Tianjin-Hebei Urban Agglomeration, China. *Transportation Research Record: Journal of the Transportation Research Board, 2678*(5), 827–840.

Chapter 5
Artificial Intelligence—Basic Notions and Applications for Reducing Emissions

Abstract This Chapter presents the AI technology and its dynamic role in promoting climate-friendly mobility. It first presents basic AI techniques that can perform tasks requiring human-like intelligence such as reasoning, decision-making, and learning. It then presents examples of specialized AI algorithms and models that are the backbone of artificial intelligence applications as they represent the mathematical and computational frameworks that enable machines to learn, reason, and solve problems. Short presentations are made to general application AI algorithms such as *Supervised Learning*, *Unsupervised Learning*, and *Reinforcement Learning (RL)* as well as to more specific ones such as for *Natural Language Processing, Computer Vision, Robotics and Automation, big data handling and data mining.* After a brief and concise presentation of the phases of developing AI software systems and platforms, the chapter presents and discusses various AI applications for reducing transport emissions. Of particular interest to the reader may be four specific AI applications that are given as examples from real-life systems and use cases that are presented and analyzed in length . These examples are presented in terms of the current state of technology, the role of AI in reducing carbon emissions, and their potential impacts. Finally, the chapter reviews the most pronounced challenges that exist for the widespread application of AI in climate friendly mobility such as its reliance on large-scale data, the need to construct robust network infrastructures for data transmission and real-time processing, the need for continuous iteration and optimization needed for AI systems, and the substantial computing and energy resources that are necessary.

Keywords AI concepts · AI tools · AI techniques · AI algorithms · Big data handling · AI systems development · Model training · AI mobility applications · China AI applications · C-ITS · High-speed railway maintenance · Public transport operation · Shared mobility

G. A. Giannopoulos and Y. Li, *Human Mobility, Artificial Intelligence and Climate Change*, Lecture Notes in Mobility, https://doi.org/10.1007/978-3-032-08171-1_5

5.1 Introduction

Artificial intelligence (AI) has undergone a remarkable 50-year journey from its early conceptual stages in the early 1990s to its current advanced stage of development and is becoming a transformative and disruptive force in practically all domains of human endeavour. The evolution of AI can be traced through several significant initial milestones, each contributing to its current state of advancement. Early milestones were characterized by the development of fundamental concepts and theoretical frameworks for machine intelligence and problem-solving abilities like the Alan Turing's machine[1] that laid the groundwork for machine intelligence. The Turing Test became known as a measure of a machine's ability to exhibit intelligent behaviour. The mid-twentieth century witnessed the birth of AI as a formal field of study. Early AI programs, such as those designed for playing chess or proving mathematical theorems, showcased the potential of machines to perform tasks traditionally associated with human intelligence (Carter & Nielsen, 2017). However, these programs were limited by the relatively modest computational power and algorithmic sophistication of the time. The subsequent decades saw the emergence of symbolic AI, where knowledge representation and logical reasoning took centre stage. The development of expert systems in the 1980s marked a significant leap, as these systems could solve specific, complex problems by emulating the decision-making processes of human experts (Waterman, 1986).

Despite their success, expert systems also highlighted the limitations of rule-based approaches, particularly in handling uncertainty and common-sense reasoning. The resurgence of *connectionism* in the 1980s and 1990s, i.e., the approach developed in the 1940's that resembled human cognition to a process of interactions between a highly interconnected network of neuron-like processing units, brought new life to AI research. The new AI models, inspired by the human brain, demonstrated remarkable capabilities in pattern recognition and data processing (Adorf, 2005). The advent of deep learning in the early twenty-first century revolutionized AI, enabling machines to learn from vast amounts of data and achieve human-like performance in tasks such as image and speech recognition (Sejnowski, 2018). The current era of AI is characterized by the development of large language models and sophisticated algorithms capable of natural language processing, autonomous decision-making, and creative problem-solving. AI systems like GPT-3 have demonstrated the ability to generate human-like text, while advancements in reinforcement learning have produced AI that can outperform humans in complex games.

The evolution and current state of AI in transportation, highlights a remarkable journey from initial theoretical concepts to practical, real-world applications which are currently taking place all over the transportation field. Early applications of AI in transportation focused on bringing intelligence and automation in basic traffic management functions such as traffic light control and route planning, but soon more advanced AI applications are taking over many transport-related tasks including—lately—the tasks of decarbonising passenger and freight transport operations and

[1] https://www.britannica.com/technology/Turing-machine.

creating gradually the climate-friendly mobility of the future. Most of these advancements are due to technical progress in machine learning, computer vision, and sensor technologies combined with genius functional algorithms that enable vehicles to interpret sensor data and navigate their environment without human intervention or control centres to be aware and organising appropriate responses to emerging traffic situations in real time. Other AI applications optimise fleet management operations from predictive maintenance of the vehicles (trains, buses, etc.), to AI-driven route optimization and smart ticketing systems for public transport and optimizing routes and enhancing supply chain efficiency in logistics and freight transportation. According to the International Union of Public Transport (UITP), nearly 90% of public transportation companies are actively engaged, currently, in developing and implementing AI into everyday operations (Besrukov et al., 2025).

From its humble beginnings in optimising traffic management systems to the emerging revolutionary applications of AI in transportation that we see today enhancing climate friendly mobility and enabling the movement of automated (autonomous) vehicles, AI is set to fundamentally transform the way we think about transportation in the future. The history of AI is a testament to human ingenuity and the relentless pursuit of knowledge. It is also a testament of the stepwise development process that has moved in phases involving small steps as well as significant strides, but each phase building on the previous one.

In this Chapter, we focus on getting a closer look at the AI technology itself, and on its dynamic role in promoting climate-friendly mobility. In doing so, we will showcase approaches and best practices for AI-enhanced climate-friendly mobility and refer to basic concepts and tools so that the reader will be able to understand better and assess the emerging new AI-empowered future in transport technologies and operating concepts.

5.2 AI Concepts and Tools

5.2.1 Basic Concepts

Artificial Intelligence (AI) is the technology that performs simulation of human intelligence in machines that are programmed to think, learn, and perform tasks that typically require human intelligence. These tasks include reasoning, learning, problem-solving, perception, language understanding, and decision-making. AI is used today in smart search engines, decision making or recommendation systems, voice interaction systems, strategy games and other creative tools, as well as smart operation systems in all fields of transportation. The rapid growth of AI raises ethical questions, impacts on job markets, and necessitates discussions on safety and regulatory policies. These issues are discussed in Chap. 9. There are basically two broad types of AI systems and applications:

A. *Narrow AI* (NAI) or Weak AI: This type of AI is designed and trained for auxiliary tasks. Examples include virtual assistants such as *Siri* (created by Apple) and *Alexa* (created by Amazon), as well as recommendation systems used by platforms like Netflix and Amazon. Image recognition systems, such as those used in Google Photos for sorting and categorizing images, are also examples of narrow AI. Narrow AI systems are highly effective at performing the tasks they are designed for, but they normally lack general cognitive abilities.
B. *General AI* (Artificial General Intelligence—AGI) or Strong AI: AGI refers to a type of AI that possesses the ability to understand, learn, and apply knowledge across a wide range of tasks at a level comparable to that of a human. It aims to replicate the broad intellectual capabilities of humans. AGI is still in its initial stages of development being basically a theoretical concept, but its technology is well advanced and raising already many ethical as well as legal questions.

Further and above these basic concepts there are many AI techniques, tools, algorithms and models that are now used which are set to revolutionise the future of the transport and mobility sector. The most well-known of these, are mentioned below.

5.2.2 AI Techniques

An AI technique is a comprehensive set of methods and algorithms that is used to build a computer-based AI model that can perform tasks requiring human-like intelligence such as reasoning, decision-making, and learning. AI techniques are used to assess large amounts of data, to identify patterns, and make logical decisions based on the collected data. The following broad categories of AI techniques are the most well-known ones.

5.2.2.1 Machine Learning

Machine learning (ML) is a subfield of artificial intelligence (AI), broadly defined as the capability of machines to replicate intelligent human behavior and perform complex tasks in ways that resemble human problem-solving. The primary goal of ML, within an AI model, is to develop the computer software that exhibits "intelligent behaviour" similar to humans based on the analysis of large amounts of data. This ability includes recognizing visual scenes, understanding text in natural language, or performing actions as in the physical world. ML is a core technology within AI, finding applications in various fields such as natural language processing, computer vision, and speech recognition.

5.2.2.2 Deep Learning

Deep Learning (DL) is a subset of machine learning that involves the full "training" of the AI algorithm, on large datasets, to recognize patterns and make decisions. In doing so, they mimic the structure and function of the human brain's neurons, enabling the AI model to learn and improve over time. The following are the three main forms of deep learning used today:

a. *Artificial Neural Networks* (ANNs). These represent the foundational architecture in deep learning. They consist of layers of interconnected nodes, or neurons, that transmit information to each other. Each node processes input data and passes the result to the next layer, ultimately producing an output. ANNs can be trained to perform various tasks, such as image recognition, speech recognition, and predictive analytics.
b. *Convolutional Neural Networks* (CNNs). These are specialized ANNs designed for processing grid-like data, such as images and videos. They use convolutional layers to automatically and adaptively learn spatial hierarchies of features from input data. CNNs have revolutionized computer vision, enabling machines to understand and interpret visual data with very high accuracy.
c. *Recurrent Neural Networks* (RNNs). They are another type of ANN designed to handle sequential data, such as text and time-series data. Unlike traditional ANNs, RNNs can retain information from previous inputs, making them ideal for tasks that require memory, like language translation and sentiment analysis. *Long, Short-Term Memory* (LSTM) networks and *Gated Recurrent Units* (GRUs) are advanced RNN architectures that mitigate the so-called vanishing gradient problem, allowing them to learn long-term dependencies more effectively.

The simplified diagram of Fig. 5.1 shows the interrelationship between ML and DL, as the basic techniques used in AI.

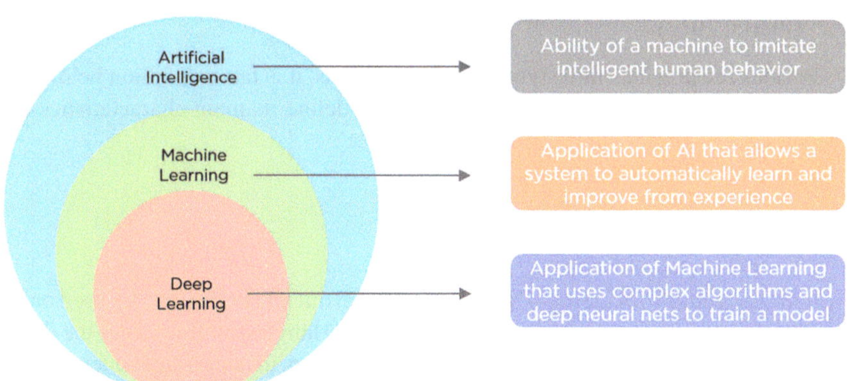

Fig. 5.1 The relationship between AI, ML and DL

5.3 AI Algorithms and Models

5.3.1 *General*

AI algorithms and models are the backbone of artificial intelligence. They represent the mathematical and computational frameworks that enable machines to learn, reason, and solve problems. A general, overall, classification of AI algorithms would distinguish three basic types:

i. *Supervised Learning Algorithms (SLA).* These involve training a model on a labeled dataset, where each training example includes input features and a corresponding target output (Suyal & Goyal, 2022). Common SLAs include linear regression, logistic regression, and support vector analysis. These algorithms are widely used for tasks such as prediction, classification, and regression (Cunningham et al., 2008).
ii. *Unsupervised Learning Algorithms (ULA).* They deal with unlabeled data, seeking to discover hidden patterns and structures within the data (Naeem et al., 2023). Clustering algorithms, such as k-means and hierarchical clustering, are examples of ULAs. These algorithms are useful for tasks like customer segmentation, anomaly detection, and dimensionality reduction (Ghahramani, 2004).
iii. *Reinforcement Learning (RL).* RLs are a paradigm of machine learning algorithms where an agent learns to make decisions by performing actions in an environment and receiving rewards or penalties based on the consequences of those actions (Wang et al., 2022a). *Deep reinforcement learning (DRL)* combines reinforcement learning with deep learning techniques, enabling agents to learn complex strategies and behaviors from high-dimensional input data, such as images and sensor readings. The DRL approach has shown promising results in applications like robotics, gaming, and autonomous systems (Wiering & Van Otterlo, 2012).

Besides the previous three types of AI algorithms, it is fair to mention below also the four basic areas of application of AI which define its main characteristics and possibilities.

5.3.1.1 Natural Language Processing

Natural Language Processing—NLP, is a field of artificial intelligence focused on the interaction between computers and human (natural) languages. It involves the development of algorithms and models that enable computers to understand, interpret, and generate human language (Chowdhary, 2020; Fanni et al., 2023).

Text Analysis is a core aspect of NLP, encompassing tasks such as sentiment analysis, entity recognition, and topic modeling. Sentiment analysis, for instance, involves determining the emotional tone of a piece of text, which is crucial for customer feedback analysis and social media monitoring (Rha & Lee, 2022). Entity recognition, on the other hand, identifies and classifies relevant entities within text, such as people, places, and organizations.

Machine Translation is another significant application of NLP, enabling computers to translate text from one language to another automatically. Neural machine translation (NMT) systems, which use deep learning techniques, have significantly improved translation quality by capturing the context and nuances of language more effectively than rule-based systems (Wang et al., 2022b).

Chatbots and *Virtual Assistants* leverage NLP to understand and respond to user queries in natural language. These systems use natural language understanding (NLU) to interpret user input and natural language generation (NLG) to produce human-like responses (Agarwal et al., 2022). They are increasingly being integrated into customer service, healthcare, and education sectors to provide personalized and efficient interactions.

5.3.1.2 Computer Vision

Computer Vision is an interdisciplinary field of AI applications that deals with how computers can gain high-level understanding from digital images or videos. It combines elements of machine learning, image processing, and artificial intelligence to enable machines to see, interpret, and respond to the visual world (Xu et al., 2021).

Object Detection and Recognition is a fundamental task in computer vision, involving identifying and classifying objects within an image or video. Techniques such as *bounding box prediction* and *region-based convolutional neural networks* (R-CNNs) have made significant strides in this area, enabling real-time object detection in various applications, including autonomous driving and surveillance systems (Kaur & Singh, 2023).

Image Segmentation goes beyond object recognition by dividing an image into multiple segments or regions, each representing a different object or part of the scene. Semantic segmentation assigns a class label to each pixel, while instance segmentation further distinguishes between individual objects of the same class (Ramesh et al., 2021). These techniques are crucial for medical imaging, autonomous navigation, and augmented reality.

Facial Recognition is another critical application of computer vision, used for biometric authentication, security surveillance, and personalized user experiences. Deep learning algorithms, particularly CNNs, have greatly enhanced the accuracy and reliability of facial recognition systems by learning to extract and match detailed facial features (Du et al., 2022).

5.3.1.3 Robotics and Automation

Robotics and Automation is the set of AI techniques that enables the design, construction, operation, and application of robots and automated systems to perform tasks traditionally carried out by humans or that are dangerous, repetitive, or impractical for human performance (Hofer et al., 2021).

Mobile Robotics focuses on the development of robots capable of navigating and interacting with their environments autonomously. These robots often rely on sensors, such as cameras, lidars, and inertial measurement units (IMUs), to perceive their surroundings and make decisions. SLAM (Simultaneous Localization and Mapping) is a key technology enabling robots to build maps of their environment while navigating them (Garaffa et al., 2021).

Industrial Robotics is concerned with the use of robots in manufacturing and assembly processes. These robots are programmed to perform specific tasks, such as welding, painting, and material handling, with high precision and efficiency (Dzedzickis et al., 2021). Collaborative robots (*cobots*) are a recent advancement, designed to work alongside humans in shared spaces, enhancing safety and productivity.

Service Robotics encompasses robots designed to assist humans in various non-industrial settings, such as healthcare, hospitality, and personal assistance. These robots often incorporate NLP and computer vision technologies to understand and respond to human commands and environments, providing services such as patient monitoring, delivery, and cleaning (Gonzalez-Aguirre et al., 2021).

5.3.1.4 Big Data Handling and Data Mining

Data Mining is a set of AI techniques that process large amounts of data with a view to discovering patterns, anomalies, and correlations within them using advanced analytical techniques. They combine elements of machine learning, statistics, and database management systems to extract valuable insights and make informed decisions (Dogan & Birant, 2021).

Classification and Prediction are fundamental tasks in data mining, involving the development of models that categorize data into predefined classes or predict future outcomes based on historical data. Techniques such as decision trees, support vector machines, and neural networks are commonly used for these purposes (Khan & Soo, 2022).

Association Rule Learning is a critical aspect of data mining, used to identify relationships between items in a transactional database. This is particularly useful in market basket analysis, where retailers can identify patterns in customer purchasing behavior to optimize product placement and promotions (Santoso, 2021).

Clustering is a data mining technique used to group similar objects together based on their attributes. Unlike classification, clustering does not require predefined labels; instead, it seeks to discover natural groupings within the data (Ren et al., 2024). Clustering is widely used in customer segmentation, anomaly detection, and image segmentation.

5.3.2 *Examples of Specialized Algorithms*

There are many specially designed algorithms and models that are used in AI software to enhance optimization and data management capabilities. They can be categorized within the above strategic categories but are also well-known by themselves as categories of their own. Some of these algorithms are mentioned below:

5.3.2.1 Evolutionary Algorithms (EA)

Inspired by biological evolution phenomena such as inheritance, mutation, natural selection, and crossover, EAs simulate these processes to solve optimization problems. They are a general family of successful optimization algorithms, the most well-known of which are the *Genetic Algorithms* (GA) which use computational mathematics for their optimization search.

In a typical GA implementation a population of candidate solutions, called "individuals", is abstractly represented as chromosomes. Traditionally, these chromosomes are encoded as binary strings (0s and 1s), although other representations can be used. The evolutionary process starts with a randomly generated population. Each "generation" involves evaluating the fitness of the entire population, selecting individuals based on their fitness, and generating a new population through crossover and mutation (Hua et al., 2021).

5.3.2.2 Simulated Annealing (SA)

Simulated Annealing (SA) is a probabilistic algorithm designed to approximate the global optimum of a given function (Aarts et al., 2005). It is classified as a metaheuristic algorithm and is commonly used to find near-optimal solutions within a large search space in a limited time frame. SA is particularly effective in generating scenarios with numerous local optima, as it has the capability to identify the global optimum (Guilmeau et al., 2017).

Simulated Annealing is often applied to problems with discrete search spaces, such as the traveling salesman problem. For cases, where finding an approximate global optimum quickly is more critical than finding a precise local optimum, SA can outperform other methods. The problems addressed by SA typically involve formulation of multivariate objective functions with several constraints, which can be incorporated into the objective function with penalties.

A key feature of SA is its "slow cooling" process, which refers to the gradual decrease in the probability of accepting worse solutions as the search progresses. This mechanism allows for a broader exploration of the search space to avoid being trapped in local optima. The algorithm operates by gradually lowering the "temperature" (which is a control parameter that governs the probability of accepting worse solutions during the search process) from an initial value to zero. At each step, a solution close to the current one is randomly selected and based on a probability related to the current "temperature", the algorithm may accept a worse solution. As the search continues, this probability approaches zero.

5.3.2.3 Ant Colony Optimization (ACO)

The Ant Colony Optimization (ACO) algorithms is a class of algorithms modeled on ant colony behaviour. It is a probabilistic technique for solving computational problems that can be framed as finding good paths through graphs (Dorigo et al., 2006). Inspired by the behaviour of real ants, artificial 'ants' (simulation agents) explore a parameter space representing all possible solutions to locate optimal solutions (Fidanova, 2021). Similar to real ants that lay down pheromones to guide each other to resources, simulated ants record their positions and the quality of their solutions. Their "pheromone trail" helps subsequent "ants" find better solutions in future iterations. ACO, combined with local search algorithms, is widely used for optimization tasks involving graphs, such as vehicle routing and internet routing.

5.3.2.4 Bee Colony Optimization (BCO) Algorithms

The BCO algorithms or simply Bees Algorithms are population-based search algorithms inspired by the food foraging behavior of honeybee colonies (Evangeline & Angel, 2023). They are applicable to both combinatorial and continuous optimization problems, provided a measure of "distance" between solutions can be defined (Teodorović, 2009). The BCO algorithm uses a combination of neighborhood (i.e. local) and global search techniques to locate optimal solutions. In this algorithm, each candidate solution is considered a food source ("flower"), and a population of agents (bees) searches for the solution space. Each "bee" evaluates the fitness (profitability) of the solutions it visits. The algorithm comprises an initialization procedure followed by a main search cycle, which repeats for a specified number of iterations (T) or until an acceptable solution is found. Each cycle includes five key procedures: recruitment, local search, neighborhood shrinking, site abandonment, and global search. These procedures collectively enable the algorithm to efficiently explore the solution space and converge on high-quality solutions by balancing exploitation of known good areas with exploration of new areas.

5.3.2.5 Fuzzy Logic Models (FLM)

Fuzzy logic is a type of many-valued logic where the truth value of variables can be any real number between 0 and 1, allowing it to handle the concept of partial truth. This contrasts with Boolean logic, where variables can only be either 0 or 1. Fuzzy logic is based on the observation that human decision-making often relies on imprecise and non-numerical information. Fuzzy models or sets provide a mathematical framework to represent and work with this vagueness and imprecision. These models can recognize, represent, manipulate, interpret, and utilize data and information that are uncertain and lack clarity (Hooda et al., 2022).

Fuzzy logic has found applications in various fields, from control theory to artificial intelligence. Unlike classical logic, which only allows for absolute true or false conclusions, fuzzy logic accommodates propositions with variable answers. For example, when a group of people is asked to identify a color, their answers might span a spectrum rather than fit into a binary true/false classification.

5.4 Phases of AI Systems Development

5.4.1 Data Collection and Preprocessing

Data is the cornerstone of any AI system, serving as the foundation upon which models are built, and decisions are made. High-quality, well-prepared data is indispensable for training effective AI models, as it directly impacts the accuracy, reliability, and generalizability of the final system.

Data collection involves gathering relevant information from a wide range of sources, including databases, sensors, user inputs, and external application programming interfaces—APIs. This process requires careful planning and execution to ensure that the collected data is comprehensive, representative, and relevant to the problem at hand.

Once the data is collected, preprocessing steps are essential to ensure that it is suitable for model training. These steps include *data cleaning*, which involves removing errors, duplicates, and irrelevant information; *normalization*, which scales the data to a common range or distribution; and *feature engineering*, which involves creating new features or transforming existing ones to improve model performance.

5.4.2 Model Training and Evaluation

After the data is prepared, the next step is to train the AI model. This involves selecting an appropriate algorithm and feeding the pre-processed data into the model. The algorithm then learns the underlying patterns or relationships within the data, enabling the model to make predictions or decisions based on new, unseen data.

Model evaluation takes place immediately after the model training phase and is a critical step in the model development process, as it assesses the performance of the trained model and helps identify areas for improvement. Techniques such as cross-validation, which involves dividing the data into multiple subsets and training and evaluating the model on each subset, provide a more accurate estimate of model performance than using a single training and test set. Confusion matrices and precision-recall curves are commonly used to evaluate model performance, providing insights into the model's ability to distinguish between different classes or outcomes (Miao & Wei, 2022).

5.4.3 Deployment and Monitoring

Once the model is trained and evaluated, it can be deployed to make predictions or decisions in real-world cases. Deployment involves integrating the model into the appropriate software or hardware platform and ensuring that it can process new data and produce accurate results in a timely manner.

Continuous monitoring is essential to ensure that the deployed AI model performs well over time and adapts to changes in the data or environment. Techniques such as *A/B testing*, which involves comparing the performance of two or more models or versions of a model, can help identify the most effective configuration. *Canary* releases, which involve deploying a new version of the model to a small subset of users before rolling it out to the entire population, provide a way to test and validate the model in a real-world environment before making it widely available. Model versioning, which involves maintaining multiple versions of the model and updating them as needed, ensures that the system can continue to operate effectively even if the underlying data or environment changes.

5.5 AI Applications for Reducing Transport Emissions

The mobility sector is a major contributor to greenhouse gas emissions, accounting for nearly one-third of global emissions. As the world grapples with the challenges of climate change, the mobility sector is faced with two major challenges:

a. Reducing its emissions to mitigate the climate change phenomenon, and
b. Adapting its infrastructure and services to achieve high resilience against climate induced extreme weather events.

The first of these two challenges is examined in Chap. 3 under the title "Measures and policies for net-zero transport emissions". We present there the most widely accepted and recognised policies for reducing greenhouse gas (GHG) emissions in the transport sector while here below we focus on the main AI applications used or that can be used. The second of the above challenges (adaptation) is the subject of Chap. 4 which examines transport sector resilience in theoretical and practical terms and of Chap. 6 where the focus is on the potential for AI applications for adaptation to climate change.

5.6 AI Application Areas for Efficiency and Climate Change Mitigation

Leveraging artificial intelligence (AI) to optimize transportation systems and reduce carbon footprints offers a promising pathway because AI has the potential to transform mobility systems by improving route planning, fleet management, autonomous vehicle development, public transit optimization, and also managing the demand for transport services as has been mentioned on several occasions in this book and in many other publications (Chen et al., 2023; Iyer, 2021; Peksen, 2022). The dynamic, transformative, role of AI in promoting climate-friendly mobility is materialised through the enabling of the development of new climate friendly transport technologies, the optimisation of the use of existing transport systems and services, and the enabling of new forms of travel demand management to make it climate aware, friendly, and sustainable.

The main application areas in which AI is expected to reshape the operation of transportation systems by improving efficiency, reducing environmental impact, and encouraging sustainable practices, are presented in the following.

5.6.1 Connected Intelligent Transportation Systems Operation

The rapid development of Connected Intelligent Transportation Systems or C-ITS using modules empowered with advanced AI software and connected via fast telecommunication links to vehicles and the various forms of infrastructure (V2X), is transforming the functionalities and efficiency of modern-day transport systems to reduce travel times, congestion and emit lower quantities of GHG. The Connected, Coordinated, and Automated Mobility (CCAM), or Intelligent Connected Vehicles (ICVs) as called in China, will demonstrate highly integrated sensors, controllers, and actuators, which combined with artificial intelligence (AI) technology and modern communication networks will enable intelligent information sharing and interaction between vehicles, humans, infrastructure, and cloud systems. Vehicle–road collaboration will be enabled through systems that typically rely on intelligent models with large-scale perception fusion and reasoning capabilities by collecting and analysing high-precision positioning data, global path planning inputs, road condition information, and vehicle operation parameters from diverse heterogeneous sensors (such as millimeter-wave radar, LiDAR, cameras, and GNSS). By leveraging this multi-modal data, the system analyzes driving scenarios, identifies and tracks key objects, and predicts their future behavioral trajectories to support accurate and timely driving decisions. Such AI-powered large-scale model systems have already been applied in practice as, for example, in the vehicle–road–cloud integrated system of the Beijing High-Level Autonomous Driving Demonstration Zone.

The application of Artificial Intelligence technology in the case of ICVs and C-ITS not only enhances driving safety and efficiency but also plays a significant role in reducing carbon emissions of the related transport activities. Traditional vehicles often experience unnecessary acceleration, deceleration, and idling due to differences in driver behaviour, delayed reactions to traffic signals, and road congestion, leading to excessive fuel consumption and carbon emissions. With AI-powered intelligent systems, vehicles can achieve more precise and smoother driving behaviour, significantly reducing energy waste. AI algorithms optimize driving behaviour to minimize energy consumption. By utilizing deep learning and reinforcement learning, these systems can analyse road conditions in real time, predict traffic signal changes, and adjust speed in advance, avoiding abrupt stops and starts, which reduces fuel consumption and carbon emissions. AI can also optimize driving routes, minimizing unnecessary distance and idling time, especially in urban traffic, by coordinating road conditions and traffic signals to avoid congestion.

In vehicle-to-everything (V2X) communication, AI interacts with traffic infrastructure, receiving real-time information to optimize driving decisions. For example, when approaching a red light, the system intelligently decelerates to avoid unnecessary stops, thus conserving fuel. Also, AI by dynamically adjusting cruising speeds, can improve fuel efficiency and reduce carbon emissions from both individual vehicles and overall traffic flow. Similarly, for freight and logistics operations, AI can analyse fleet data and intelligently schedule tasks to reduce empty runs and waiting times, thus cutting fuel consumption. With Advanced Driver Assistance Systems (ADAS), AI can help vehicles maintain appropriate speeds in congested traffic and further reduce unnecessary acceleration and carbon emissions.

The application of AI-driven ICVs and C-ITS systems offer broad expansion potential with its core advantage lies in using *large-scale model* technology to transform complex driving behaviour decisions into an "end-to-end" mapping, which greatly enhances the system's generalization capability and adaptability across different scenarios.

5.6.2 Route Optimisation

The optimisation of transportation routeing through AI-powered algorithms that enable eco-friendly route selection is performed by considering factors such as traffic patterns, road and weather conditions, fuel consumption rates, etc. (Chavhan et al., 2020). These algorithms can suggest routes that minimize fuel consumption and emissions while reducing travel times. They also integrate multi-modal options, facilitating seamless transitions between transport modes. For example, with deep reinforcement learning, the system combines the perception abilities of *deep learning* (akin to human vision) with the intelligent decision-making capabilities of *reinforcement learning* (like human brain reasoning and judgment). This combination allows the system to automatically adapt to different driving environments, improving the efficiency of perception and decision-making. The introduction of *large-scale model* technology accelerates the progress of autonomous driving, especially in complex traffic scenarios such as jaywalking pedestrians, unprotected left turns, irregular obstacles, and abnormal traffic conditions. Through extensive simulations and real-world tests, AI-powered vehicles and vehicle–road collaborative systems have not only improved traffic safety but also reduced collision rates. The widespread application of this technology offers valuable lessons for the development of intelligent transportation systems and will continue to drive the evolution of smart and sustainable transportation.

5.6.3 *Vehicle Fleet Management—Predictive Maintenance*

For the fleet managers, AI will play a critical role in fleet management by improving maintenance schedules and refuelling logistics through appropriate predictive analytics (Alexandru et al., 2022). It will also anticipate maintenance needs and will be planning refuelling stops to minimise vehicle downtime and reducing fuel consumption. These benefits are expected to be greater with the development and operation of automated vehicles (at levels L4 and L5) where AI will have the full potential to manage the trip and significantly improve fuel efficiency and lower emissions (Tyagi & Aswathy, 2021).

Related to fleet management, AI systems for predictive maintenance can analyse sensor data from vehicle fleets to detect potential technical issues before breakdowns occur. By optimizing maintenance schedules and reducing downtime, these systems enhance fuel efficiency and extend vehicle lifespans. This results in cost savings and a reduction in waste and emissions, making fleet operations more sustainable.

5.6.4 *Dynamic Public Transit Optimisation and Management*

Public transportation systems are also set to greatly benefit from AI. By leveraging data on passenger demand and traffic flow, AI improves scheduling and route planning, reduces inefficiencies, such as empty runs, and ensures optimal utilization of resources, ultimately leading to lower emissions (Nikitas et al., 2020). It can also be employed to encourage users to prefer public transportation modes by analysing user behaviour and preferences and designing incentives and campaigns to promote the shift towards public transit and other green mobility options (Olayode et al., 2020). AI applications in public transit operations are many and varied. They optimize schedules, routes, and resource allocation based on real-time data and can analyse passenger demand and adapt services to match this demand. Furthermore, public transit AI can monitor traffic and weather conditions and adapt transit schedules to improve efficiency, avoid congestion, and reduce emissions by reducing empty bus trips. Perhaps the greatest contribution of AI in public transit operation is its application to enhance the integration to public transit operations of micro-mobility and shared transport options (with or without automated vehicles) with a view to providing users with a door-to-door service.

5.6.5 Automated (Autonomous) Vehicles Operation

All automated (autonomous) vehicles (AVs) will be using a fully empowered AI software whose sophistication will be of a revolutionary nature. These vehicles rely on advanced AI algorithms to navigate safely and efficiently, minimizing unnecessary stops and accelerations and driving precisely according to the rules of eco-driving. When AVs are deployed in shared mobility mode, the role of AI will be crucial in selecting the appropriate persons to share a vehicle (pairing), drawing the route to be followed so as to minimise travel times for all passengers, cater for any privacy issues that may arise, provide travel information and advice to each and every one of the persons sharing the vehicle, and so on. It can be said that AVs and AI will be inextricably tied together in their future development.

5.6.6 Demand Management

As it is described analytically in Chap. 8 (Adapting the human mobility behaviour), the management of the demand for travel is of crucial importance for achieving sustainability in the future urban mobility systems. Through travel demand management, strategic as well as operational objectives can be achieved. The strategic objectives include reduction in car ownership, acceptance of car sharing, adoption of teleworking, use of environmentally friendly energy vectors, and so on. The more operational objectives include selection of environmentally friendly transport modes, selection of an off-peak hour for the travel, and so on. In all these tasks, AI will be a crucial enabler as it can fully utilise its ability to analyse large amounts of data in real time and synthesise optimal solutions based on its learning and optimisation algorithms.

5.6.7 Smart EV Charging Networks

AI enhances electric vehicle (EV) adoption through managing intelligent charging networks in that by predicting energy demand and availability can ease the energy loads on the carrying networks. These systems integrate renewable energy sources, dynamically allocate charging slots, and prioritize energy distribution during peak usage. By reducing grid strain and ensuring renewable energy usage, AI helps make EVs a truly sustainable mobility choice.

5.6.8 *Traffic Flow Optimization*

AI-powered road traffic management systems will, by analysing live data from cameras, sensors, and IoT devices, optimize traffic light timings and road usage. By reducing congestion and idle times, these systems will lower vehicle emissions and enhance overall mobility efficiency, particularly in urban environments.

5.6.9 *AI-Powered Freight And Logistics*

Finally, an important area in which AI can play a substantial role is freight transport and logistics. Here, AI by determining the most efficient delivery routes, consolidating the shipments, and predicting delays can reduce fuel consumption and operational costs while at the same time also facilitating the use of low-emission transport modes, such as rail or electric trucks and integrating them into the supply chains.

5.7 Examples of AI Applications for Efficiency and Reduced Emissions in China

5.7.1 *Connected Intelligent Transport Systems, C-ITS*

5.7.1.1 Current State of the Technology

The rapid development of Connected Intelligent Transportation Systems (C-ITS) or Intelligent Connected Vehicles (ICVs) as called in China, using modules empowered with advanced AI software and connected via fast telecommunication links to other vehicles and the various forms of infrastructure (V2X), is transforming the functionalities and efficiency of modern-day transport systems. The intelligent connected vehicles are equipped with integrated sensors, controllers, and actuators, which combined with artificial intelligence (AI) technology and modern communication networks enable intelligent information sharing and interaction between vehicles, humans, infrastructure, and cloud systems. The application of AI not only empowers the perception, decision-making, and control of intelligent transport systems but also significantly improves overall traffic safety, efficiency, and comfort, ultimately leading to reduced emissions.

Vehicle–road collaboration involves shared perception, decision-making, and control via real-time communication. These systems typically rely on intelligent models with large-scale perception fusion and reasoning capabilities. They collect high-precision positioning data, global path planning inputs, road condition information, and vehicle operation parameters from diverse heterogeneous sensors (such as millimeter-wave radar, LiDAR, cameras, and GNSS). By leveraging this multi-modal data, the system analyzes driving scenarios, identifies and tracks key objects, and predicts their future behavioral trajectories to support accurate and timely driving decisions.

5.7.1.2 Use Case Example

AI-powered large-scale model systems for C-ITS have been applied in the case of the vehicle–road-cloud integrated system deployed in the Beijing High-Level Autonomous Driving Demonstration Zone located in Southeast Beijing.[2] In this application, when automated vehicles encounter complex intersections or congested urban roads, the onboard AI algorithms work in coordination with roadside intelligent sensing units. They fuse real-time environmental information with historical traffic data and incorporate prior knowledge such as traffic rules and driving behavior patterns to generate optimal driving strategies. These strategies are computed and verified on the vehicle side or at a cloud control center, and then rapidly disseminated and updated via V2X communication, enabling a closed loop of perception, decision-making, and execution. By continuously collaborating, the vehicle-side and roadside systems not only extend the perception and reasoning range of individual vehicles but also optimize road resource allocation and traffic flow organization at a network level. This architecture significantly enhances the safety, operational efficiency, and sustainability of the transportation system and represents a key direction for the future development of intelligent mobility.

On the roadside the system acquires necessary data inputs, such as traffic flow and road infrastructure status, via cloud data interfaces. These data are processed by a road-cloud decision module that integrates historical information and traffic rules to identify obstacles and provide collision avoidance strategies. Simultaneously, the vehicle-side system processes task demands, vehicle status, and environmental perception data, forming a vehicle-roadside state space. Through decision logic modules, the AI system makes final behaviour decisions, which are then fed back to the vehicle's control module. This collaborative perception and decision-making mechanism allows ICVs to make complex driving decisions within a C-ITS at very short timeframes, ensuring driving safety and efficiency.

The use of AI-powered large-scale models in the Beijing High-Level Autonomous Driving Demonstration Zone is a fundamental break with traditional vehicle–road collaboration algorithms which relied on small data sets and models, which, though effective for specific tasks, required large amounts of manual data annotation, leading to high development costs, slow iteration, and limited generalization capability. Using AI and the introduction of large-scale models, which offer superior generalization and performance, will lower the barriers for development and use of C-ITS, ushering in a new paradigm for vehicle–road cooperation mobility.

[2] See: https://english.beijing.gov.cn/beijinginfo/sci/latesttrend/latesttrendlist.

5.7.1.3 AI's Role in Reducing Carbon Emissions

The application of Artificial Intelligence technology in the case of ICVs and C-ITS not only enhances driving safety and efficiency but also plays a significant role in reducing carbon emissions of the related transport activities. Traditional vehicles often experience unnecessary acceleration, deceleration, and idling due to differences in driver behaviour, delayed reactions to traffic signals, and road congestion, leading to excessive fuel consumption and carbon emissions. With AI-powered intelligent systems, vehicles can achieve more precise and smoother driving behaviour, significantly reducing energy waste. AI algorithms optimize driving behaviour to minimize energy consumption. By utilizing deep learning and reinforcement learning, these systems can analyse road conditions in real time, predict traffic signal changes, and adjust speed in advance, avoiding abrupt stops and starts, which reduces fuel consumption and carbon emissions. AI can also optimize driving routes, minimizing unnecessary distance and idling time, especially in urban traffic, by coordinating road conditions and traffic signals to avoid congestion.

In vehicle-to-everything (V2X) communication, AI interacts with traffic infrastructure, receiving real-time information to optimize driving decisions. For example, when approaching a red light, the system intelligently decelerates to avoid unnecessary stops, thus conserving fuel. Also, AI by dynamically adjusting cruising speeds, can improve fuel efficiency and reduce carbon emissions from both individual vehicles and overall traffic flow. Similarly, for freight and logistics operations, AI can analyse fleet data and intelligently schedule tasks to reduce empty runs and waiting times, thus cutting fuel consumption.

5.7.1.4 Potential Future Developments

The application of AI-driven ICVs and C-ITS systems in the transportation industry has broad expansion potential. Its core advantage lies in using *large-scale model* technology to transform complex driving behaviour decisions into an "end-to-end" mapping, which greatly enhances the system's generalization capability and adaptability across different scenarios. For example, with deep reinforcement learning, the system combines the perception abilities of *deep learning* (akin to human vision) with the intelligent decision-making capabilities of *reinforcement learning* (like human brain reasoning and judgment). This combination allows the system to automatically adapt to different driving environments, improving the efficiency of perception and decision-making.

The introduction of *large-scale* C-ITS technology will accelerate the progress of automated driving, especially in complex traffic scenarios such as jaywalking pedestrians, unprotected left turns, irregular obstacles, and abnormal traffic conditions. Through extensive simulations and real-world tests, AI-powered vehicles and vehicle–road collaborative systems will be developed for improved traffic safety and reduced collision rates. The widespread application of this technology will offer valuable lessons for the development of intelligent transportation systems and will continue to drive the evolution of smart and sustainable transportation.

5.7.2 *High-Speed Railway Maintenance*

5.7.2.1 Current State of the Technology

As high-speed railway networks continue to expand, the complexity and need for continuous, real-time, railway track maintenance and inspection have intensified. Traditional manual maintenance methods are not only inefficient but also fraught with risks of human error, potentially undermining the safety and stability of railway operations. To address these challenges, AI-driven intelligent railway inspection systems have gradually become the cornerstone of modern railway maintenance regimes especially in high-speed railway systems. Traditional inspection systems require line shutdowns for manual checks, using specially equipped vehicles, causing delays and extra CO_2 emissions. AI systems can perform round-the-clock monitoring while capturing and analysing environmental data in real-time during train operations. These features not only enhance maintenance efficiency but also significantly reduce energy consumption and economic losses caused by train stoppages.

5.7.2.2 Use Case Example

The system reviewed here was developed by the China Academy of Railway Sciences and applied in inspection vehicles on multiple railway sections, including the Beijing-Shanghai High-Speed Railway, Beijing-Guangzhou High-Speed Railway, Beijing-Shenyang Line, and Chengdu-Kunming Line. The practical application of this intelligent system has already covered more than 1700 kms of operational rail lines in China significantly reducing manual maintenance costs. The deep learning algorithms employed, not only enhance detection efficiency but also alleviate the excessive reliance on manual intervention inherent in traditional inspection methods, advancing the automation and intelligence of high-speed rail maintenance.

The system relies heavily on deep learning algorithms and computer vision technology to automatically analyse images and video data, captured by cameras along the high-speed railway network. Once pre-processed, this data undergoes AI-driven semantic segmentation and feature extraction, allowing for the detection of key issues such as rail surface damage and foreign object intrusion (e.g., balloons, kites, bird nests) and abnormalities in overhead line equipment. The system developed is capable of accurately identifying various types of railway defects through diversified training datasets. By adopting transfer learning and online learning strategies, the AI models developed can adapt to image data from varying lighting conditions, weather patterns, and track environments, overcoming the low detection accuracy of traditional systems in complex settings. AI detection now achieves high levels of accuracy in complex scenarios, such as ballast foreign objects or track surface damage. Particularly in foreign object detection, multiple deep learning models are employed to automatically classify object types and trigger alerts, greatly improving response times in maintenance.

5.7.2.3 AI's Role in Reducing Carbon Emissions

Optimizing maintenance efficiency, thereby reducing redundant energy consumption and emissions, is primarily how AI technology facilitates carbon emissions reduction in high-speed railway maintenance systems. Traditional railway maintenance practices often involve the frequent mobilization of manpower and equipment for large-scale inspections, leading to high energy consumption and additional carbon emissions. AI systems, can achieve significant energy savings in railway maintenance through the following mechanisms:

1. Automated detection and analysis of track data during train operations, eliminating the need for frequent shutdowns, as required in traditional methods. This dramatically cuts the energy consumption associated with idling and halted trains. In high-density transportation environments like high-speed rail, the energy savings from reduced downtime are considerable, so AI technology indirectly contributes to carbon emissions reduction.
2. Traditional railway equipment maintenance follows fixed schedules and is susceptible to human error, often leading to issues being discovered only after they have worsened, necessitating more frequent repairs and replacements. AI systems, by enabling precise predictions and early detection, can effectively extend the service life of equipment, thereby avoiding the resource wastage and increased carbon emissions associated with frequent replacements and repairs.
3. The application of AI technology has greatly reduced the need for maintenance equipment mobilization, lowering the related fuel consumption and carbon emissions. AI systems effectively minimize redundant equipment usage and ensure that maintenance is performed precisely when needed.

According to existing research data, the application of AI technology in the use case considered has reduced carbon emissions related to railway maintenance by approximately 5 to 7% per kilometre of track. For large-scale high-speed rail systems, this translates into hundreds of thousands of tons of CO_2 emissions reductions annually.

5.7.2.4 Potential Future Developments

Looking ahead, the application of AI in high-speed railway maintenance systems is expected to bring a wide range of potential future impacts. Economically, AI systems will significantly lower long-term operational and maintenance costs, reducing reliance on high-cost manual labour. As AI technology continues to advance, railway operators will be able to optimize maintenance schedules and improve operational efficiency through more precise data analysis and prediction.

In terms of environmental protection, the continued application of AI will greatly promote the development of low-carbon railways. Against the backdrop of growing global demand for sustainable transportation, AI in high-speed railway maintenance not only will help reduce carbon emissions but also will enhance overall resource efficiency, reducing material waste and excessive energy consumption.

5.7.3 Public Transport Operation (Ticket Payment and Verification)

5.7.3.1 Current State of the Technology

AI is set to optimize public transportation routes and schedules, making it more efficient and attractive to users, and help reduce the number of private vehicles on the road. As urban public transportation systems continue to evolve, many cities are facing the challenge of a mismatch between public transport capacity and passenger demand, particularly at hub stations. During peak hours or special periods, passenger pressure often leads to station congestion and inefficient entry and exit processes, further exacerbating overall traffic inefficiency. To address these issues, more and more cities are turning to artificial intelligence (AI) and big data technologies to optimize rail transit operations and management.

The application of AI in optimizing capacity utilization across public transport systems leads to better operational efficiency. AI enables such systems to adjust schedules and capacity in real-time, based on predictive analytics and real-time data. This allows for a more dynamic response to passenger demand, reducing inefficiencies and improving service during both peak and off-peak periods. Through optimized scheduling, the system can reduce energy consumption during low-demand hours while ensuring that resources are allocated effectively during high-demand periods, leading to a more efficient and cost-effective operation.

Furthermore, AI brings significant improvements to safety and emergency management. Real-time data analysis allows for quicker identification of potential security threats or operational disruptions. AI-powered systems can instantly detect anomalies or unauthorized access, ensuring that only authorized individuals can enter stations and travel on the metro network. Furthermore, AI enables rapid emergency response by automatically adjusting operations during disruptions or crises, helping to maintain smooth system operations and ensure passenger safety. This ability to respond swiftly to unexpected events also contributes to the stability and reliability of the metro system.

5.7.3.2 Use Case Example

The case examined, is the application of AI for ticket payment verification namely the *SenseMeteor Intelligent Rail Transit Platform*, a system developed by Hong Kong based *Sensetime* company (https://www.sensetime.com/en) in collaboration with partners. The *SenseMeteor* platform includes the *SenseMeteor Smart Commute System* that is leveraging advanced facial recognition and QR code-based access for station entry and exit. Passengers can choose to use either facial recognition or QR codes, bypassing the need to queue for tickets or pass through traditional ticket gates, thus significantly enhancing efficiency.

The *SenseMeteor* system is powered by deep learning technologies that allow for fast and accurate facial recognition, enabling passengers to travel by simply scanning their faces, with post-trip payment facilitated. By eliminating the need for ticket purchases or waiting in line at ticket gates, this solution greatly improves entry and exit efficiency. Additionally, the *SenseMeteor* platform is deeply integrated with the rest of the public transport (in this case metro) system operation, ensuring a seamless experience from entry and exit to fare settlement. This intelligent solution not only optimizes passenger flow but also alleviates congestion during peak periods. The system is based on a specially optimized deep learning algorithm that ensures high accuracy even in high-traffic environments. With a misrecognition rate of less than one in a million, the platform efficiently handles millions of daily passengers while maintaining high precision. The success of this technology is evident in its implementation across metro systems in Chinese cities such as Zhengzhou, Xi'an, and Harbin, while it is currently being deployed on other newly constructed metro lines elsewhere. This marks a significant step toward smart management in urban rail public transit.

5.7.3.3 AI's Role in Reducing Carbon Emissions

The integration of AI into urban public transport systems improves operational efficiency and plays a crucial role in reducing carbon emissions. AI achieves this by optimizing train schedules and operations to reduce energy consumption during non-peak periods. AI can adjust train frequencies and departure times during off-peak hours when demand is lower, ensuring that energy is used more efficiently and avoiding unnecessary energy expenditure. Furthermore, AI systems powered by big data can monitor real-time passenger flows and adjust train frequencies, accordingly, reducing waste in the system.

Optimization of resources leads to a decrease in energy consumption and emissions. By reducing overcrowding and improving the flow of passengers, AI contributes to lowering the overall carbon footprint of the public transport system. The increased energy efficiency and reduced resource wastage contributes to more sustainable urban transportation systems, providing significant environmental benefits. AI-based solutions like the *SenseMeteor Smart Commute System* help reduce passenger waiting times and streamline the entry and exit processes. By minimizing delays and idle time, both in stations and during travel, AI technology helps reduce overall energy use and improves the environmental performance of the rail network..

5.7.3.4 Potential Future Developments

The integration of AI and big data into urban public transport systems operation will bring profound and far-reaching impacts across all dimensions of the urban public transportation system. The adoption of AI will significantly improve the passenger experience by reducing waiting times, eliminating long queues, and providing a more efficient and streamlined travel process while for the operator will enhance efficiency, and sustainability. Passengers will be able to pass through turnstiles or checkpoints for ticket validation without physically holding a ticket quickly and efficiently through face recognition or other biometric checks thus reducing the stress and annoying delays that often occur during peak hours. With AI's ability to predict and manage passenger flow in real-time, congestion can be alleviated, and the overall experience will become far more user friendly and convenient.

5.7.4 Shared Mobility

5.7.4.1 Current State of the Technology

Shared mobility is hailed by many as a key to solving the future mobility problems by helping reduce the continuous growth of car ownership and use that creates more travel demands, traffic congestion and ever-increasing travel times, especially during peak hours. Its characteristics and technical data are detailed in other chapters namely Chaps. 3 and 10. Currently, shared mobility is used in ever increasing numbers while the advent of automated mobility is expected to exponentially increase its use and hopefully reduce car ownership and use very considerably mainly in urban areas. Optimizing the operational characteristics of shared mobility in terms of scheduling the vehicles and matching passenger travel demands to improve vehicle utilization and reduce the number of vehicles on the road, is of crucial importance and one of the most suitable applications of AI in this area.

5.7.4.2 Use Case Example

The case to consider here is the leading mobility technology platform of *DiDi Global Inc* (https://web.didiglobal.com/). This company operates (mainly) in China and several other countries serving more than 550 million users around the world. It compares in terms of size and diversity to Uber serving not only daily mobility needs but also the needs in food delivery, financial services and more (Yuan, 2025). The Didi platform offers a wide range of AI app-based services, including ride hailing, taxi hailing, designated driving, carpooling and other forms of shared mobility as well as certain energy and vehicle services, food delivery, and intra-city freight services. It leverages massive data and uses advanced AI technologies to develop deep learning algorithms, that provide precise ETA (Estimated Time of Arrival), intelligent dispatching, and demand forecasting thereby achieving efficient vehicle matching and dynamic scheduling.

The platform's intelligent dispatching system automatically adjusts vehicle allocation and routes based on user demand, vehicle location, and real-time traffic conditions, optimizing vehicle utilization and avoiding idle time, thus improving travel efficiency and reducing unnecessary fuel consumption. It also uses demand forecasting to anticipate travel demand at different times, ensuring that vehicles are allocated to high-demand areas promptly and efficiently preventing congestion during peak hours. This combination of technologies has significantly improved the efficiency of shared mobility, reduced urban traffic pressure, and promoted more sustainable and environmentally friendly mobility.

5.7.4.3 AI's Role in Reducing Carbon Emissions

The application of AI technologies in shared mobility has not only significantly enhanced travel efficiency but also played a significant role in reducing carbon emissions. Through intelligent dispatching and accurate ETA predictions, shared mobility platforms can dynamically adjust vehicle routes and speeds based on real-time traffic and weather data, thereby minimizing idle and waiting times and reducing fuel consumption. The intelligent dispatching system effectively avoids idle time during low-demand periods, reducing unnecessary carbon emissions. The ETA prediction system, by optimizing travel paths, also helps avoid traffic congestion during peak hours, thus lowering fuel consumption and emissions. In addition, demand forecasting allows platforms to understand travel needs at different times, efficiently allocate vehicle resources, and prevent congestion, further reducing unnecessary traffic burdens. The integration of these technologies enables shared mobility platforms to not only improve travel efficiency but also significantly cut down on carbon emissions, advancing green travel development.

5.7.4.4 Potential Future Developments

The technological innovations introduced by shared mobility platforms can have a profound impact on urban transportation systems operation and the environment. Through precise ETA predictions and intelligent dispatching of shared mobility vehicles, such platforms will greatly improve vehicle utilization, optimize resource allocation, reduce idle times, and alleviate congestion. Through environmentally sensitive route selection optimization AI can help reduce congestion during peak hours and avoid unnecessary idle time thus lowering pollutant emissions. Real-time scheduling of the shared vehicles will also minimize unnecessary vehicle operation and improve the system's energy efficiency. The successful application of AI-assisted shared mobility technologies will provide valuable incentives to abandon car ownership and use which in itself will be the most visible positive effect that will reduce congestion and minimise the vehicle-miles on the roads. Reducing the car ownership and use in urban areas will probably be the greatest benefit of the use of AI in transportation. By leveraging intelligent scheduling and prediction systems and making shared mobility a user accepted and preferred mode of transport, AI shared mobility platforms will not only improve operational efficiency but will also support the realization of a more sustainable and environmentally friendly mobility driving the global development of green and climate friendly transportation.

5.8 Challenges and Weaknesses

The widespread application of AI in climate friendly mobility comes with certain risks and challenges which must be properly addressed if we want to achieve the maximum of the potential. The system's reliance on large-scale data, necessitates the construction of more robust network infrastructure to ensure smooth data transmission and real-time processing of the gathered data. Additionally, the continuous iteration and optimization of AI systems requires substantial computing resources, raising questions about how to minimize the carbon footprint of these computational processes. Overall, AI technology offers considerable potential for improving the efficiency of new mobility systems and operations while playing a critical role in reducing the environmental impacts of these activities, but key challenges exist, and these are discussed below.

5.8.1 Achieving Data Interoperability

Interoperability is a critical challenge in any large-scale system that consists of different modules or subsystems. The data sets that are to be used by AI algorithms for the various climate friendly mobility applications are many and varied and coming from different sources with different methods of collection and storage. This fragmentation makes it difficult and is indeed risky for AI to utilise fully i.e. collect, analyse, and interpret such data comprehensively. For example, public transport data may come from different public transport organisations which often have proprietary data silos and thus cannot be utilised in a straightforward way. Similarly, data on private vehicles may rely on different GPS systems and communication protocols which makes integrating them in one data base difficult and risky.

To achieve data interoperability, AI systems must be designed to recognise and bridge existing discrepancies and gaps and ensure—at the end—a uniform data base on which to perform its analyses. Achieving data interoperability requires the development of robust data standards and protocols that can facilitate data sharing and integration across different platforms. Additionally, advanced algorithms and sophisticated software architectures are needed to aggregate and analyse data from multiple sources in real-time.

Achieving data interoperability also necessitates overcoming technical and regulatory barriers. Different countries and regions may have varying data privacy laws and standards, which can complicate data sharing and integration. Collaboration between governments, private sector companies, and academic institutions is essential to develop a comprehensive framework that supports such interoperability while safeguarding other criteria such as data integrity, representativeness, accuracy, compatibility, and so on. With the establishment of clear guidelines and standards for data integration and sharing, we can ensure data interoperability, and data availability for extensive AI application while at the same time protecting sensitive information.

5.8.2 Achieving Interoperability of AI Applications

Establishing interoperability of the various AI applications in different countries and jurisdictions, is another necessity which calls for international standards for AI applications in climate-friendly mobility. The issue is not only to establish such standards but to establish uniform standards applicable (ideally) at a global scale. There is no use for standards that are valid and have been developed for one country or region but standards that are accepted and honoured by many and different countries and regions. These standards should encompass data privacy, security, and other ethical considerations as reported in detail in Chap. 9. Another version of this interoperability requirement would be to develop AI systems that are designed to comply with evolving regulatory requirements and ethical guidelines in different countries and regions.

Developing standards for AI in transportation applications requires a collaborative effort between stakeholders, including governments, private sector companies, and academic institutions. Governments can play a pivotal role in setting the groundwork for AI standards and their wider adoption by providing clear regulatory frameworks and guidelines. This includes establishing data privacy laws, setting ethical guidelines for AI use, and providing incentives for companies to adopt sustainable practices. The private sector can contribute by investing in research and development, adopting best practices, and collaborating with other stakeholders to develop the corresponding AI development standards at industry or company level (see also Chap. 9). Academic institutions can offer expertise and insights into ethical considerations and societal impacts, ensuring that AI systems are designed to promote fairness, transparency, and accountability.

5.8.3 Adaptability of AI Applications

AI applications for planning and executing human travel in climate-friendly mobility must be able to adapt to changing circumstances, such as unexpected delays or cancellations of mobility services, adapting to alternative routes or solutions in a timely and effective manner and so on. This requires that AI systems are designed to monitor and adapt their recommendations so as to always reflect the existing transportation system's operating conditions and influencing factors ensuring a realistic and user-friendly experience for travellers.

To achieve this, AI apps and platforms need to be equipped with advanced machine learning algorithms that can learn from historical data and adapt to new patterns. They must also be capable of adapting with the varying transportation system operating conditions and technologies used especially in areas such as public transport, vehicle-sharing, or autonomous vehicles operation to provide comprehensive travel solutions. Collaboration between AI developers and transportation planners is essential to ensure that future AI systems are designed to promote sustainable travel practices and reduce greenhouse gas emissions.

5.8.4 Data Privacy Issues

The utilization of artificial intelligence technologies in transportation systems heavily depends on the aggregation and interpretation of vast amounts of data, including sensitive personal information such as user locations, travel patterns, and preferences. Ensuring the confidentiality and security of this data is paramount for building trust in AI-driven transportation systems. Any lapse in data privacy protection can lead to privacy breaches, unauthorized access, and misuse of personal information, which can undermine public confidence and deter the adoption of climate-friendly transportation solutions. This is particularly important since AI systems often require sharing data across multiple stakeholders, including service providers, vehicle manufacturers, and infrastructure operators, which increases the complexity of maintaining robust data privacy and security protocols. Ensuring compliance with data protection regulations, such as the GDPR or CCPA, adds another layer of responsibility and complexity for transportation operators.

Advanced cyberattacks, targeting interconnected transportation data networks can result in unauthorized access to centralized databases or manipulation of AI algorithms and could disrupt transportation services or compromise user safety. Addressing these challenges necessitates the implementation of cutting-edge encryption techniques, decentralized data storage, rigorous access controls, and continuous monitoring systems to safeguard data and maintain system integrity. Balancing data privacy with the need for extensive data sharing remains a critical challenge in fostering trust and adoption of AI-enabled, climate-friendly mobility solutions.

5.8.5 *Development of the Necessary Infrastructure*

Implementing artificial intelligence technologies for climate-friendly mobility requires substantial investment in advanced infrastructure and technology. Key components like high-resolution sensors, cameras, and edge computing devices are essential for collecting and processing real-time data to enable AI applications like route optimization, autonomous driving, and predictive maintenance. In addition, robust communication networks, such as 5G or 6G for vehicle-to-everything (V2X) connectivity, are necessary to facilitate seamless data exchange between vehicles, infrastructure, and centralized systems.

For many smaller transportation systems or those in developing countries, these requirements pose significant financial and logistical challenges. Limited budgets may prevent the acquisition and deployment of cutting-edge hardware, while inadequate technological expertise and workforce training hinder the integration and maintenance of such systems. Additionally, outdated infrastructure, such as poorly maintained roads and insufficient charging networks for electric vehicles, can impede the successful adoption of AI-driven climate-friendly transportation solutions.

Even in developed countries, upgrading legacy systems to accommodate AI technologies can be resource-intensive and time-consuming. The need for large-scale sensor deployment and infrastructure adaptation often conflicts with budget constraints and urban planning priorities. Governments and private sectors must collaborate to bridge these gaps, ensuring equitable access to advanced transportation technologies while avoiding an overreliance on centralized systems that may exacerbate existing disparities.

Finally, the rapid pace of AI innovation often outstrips the rate at which infrastructure can be upgraded, creating a mismatch between technological capabilities and practical implementation. Overcoming these infrastructure challenges will require strategic investments, international cooperation, and the development of scalable, cost-effective solutions tailored to diverse contexts and resource levels.

References

Aarts, E., Korst, J., & Michiels, W. (2005). Search methodologies: introductory tutorials in optimization and decision support techniques. In E. K. Burke, & G. Kendall (Eds.), *Simulated annealing* (Chap. 7, pp. 187–210). Springer.

Adorf, H. M. (2005). Connectionism and neural networks. In A. Heck, & F. Murtagh (Eds.), *Knowledge-based systems in astronomy. Lecture notes in Physics* (Vol. 329, pp. 213–245). Springer. https://doi.org/10.1007/3-540-51044-3_25

Agarwal, Sh., Agarwal, Bh., & Gupta, R. (2022). Chatbots and virtual assistants: A bibliometric analysis. *Library Hi Tech, 40*(8), 1013–1030. https://doi.org/10.1108/lht-09-2021-0330

Alexandru, M., Dragoș, C., & Bălă-Constantin, Z. (2022). Digital Twin for automated guided vehicles fleet management. *Proceedings Computer Science, 199*, 1363–1369. https://doi.org/10.1016/j.procs.2022.01.172

Besrukov, O. (2025). Navigating the Future: Ethical Considerations in AI-Driven Transportation. Techstack blog 28 May 2025. In: https://tech-stack.com/blog/ai-in-transportation/

Carter, S., & Nielsen, M. (2017). Using artificial intelligence to augment human intelligence. *Distill, 2*(12). https://doi.org/10.23915/distill.00009

Chavhan, S., Gupta, D., Chandana, B. N., Khanna, A., & Rodrigues, J. P. C. (2020). IoT-based context—aware intelligent public transport system in a metropolitan area. *IEEE Internet Things*. https://doi.org/10.1109/JIOT.2019.2955102

Chen, L., Chen, Z., Zhang, Y., Liu, Y., Osman, A. I., Farghali, M., Hua, J., Al-Fatesh, A., Ihara, I., Rooney, D. W., & Yap, P.-S. (2023). Artificial intelligence-based solutions for climate change: A review. *Environmental Chemistry Letters, 21*(5), 2525–2557.

Chowdhary, K. R. (2020). Natural language processing. In K. R. Chowdhary (Ed.), *Fundamentals of artificial intelligence* (pp. 603–649). Springer Nature. https://doi.org/10.1007/978-81-322-3972-7_19

Cunningham, P., Cord, M., & Delany, S. J. (2008). Supervised learning. *Machine learning techniques for multimedia: Case studies on organization and retrieval* (pp. 21–49). Springer.

Dogan, A., & Birant, D. (2021). Machine learning and data mining in manufacturing. *Elsevier, Expert Systems with Applications, 166*. https://doi.org/10.1016/j.eswa.2020.114060

Dorigo, M., Birattari, M., & Stutzle, T. (2006). Ant colony optimization. *IEEE Computational Intelligence Magazine, 1*(4), 28–39.

Du, H., Shi, H., Zeng, D., Zhang, X.-P., & Mei, T. (2022). The elements of end-to-end deep face recognition: A survey of recent advances. *ACM Computing Surveys (CSUR) 54*(10s), 1–42. https://doi.org/10.48550/arXiv.2009.13290

Dzedzickis, A., Subačiūtė-Žemaitienė, J., Šutinys, E., Samukaitė-Bubnienė, U., & Bučinskas, V. (2021). Advanced applications of industrial robotics: New trends and possibilities. *Applied Sciences, 12*(1). https://doi.org/10.3390/app12010135

Evangeline, R. C., & Angel, R. E. P. (2023). A novel approach for flexible nurse rostering allotment using the nature inspiration Bee Colony Optimization (BCO) algorithm. *SN Computer Science, 4*(5), 612.

Fanni, S. C., Febi, M., Aghakhanyan, G., & Neri, E. (2023). Natural language processing. In M. E. Klontzas, S. C. Fanni, & E. Neri (Eds.), *Introduction to artificial intelligence* (pp. 87–99). Springer Nature. https://doi.org/10.1007/978-3-031-25928-9_5

Fidanova, St. (2021). *Ant Colony optimization and applications* (vol. 947). Springer Nature. ISBN: 978-3-030-67379-6.

Garaffa, L. C., Basso, M., Konzen, A., & de Freitas, P. E. (2021). Reinforcement learning for mobile robotics exploration: A survey. *IEEE Transactions on Neural Networks and Learning Systems, 34*(8), 3796–3810. https://doi.org/10.1109/TNNLS.2021.3124466

Ghahramani, Z. (2004). Unsupervised learning. In O. Bousquet, U. von Luxburg, & G. Rätsch (Eds.), *Advanced lectures on machine learning. ML 2003. Lecture notes in computer science* (vol. 3176, pp. 72–112). Springer. https://doi.org/10.1007/978-3-540-28650-9_5

Gonzalez-Aguirre, J. A., Osorio-Oliveros, R., Rodriguez-Hernandez, K. L., Lizarraga-Iturralde, J., Morales Menendez, R., Ramirez-Mendoza, R. A., Ramirez-Moreno, M. A., & Lozoya-Santos, J. de J. (2021). Service robots: Trends and technology. *Applied Sciences, 11*(22). https://doi.org/10.3390/app112210702

Guilmeau, Th., Chouzenoux, E., & Elvira, V. (2017). Simulated annealing: A review and a new scheme. In *SSP 2021—IEEE statistical signal processing workshop, July 2021, Rio de Janeiro, Brazil*. ffhal-03275401f.

Hofer, S., Bekris, K., Handa, A., Gamboa, J. C., Mozifian, M., Golemo, Fl., Atkeson, Ch., Fox, D., Goldberg, K., & Leonard, J. (2021). Sim2real in robotics and automation: Applications and challenges. *IEEE Transactions on Automation Science and Engineering, 18*(2), 398–400. https://doi.org/10.1109/TASE.2021.3064065

Hooda, D. S., Raich, V., & Cahyomo, St. (2022). Fuzzy logic models and fuzzy control: An introduction. *Alpha Science International*. ISBN: 978-1-78322-326-5.

Hua, Y., Liu, Q., Hao, K., & Jin, Y. (2021). A survey of evolutionary algorithms for multi-objective optimization problems with irregular Pareto fronts. *IEEE/CAA Journal of Automatica Sinica, 8*(2), 303–318.

Iyer, L. S. (2021). AI enabled applications towards intelligent transportation. *Transportation Engineering, 5*. https://doi.org/10.1016/j.treng.2021.100083

Kaur, R., & Singh, S. (2023). A comprehensive review of object detection with deep learning. *Digital Signal Processing, 132*. https://doi.org/10.1016/j.dsp.2022.103812

Khan, A., & Soo Kim, H. (2022). Classification and prediction of multidamages in smart composite laminates using discriminant analysis. *Mechanics of Advanced Materials and Structures, 29*(2), 230–240. https://doi.org/10.1080/15376494.2020.1759164

Miao, J., & Wei, Z. (2022). Precision–recall curve (PRC) classification trees. *Evolutionary Intelligence, 15*(3), 1545–1569.

Naeem, S., Ali, A., Anam, S., & Ahmed, M. M. (2023). An unsupervised machine learning algorithms: Comprehensive review. *International Journal of Computing and Digital Systems IJCDS, 13*(1), 911–921. https://doi.org/10.12785/ijcds/130172

Nikitas, A., Michalakopoulou, K., Njoya, E. T., & Karampatzakis, D. (2020). Artificial Intelligence, transport and the smart city: Definitions and dimensions of a new mobility era. *Sustainability*. https://doi.org/10.3390/su12072789

Olayode, O. I., Tartibu, L. K., & Okwu, M. O. (2020). Application of artificial intelligence in traffic control system of non-autonomous vehicles at signalized road intersection. *Procedia CIRP, 91*, 194–200. https://doi.org/10.1016/j.procir.2020.02.167

Peksen, M. M. (2022). Artificial intelligence-based machine learning toward the solution of climate-friendly hydrogen fuel cell electric vehicles. *Vehicles, 4*(3), 663–680.

Ramesh, K. K. D., Kiran Kumar, G., Swapna, K., Datta, D., & Suman Rajest, S. (2021). A review of medical image segmentation algorithms. *European Alliance for Innovation (EAI) Endorsed Transactions on Pervasive Health and Technology, 7*(27). https://doi.org/10.4108/eai.12-4-2021.169184

Ren, Y., Pu, J., Yang, Zh., Xu, J., Li, G., Pu, X., Yu, P. S., & He, L. (2024). Deep clustering: A comprehensive survey. *IEEE Transactions on Neural Networks and Learning Systems*, 1–25. https://doi.org/10.1109/TNNLS.2024.3403155

Rha, J. S., & Lee, H.-H. (2022). Research trends in digital transformation in the service sector: A review based on network text analysis. *Service Business, 16*, 77–98. https://doi.org/10.1007/s11628-022-00481-0

Santoso, M. H. (2021). Application of association rule method using a priori algorithm method to find sales patterns case study of Indomaret Tanjung Anom. *Brilliance: Research of Artificial Intelligence, 1*(2), 54–66.

Sejnowski, T. J. (2018). *The deep learning revolution*. MIT Press. https://doi.org/10.7551/mitpress/11474.001.0001. ISBN electronic: 9780262346825.

Suyal, M., & Goyal, P. (2022). A review on analysis of k-nearest neighbour classification machine learning algorithms based on supervised learning. *International Journal of Engineering Trends and Technology, 70*(7), 43–48. https://doi.org/10.14445/22315381/IJETT-V70I7P205

Teodorović, D. (2009). Bee Colony Optimization (BCO). In C. P. Lim, L. C. Jain, & S. Dehuri (Eds.), *Innovations in Swarm intelligence. Studies in computational intelligence* (vol. 248, pp. 39–60). Springer. https://doi.org/10.1007/978-3-642-04225-6_3

Tyagi, A. K., & Aswathy, S. U. (2021). Autonomous intelligent vehicles (AIV): Research statements, open issues, challenges and road for future. *International Journal of Intelligent Networks (IJIN)*, (2), 83–102. https://doi.org/10.1016/j.ijin.2021.07.002

Wang, H., Wu, H., He, Zh., Huang, L., & Church, K. W. (2022a). Progress in machine translation. *Engineering, 18*, 143–153. https://doi.org/10.1016/j.eng.2021.03.023

Wang, X., Wang, S., Liang, X., Zhao, D., Huang, J., Xu, X., Dai, B., & Miao, Q. (2022b). Deep reinforcement learning: A survey. *IEEE Transactions on Neural Networks and Learning Systems, 35*(4), 5064–5078. https://doi.org/10.1109/TNNLS.2022.3207346.

Waterman, D. A. (1986). A guide to expert systems. Addison-Wesley Longman Publishing Co., Inc., July 1st, 1986. ISBN: 978-0201083132.

Wiering, M. A., & Van Otterlo, M. (Eds.). (2012). Reinforcement learning: State-of-the-art. In *Adaptation learning, and optimization series* (vol. 12). Springer Nature. https://doi.org/10.1007/978-3-642-27645-3

Xu, S., Wang, J., Shou, W., Ngo, T., Sadick, A.-M., & Wang, X. (2021). Computer vision techniques in construction: A critical review. *Archives of Computational Methods in Engineering, 28*, 3383–3397. https://doi.org/10.1007/s11831-020-09504-3

Yuan, J. (2025). The competition and comparison among travel service giants—A case study of Didi and Uber. In *Global dialogue on media dynamics, trends and perspectives on public relations and communication* (1st ed.). CRC Press, Taylor & Francis. ISBN: 9781003564966. https://doi.org/10.1201/9781003564966

Chapter 6
Artificial Intelligence Applications for Climate Friendly Mobility

Abstract This chapter investigates the use of Artificial Intelligence (AI) as an emerging and powerful tool in addressing the complex challenges imposed by adaptation of the transport sector to climate change. There is also a discussion on the potential of AI to help optimise the operation of the transport system in terms of traffic management, traffic safety, public transport operation, and the like, as well as in terms of aspects like social equity and accessibility, reducing mobility disparities, reducing carbon footprints and other. It comes as an additional examination of the issues of climate change adaptation discussed in Chap. 4 focused in more detail on the AI applications that can be used. As climate impacts intensify, the transportation system infrastructure as well as the operational elements of it will need to be protected against increasing risks from the impacts of extreme weather events. This chapter follows a holistic approach in presenting the need for a proactive climate adaptation while maintaining operational efficiency across the transport sector. It is based on the imperative to adapt transportation systems to increase resilience to climate change since failure to adapt and make those systems more resilient will result in large economic losses, grave safety risks, and disruptions to vital supply chains and mobility services. The chapter proposes an integrated framework for Artificial Intelligence applications for adaptation in the transport sector and discusses applications in predictive infrastructure maintenance, real-time monitoring and risk assessment, adaptive traffic management, optimised and inclusive transportation planning and design. It also makes detailed reference to the need for reducing mobility disparities, ensuring social equity and accessibility, establishing a viable workforce planning process, and to life-cycle analysis of mobility options.

Keywords AI advantages · Infrastructure monitoring · AI assisted transportation planning · AI backstage applications · Workforce planning · Energy consumption · Carbon footprints · Economic sustainability · Financial sustainability · AI challenges · AI weaknesses

G. A. Giannopoulos and Y. Li, *Human Mobility, Artificial Intelligence and Climate Change*, Lecture Notes in Mobility, https://doi.org/10.1007/978-3-032-08171-1_6

6.1 Introduction

The main policies for the transport sector's contribution to mitigating the climate change phenomenon and the expected global warming, were addressed in Chap. 3. The issues involved in developing the transport system's resilience and adaptation to climate change were addressed in Chap. 4. This chapter further investigates the use of Artificial Intelligence (AI) as an emerging and powerful tool in addressing the complex challenges imposed by climate change on the transport sector's adaptation to climate change. The focus is not only on adaptation but also on the great potential of AI to help optimise the operation of the transport system in terms of traffic management, traffic safety, public transport operation, and the like, and also in terms of aspects like social equity and accessibility, reducing mobility disparities, reducing carbon footprints and so on. As climate impacts intensify, the transportation system infrastructure as well as the operational elements will need to be protected against increasing risks from the impacts of extreme weather events (Jha & Dev, 2024). AI technologies offer innovative solutions and can help in facing these risks in a multitude of ways:

1. *Predictive maintenance*: Systems, powered by AI, that predict when infrastructure components are likely to fail, especially under climate stress. This is enabled by AI's potential to analyse vast amounts of data from sensors and historical records and so predict failure accurately. Then proactive maintenance can reduce the risk of catastrophic failures during extreme weather events.
2. *Real-time monitoring and early warning systems*: Machine learning algorithms can process data from multiple sources (e.g., weather stations, satellites, road sensors). This provides real-time monitoring and insights on all aspects of the transport system's operation as well as its infrastructure conditions and so detect potential hazards. This allows the establishment of early warning systems for rapid response and re-routing of traffic during climate-related (or other) disruptions.
3. *Optimised planning and design*: Planning and designing operations are the first step in any new transport and mobility system (project of construction, operation or maintenance). AI can analyse climate projections and system performance data to inform and enable more resilient designs for new transport infrastructure and/or identify priority areas for upgrading existing assets.
4. *Adaptive traffic management*: AI-driven traffic management systems can dynamically adjust and manage traffic flows based on real-time conditions, with the aim of reducing congestion and emissions while improving system efficiency under varying weather conditions.
5. *Enhanced disaster response*: This will bring into the system the experience from previous disasters and the assessment of the response made then. In the aftermath of a climate-related (or other) disaster, AI can assist in rapid damage assessment, prioritization of repair efforts, and optimization of emergency response logistics.

Figure 6.1 presents the interplay between the five above areas within an integrated framework for AI-powered climate change adaptation of the transportation sector.

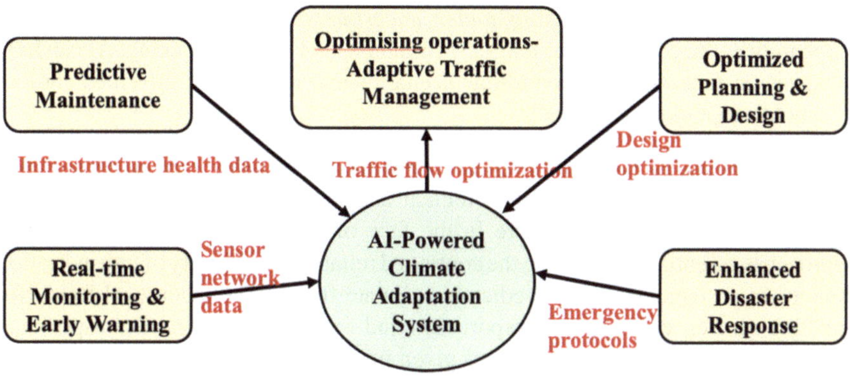

Fig. 6.1 A framework for climate change adaptation in the transport sector, supported by AI

The framework demonstrates the flow of data, information and control that feed into a central AI system, which in turn generates design guidelines, optimization suggestions and emergency response protocols. Such a holistic approach enables proactive climate adaptation while maintaining operational efficiency across the transport sector. The imperative to adapt transportation systems to climate change is clear and urgent. Transportation systems including roads, railways, ports, and airports, represent massive long-term investments that are all vulnerable to climate impacts. Failure to adapt these systems and make them more resilient could result in large economic losses, grave safety risks, and disruptions to vital supply chains and mobility services. AI technologies offer compelling advantages in addressing these challenges and in particular they offer:

1. *Ability to handle complexity*: Climate change impacts on transportation are multifaceted and involve complex interactions between weather patterns, infrastructure performance, and human behaviour. AI's capacity to process and find patterns in large, complex datasets makes it well-suited to tackle these intricate problems.
2. *Continuous learning and improvement*: Machine learning algorithms can continuously update and refine their models based on new data, allowing adaptation strategies to evolve as climate conditions change and new information becomes available.
3. *Cost-effective solutions*: While initial investments in AI systems may be substantial, they can lead to significant long-term cost savings by optimizing maintenance schedules, extending infrastructure lifespans, and reducing the impact of climate-related disruptions.
4. *Scalability*: Once developed, AI solutions can often be scaled (up or down) and adapted to different geographic contexts and types of transportation infrastructure, allowing for widespread implementation of best practices.
5. *Integration with existing systems*: AI can be integrated with existing transportation management and information systems, enhancing their capabilities without requiring complete overhauls of current infrastructure.

6. *Support for decision-making under uncertainty*: Climate change introduces significant uncertainty into long-term planning and decision making. AI can help decision-makers evaluate multiple scenarios and make more informed choices about adaptation strategies.

By leveraging AI technologies, transportation authorities and planners can thus develop more robust, flexible, and efficient adaptation strategies that enhance the resilience of critical infrastructure in the face of climate change. This proactive approach is essential for ensuring the continued reliability and safety of transportation systems in an increasingly unpredictable climate future. The theoretical frame for handling resilience issues in transportation and for the adaptation of transport and mobility systems to climate change, is given in Chap. 4.

The basic contents and characteristics of the AI-powered climate change adaptation integrated framework suggested in Fig. 6.1, are presented in more detail in the following.

6.2 Towards an AI-Powered Climate Adapted Transport System

The integrated framework for climate change adaptation, depicted in Fig. 6.1, is further discussed below in terms of its main elements shown in that Figure. Artificial Intelligence (AI) will emerge as a powerful tool in materialising this framework, offering a range of applications that can significantly improve our ability to anticipate, respond to, and mitigate the impacts of climate change on transport infrastructure and operations. This has been a desirable objective mentioned by other authors as well (Jain et al., 2023; Alqasi et al, 2024).

6.2.1 Predictive Infrastructure Maintenance

One of the most critical applications of AI in the domain of transport system adaptation, is in infrastructure maintenance. Climate change is increasingly impacting transportation infrastructure, with more frequent extreme weather events accelerating the wear and tear on critical assets such as bridges, roads, and railways (Haseli et al., 2024). In this area, AI can revolutionize traditional approaches to ensure the longevity and safety of transportation assets (Sinha et al., 2017). Traditional infrastructure maintenance approaches often rely on periodic inspections or reactive repairs, which can be time-consuming, labour-intensive, and may miss early warning signs, leading to costly repairs and potential catastrophic safety failures.

Artificial Intelligence (AI) technologies offer a novel approach for infrastructure maintenance i.e., one that based on networks of sensors, historical data, and machine learning algorithms, can perform a reliable task in which all potential problems in existing transportation infrastructure are continuously evaluated, assessed and categorized as to their gravity (Bharadiya, 2023; Paramesha et al., 2024; Patil et al., 2024; Rane et al., 2024). AI systems can analyse real-time data collected from bridges, roads, tunnels, and other facilities to predict their risk of damage under extreme climate conditions and automatically generate maintenance schedules. A real life example of an AI based infrastructure maintenance system is presented for the case of the high speed railway network of China, in Chap. 5.

AI-powered predictive maintenance not only extends the lifespan of infrastructure but also reduces economic and social losses due to sudden failures. More importantly, these systems can continuously be updated and optimised based on climate trends and facility usage, ensuring dynamic adjustment of maintenance strategies and enhancing adaptive capacity to climate change.

6.2.2 Real-Time Monitoring and Risk Assessment

As climate change brings increased uncertainty, real-time monitoring of transport networks and continuous risk assessment will become crucial tools for adapting transport networks to climate change. Transport networks are complex systems, and real-time information is particularly important for ensuring their normal operation under extreme climate conditions. AI, by processing data from transportation infrastructure, weather forecasts, satellite imagery, weather stations, etc., can provide real-time monitoring and assessment of the conditions including a comprehensive risk assessment for all types of transport networks. This real time monitoring and dynamic risk assessment capability has already been discussed in the literature and can significantly improve the resilience of transportation systems, enabling them to better respond to the challenges posed by climate change (Alqasi et al., 2024; McMillan & Varga, 2022).

6.2.3 Optimizing Operations—Adaptive Traffic Management

Extreme climate events not only threaten transportation infrastructure but also severely disrupt transport operations, too. They can lead to traffic congestion, transport delays, and interruptions to public transport services as well as disruption and delays to freight transport and logistics services. To address these challenges, AI technologies can provide powerful support for optimizing transport operations through real-time data analysis and machine learning algorithms (Yazdani et al., 2020). For example, under extreme weather conditions, AI can automatically adjust traffic signals, guide vehicle detours, and optimise public transport routes based on real-time monitored traffic flow, weather conditions, and road conditions, thereby reducing congestion and ensuring transport efficiency. Additionally, AI can assist in

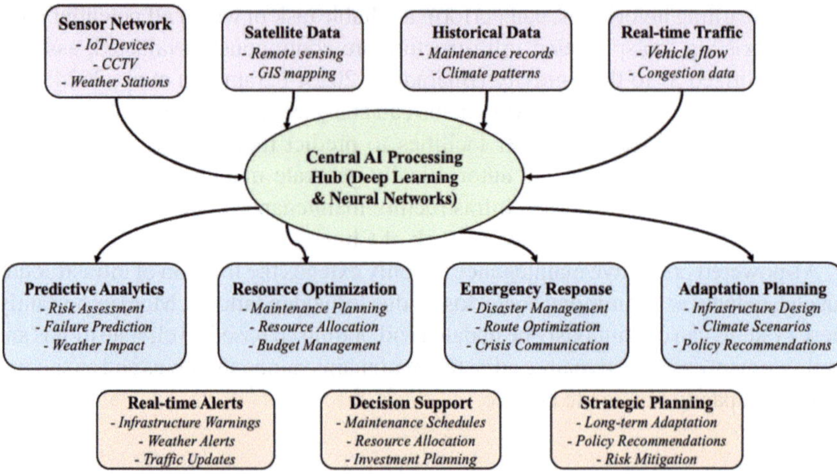

Fig. 6.2 A multi-layered AI-driven framework for transportation infrastructure monitoring, response and planning system

optimizing logistics networks, helping transport companies adjust transport routes in response to climate change, reducing delays and carbon emissions.

Figure 6.2 depicts a possible, multi-layered architecture for such an AI-enabled transport operations optimisation as well as infrastructure monitoring and response system. The framework consists of three distinct but interconnected layers:

a. A comprehensive data collection layer incorporating sensor networks, satellite data, historical records, and real-time traffic information (top line in Fig. 6.2);
b. A central AI processing hub that serves as the system's cognitive core. This processing layer contains the system's analytical capabilities, including predictive analytics, resource optimization, emergency response coordination, and adaptive planning.
c. An output layer (bottom line) delivering actionable insights through real-time alerts, decision support tools, and strategic planning recommendations. At the short-term operational level, the outputs of this process can be used as input for operational optimization applications including adaptive traffic management.

There will of course exist continuous feedback loops between the different components (not shown in the Figure for clarity), enabling dynamic system adaptation and performance optimization. This integrated approach should ensure efficient data utilization and facilitation of informed decision-making across various temporal and spatial scales, from immediate operational responses to long-term strategic planning for climate resilience.

AI-driven monitoring systems will identify potential risks to transport infrastructures from climate change in real-time such as storms, floods, and high temperatures, and at the same time provide up-to-date condition updates. In this way, traffic management can quickly be adjusted to avoid or reduce congestion and even close affected roads and bridges to ensure public safety. AI systems can predict future risk areas by analyzing historical data and current trends, helping traffic planners develop emergency plans and long-term adaptation strategies (Singh et al., 2024).

For modes of transport such as aviation and shipping, which are particularly sensitive to weather, AI can optimise take-off and landing times and adjust routes through predictive analytics, reducing operational risks due to climate uncertainties (Damaj et al., 2022; Yuan et al., 2022). By applying AI technologies, transportation systems can maintain greater flexibility and adaptability under extreme climate conditions, thereby reducing the impact of climate change on transport operations. This not only will help reduce greenhouse gas emissions but also will improve the overall resilience of transportation, ensuring the continuity and stability of socio-economic activities.

6.2.4 Optimised and Inclusive Transportation Planning and Design

There are several ways in which AI can support and enable a more effective and inclusive transportation system planning and operation. These can be grouped as follows:

- *AI assisted predictive modelling.* AI-assisted modelling can enhance the planning process by allowing planners to anticipate future mobility needs based on evolving demographic trends, urbanization, and climate migration patterns. In cities experiencing rapid population growth or shifts due to climate displacement, AI can forecast transport demand surges and suggest early interventions, such as expanding bus routes or introducing micro-mobility options in underserved areas. In this way, AI shifts transportation planning from reactive to proactive, ensuring that resources are allocated before disparities deepen.
- *AI-driven demand forecasting.* This enables a more nuanced understanding of how different communities will use transport systems. By analyzing real-time and historical travel data, AI can detect patterns in public transport usage, revealing gaps in service that disproportionately affect certain populations. For instance, in low-income neighbourhoods, AI might highlight the need for more frequent off-peak services to accommodate shift workers, or in rural areas, it could identify the need for on-demand transportation solutions. These insights allow for a more targeted deployment of resources, ensuring that mobility investments are both efficient and equitable.
- *AI-assisted equitable route optimisation.* AI goes beyond improving system efficiency and helps balance efficiency with equity. While traditional route planning often prioritizes high-traffic corridors to maximize operational performance, AI can design routes more equitably i.e. that better serve marginalized communities without sacrificing overall system viability. By analyzing the travel behaviour of underserved groups, such as women, seniors, or low-income individuals, AI can create routes that reduce travel times and improve connectivity for those who need

it most. This capability is particularly valuable in regions where climate change is altering daily travel patterns, forcing greater reliance on public transport systems.

- *Dynamic pricing models* driven by AI ensure that affordability remains central to sustainable transport initiatives. These models can adjust fares based on real-time economic conditions, making climate-friendly transport more accessible to low-income groups. AI can analyse income data, local economic trends, and household spending patterns, creating fare structures that are both fair and sustainable. This adaptability allows transport systems to remain affordable even in times of economic fluctuation, ensuring that lower-income populations do not bear the brunt of transport costs during periods of inflation or economic downturns.
- Lastly, AI-powered platforms can be increasingly used in *participatory transport planning* to enable communities to have a greater say in the planning process. By analyzing and incorporating feedback from diverse stakeholders, AI helps create more inclusive and responsive transport systems. In regions where communities have historically been excluded from decision-making due to political or economic marginalization, AI can facilitate a more democratic planning process, ensuring that the voices of underserved populations are heard and acted upon.

Over and above the previous points, which represent the "traditional"—so to speak—elements of transportation planning, there is another important potential role of AI in optimising planning and design. This refers to its ability to ensure an equitable distribution of mobility resources across different demographics and in allowing transportation planners to move beyond traditional, often biased, methods of resource allocation and instead develop systems that respond to the specific needs of diverse communities in real time (Porter & Dungey, 2021). AI can also drive systems that diagnose disparities in mobility access across different socioeconomic groups and geographies. One of the most significant contributions of AI in inclusive transport planning, also lies in equity impact assessments, where it simulates the effects of proposed transportation policies and projects on diverse demographic groups. These simulations provide planners with a clear picture of how different communities, particularly those traditionally underserved, will be impacted by new transport developments (Cirianni et al., 2023). For example, AI can identify whether a new public transportation line will truly benefit lower-income neighbourhoods or simply reinforce existing inequalities by primarily servicing affluent areas.

All these are novel capabilities in planning and designing of transport systems and facilities that permit a granular level of analysis that will allow for the adjustment of plans to ensure that even marginalized communities receive tangible benefits from climate-adaptive transport investments.

6.3 Other Important Applications

Besides the role of AI in supporting the framework for transportation infrastructure monitoring, operation, and planning outlined in Fig. 6.2, there are several other potential applications of AI, which are equally important but not so noticeable. To these we turn our attention now.

6.3.1 Reducing Mobility Disparities

This refers to the mobility gaps that exist between urban and rural areas, as well as between affluent and low-income communities. These gaps and disparities are often stark, with rural areas and economically disadvantaged groups facing limited access to reliable, affordable, and climate-friendly transportation. By utilizing AI's capacity for real-time data analysis and predictive modelling, one can develop adaptive transport systems that cater to the specific needs of these underserved populations. For example, in rural areas where public transport is often sparse and disconnected, AI-driven systems can optimise last-mile connectivity, ensuring that residents have reliable access to essential public transport services. AI can analyse real-time demand data, weather conditions, and geographic constraints to design on-demand transportation services that adapt to the unique needs of rural populations. For example, in agriculture-dependent regions, AI can help optimise transport routes during harvest seasons or periods of peak activity, ensuring that farmers and workers have timely access to markets and resources.

AI plays a critical role in multimodal transport integration, particularly in areas where single-mode solutions are insufficient to meet the needs of low-income or geographically isolated populations (Bharadiya, 2023). By integrating buses, trains, bike-sharing, and ride-hailing services into a cohesive system, AI ensures that rural and low-income residents can access a variety of transport modes, seamlessly transitioning between them. This integrated approach is especially important in areas where climate change is disrupting traditional transport systems, leading to increasingly complex and multi-stage journeys. AI's ability to optimise these multimodal networks ensures that even those in remote or underserved areas can efficiently move between modes of transport, reducing overall travel times and improving access to essential services.

One of AI's most impactful contributions in addressing mobility disparities is in infrastructure prioritization (Nikitas et al., 2020). By analyzing socioeconomic data, travel patterns, and geographic constraints, AI can identify critical areas where infrastructure investments will have the greatest impact on reducing mobility gaps. For instance, AI might highlight the need for more frequent bus services in low-income neighbourhoods or suggest the construction of pedestrian-friendly infrastructure in areas where foot traffic is high, but safety is a concern. In regions where financial resources are limited or where geopolitical tensions may curtail large-scale infrastructure projects, AI helps ensure that investments are made strategically, maximizing their impact on the most vulnerable populations.

Fare optimization is another area where AI helps bridge the gap between rich and poor communities. By analyzing data on income distribution and travel patterns, AI can design dynamic fare structures that adjust based on users' economic capabilities. For instance, AI could implement lower fares for low-income riders or offer subsidies for rural residents who face long commutes to urban centres. This ensures that sustainable, climate-friendly transport remains affordable for everyone, regardless of their economic status. Along the same lines of thinking, AI can help mitigate the digital divide by powering offline-capable mobile applications that provide transportation information and services in areas with limited internet connectivity. In many rural areas, internet access is unreliable or non-existent, which can limit the ability of residents to use modern digital tools for mobility. By enabling offline functionality, AI ensures that even in regions with poor digital infrastructure, residents can still access real-time transport information, schedule rides, and navigate public transit systems.

Finally, AI enables the efficient implementation of shared mobility services in underserved areas by optimizing vehicle deployment and route planning based on community needs. In both urban and rural low-income areas, AI can analyse travel demand to determine the best locations for shared mobility hubs, ensuring that residents have access to electric vehicles, carpooling services, and bike-sharing programmes. In doing so, AI helps reduce the dependency on private vehicle ownership, which is often economically and environmentally unsustainable in the context of climate change.

As we harness AI to reduce mobility disparities, it is essential to address potential challenges, such as algorithmic bias, which might inadvertently reinforce existing inequalities. AI models must be trained on representative data sets to avoid skewed outcomes, and privacy concerns must be rigorously addressed to protect vulnerable populations from data misuse. Additionally, digital literacy must be considered, as varying levels of technological familiarity could create new barriers for accessing AI-enhanced transport services. By addressing these challenges, AI can help build more resilient and equitable transportation systems that serve all communities, regardless of their location or economic status. We duel more in these issues below.

6.3.2 Ensuring Social Equity and Accessibility

In the face of accelerating climate change, the need for ensuring equitable access to sustainable transportation is becoming a central issue of the systems' adaptation. The ability of Artificial Intelligence (AI) to support underserved and vulnerable communities in accessing climate-friendly transport is a moral imperative. AI offers a means to transcend social barriers, ensuring that no community is left behind. The main areas in which AI supports social equity and accessibility are the following:

a. Identifying vulnerable populations with precision.
b. Predicting future needs.
c. Tailoring solutions to specific communities.
d. Breaking down barriers with multilingual and non-accessible systems.
e. Allocating resources dynamically to meet real-time demands.

Identifying vulnerable populations with precision: By integrating data on income levels, geographic isolation, demographic trends, and existing transportation infrastructure, AI can point to "transport deserts" i.e., areas where residents lack access to sustainable, reliable, and affordable transportation (Jackson et al., 2019). This data-driven identification becomes particularly crucial in regions where governments may be preoccupied with short-term economic gains or geopolitical competition, often sidelining the needs of the most vulnerable. While priorities may shift, AI can consistently highlight the socio-economic and environmental benefits of investing in underserved communities, offering a pragmatic argument for action even in complex political landscapes. For example, in regions where climate adaptation is deprioritized due to competing economic or security concerns, AI can provide clear evidence that investing in climate-friendly transport for marginalized populations leads to long-term economic resilience. This creates opportunities for governments to align their climate and economic strategies, even when global cooperation is limited.

Predicting future needs: AI's predictive analytics go beyond merely identifying current transport inequities; they also allow for the anticipation of future needs. As climate change accelerates, migration patterns, demographic shifts, and urbanization trends are likely to change dramatically. AI can model these shifts, enabling transportation planners to design systems that will meet the evolving needs of vulnerable populations quite far into the future. In regions at high risk of climate-induced displacement, AI can predict where populations are likely to relocate and what their transportation needs will be. This foresight allows for the proactive development of infrastructure that not only addresses current disparities but also anticipates future challenges. In doing so, AI helps stabilize regions that might otherwise face increased social and economic strain due to inadequate planning.

Tailoring solutions to specific communities: AI can provide valuable support in tailoring transportation solutions to the specific needs of different communities. By analyzing localized data, AI can recommend transport interventions that are not only sustainable but also culturally and demographically appropriate. For instance, AI can suggest on-demand electric shuttles for elderly populations in rural areas, while recommending shared e-bikes or micro-mobility options for younger, urban populations. These "personalized" solutions are critical in a world where geopolitical priorities vary widely. In regions where national governments may not prioritize climate adaptation, AI-driven local solutions can still meet the needs of vulnerable populations. Local governments, NGOs, and private sector actors can leverage AI to implement targeted projects that are both feasible and effective, even in the absence of broader political will.

Breaking down barriers with multilingual and non-accessible systems: AI's Natural Language Processing (NLP) capabilities open new horizons for inclusivity, particularly in multilingual or highly diverse regions. By providing real-time translation of transport information, AI ensures that non-native speakers, refugees, and migrants can access critical mobility services. This is particularly important in border areas or regions experiencing high levels of migration due to geopolitical instability or climate-related displacement. Moreover, AI-powered navigation systems can enhance accessibility for people with disabilities, offering real-time information about the suitability of transport options for individuals with mobility, auditory, or visual impairments. This, not only improves transport equity but also elevates public awareness of the need for inclusive, climate-resilient infrastructure.

Dynamic resource allocation to meet real-time demands: AI's ability to monitor transportation systems in real time allows for dynamic resource allocation that can respond to the shifting needs of underserved populations. For example, AI can analyse real-time data to identify where additional buses or ride-sharing vehicles are needed the most, ensuring that resources are deployed efficiently and equitably. In regions where financial and infrastructure resources are limited, this dynamic allocation becomes even more critical. AI can also help prioritize investments in areas where they will have the greatest impact, ensuring that even in resource-constrained environments, the most vulnerable populations receive the support they need.

6.3.3 AI in Workforce Planning

Utilizing AI to identify future workforce needs in climate-adaptive transport sectors is seen as a tool of a multitude of uses and benefits because the transition towards climate-adaptive transportation infrastructure is not just a matter of physical assets; it is also an issue that results in significant shifts in workforce composition and skills. AI is emerging as a powerful tool in anticipating and planning for these evolving workforce needs. AI-driven workforce planning in climate-adaptive transport sectors encompasses several key areas such as the following (Chhetri et al., 2020; Ogungbire & Mitra, 2024):

1. *Skills gap analysis*: AI algorithms can analyse current workforce capabilities against projected future needs, identifying potential skills gaps. This analysis considers not only technical skills related to new climate-adaptive technologies but also soft skills like adaptability and systems thinking that are crucial in a rapidly changing environment.
2. *Predictive modelling of job market trends*: By processing vast amounts of data from various sources—including job postings, academic publications, and industry reports—AI can forecast emerging job roles and declining positions in the climate-adaptive transport sector. This foresight allows for proactive workforce development strategies.

3. *Personalized training recommendations*: AI systems can assess individual employee profiles and recommend tailored training programmes to prepare the workforce for future roles in climate-adaptive infrastructure management.
4. *Scenario planning*: AI-powered simulations can model various climate scenarios and their implications for workforce needs, enabling organisations to develop flexible staffing strategies.
5. *Automated talent matching*: As new roles emerge in climate-adaptive transport sectors, AI can match job requirements with candidate profiles more efficiently, considering both traditional qualifications and adaptability to new challenges.
6. *Diversity and inclusion in workforce planning*: AI tools can help identify and address potential biases in hiring and promotion processes, ensuring a diverse workforce that brings varied perspectives to climate adaptation challenges.
7. *Continuous learning systems*: AI can facilitate the creation of adaptive learning systems that continuously update training content based on the latest developments in climate science and adaptation technologies.

AI applications in workforce planning for climate-adaptive transport sectors have been exemplified by the European Union's SKILLFUL project and the AI-powered human resource (HR) platforms developed there by forward-thinking transportation authorities (Loukea et al., 2016). The use of AI in workforce planning faces several challenges, including data privacy concerns, potential reinforcement of existing biases, and the risk of over-relying on AI predictions that may not fully capture complex social and economic factors. To address these issues, it is crucial to develop AI workforce planning tools with robust ethical guidelines and human oversight, fostering collaboration between AI developers, HR professionals, climate scientists, and transportation experts. As the sector evolves, AI-driven workforce planning will become increasingly vital in ensuring that organisations have the right talent in the right roles, contributing to the development of resilient, adaptable, and sustainable transportation systems capable of meeting climate change challenges.

6.3.4 Life-Cycle Analysis of Mobility Options

The ability to estimate the full environmental impact of different mobility options, (new transportation routes or modes, or new technologies for transport, etc.), depends on the ability to collect data and make the necessary cost–benefit analyses for all the stages of development from the initial extraction of raw materials to produce one unit to the end-of-life disposal. AI-driven *Life-Cycle Analysis* (LCA) is doing exactly that and can be a very powerful tool in assessing the full environmental footprint of a given mobility option. This holistic approach allows for more informed decision-making in transport planning, ensuring that the environmental costs of different transport options are fully accounted for and fully taken into account in determining political and economic priorities (Qiao et al., 2019).

A very illustrative example of the value of LCA is when evaluating electric vehicles (EVs). The EV's LCA will assess the environmental impact of battery production, including the mining of rare earth minerals, energy consumption during manufacturing of the battery and the vehicle, and the carbon emissions associated with electricity generation for battery charging. By comparing this data with corresponding internal combustion engine (ICE) vehicles data, AI can provide a more nuanced understanding of the trade-offs between the different options. An example of this, is depicted in Fig. 3.3 of Chap. 3. AI-driven LCA tools are essential in evaluating alternative mobility solutions, such as shared mobility services, automated vehicles, or micro-mobility options like electric scooters and bikes. By analyzing usage patterns, energy consumption, and vehicle lifespans, AI can estimate the cumulative environmental impact of these newer transport modes. For instance, AI might reveal that while shared electric scooters have a low carbon footprint during operation, their frequent replacement due to wear and tear, combined with the energy-intensive materials used in their production, results in a higher overall environmental cost than initially anticipated. This level of insight allows policymakers to make more informed decisions about which mobility options are genuinely sustainable in the long term, especially in geopolitically unstable regions where resource availability may fluctuate.

AI's predictive capabilities enhance the accuracy of life-cycle assessments, and this may alter the environmental impact assessment of different transport modes. This foresight is particularly valuable in regions where climate policy may be subject to rapid change due to shifting political landscapes. By providing a dynamic LCA that evolves with technological and policy developments, AI ensures that transport systems can adapt to future environmental challenges, even as geopolitical conditions remain uncertain. In regions where climate adaptation is deprioritized due to economic or security concerns, AI-driven life-cycle analysis can help make a compelling economic case for sustainable transportation. By demonstrating the long-term environmental and economic benefits of investing in low-carbon transport options, AI can influence decision-makers to prioritize sustainable infrastructure, even when short-term geopolitical priorities appear to conflict with climate goals.

6.3.5 Optimizing Energy Consumption

AI's role in minimizing energy use in transport systems, particularly for electric and automated vehicles will become increasingly important as climate change intensifies and geopolitical tensions create unpredictable energy markets. AI's role in minimizing energy use, can be pivotal in creating sustainable, efficient, and resilient transportation networks that can adapt to varying geopolitical contexts (Chen et al., 2022; Olatunde et al., 2024). By processing vast streams of real-time data including traffic patterns, vehicle energy consumption, and weather conditions data, AI can predict and adjust energy use in real time, ensuring that vehicles operate at maximum efficiency. For electric vehicles (EVs), AI can optimise energy use by helping to adjust driving patterns, such as acceleration and braking, based on traffic conditions or road topography. This dynamic optimisation reduces energy waste and extends the range of EVs, making them more viable in regions with limited charging infrastructure a critical issue in geopolitically unstable areas where energy resources may be scarce or inconsistent. Moreover, AI is instrumental in energy grid optimization for EV charging infrastructure. By analyzing real-time data on grid demand, energy prices, and weather conditions, it can optimise when and where EVs are charged, ensuring that they draw power during periods of low demand or when renewable energy sources, such as solar or wind, are most abundant. This not only reduces the carbon footprint of EVs but also minimizes the strain on national grid systems.

In the context of automated vehicles (AVs), AI will play an even more critical role. AVs rely on a network of sensors, cameras, and algorithms to navigate roads, all of which consume significant amounts of energy. AI-driven systems can optimise this energy use by adjusting sensor activity based on immediate environmental conditions. For example, AI may reduce the power consumption of certain sensors in low-traffic areas or during favourable weather conditions, extending the vehicle's operational efficiency. AI can also optimise route planning for AVs, selecting routes that minimize energy consumption by avoiding traffic congestion or hilly terrain. This level of optimization is particularly valuable in regions where energy costs are high or where geopolitical instability creates fluctuations in fuel prices or electricity supply (Choksi & Shah, 2024).

AI is also central in enhancing energy efficiency in public transportation systems (electric buses, trains, and metro systems). By analysing ridership data, it can adjust schedules and routes to minimize energy use, especially during off-peak hours, ensuring that vehicles do not waste energy running empty or partially full. Additionally, it can optimise regenerative braking systems, which convert a vehicle's kinetic energy into electric energy that can be stored and reused. In regions where energy prices are volatile, these AI-driven optimizations can make public transport systems more resilient and cost-effective, while also reducing their environmental footprint.

AI's ability to predict future energy needs is another crucial factor in optimizing the energy requirements of transportation systems. By forecasting how changes in population density, urban development, and climate-induced migration will affect energy demand, it enables policymakers and transport operators to plan for future energy requirements proactively. For example, AI can predict where additional EV charging stations will be needed in the coming years, ensuring that infrastructure investments are made in advance of demand surges. This foresight helps mitigate the risk of energy shortages or system overloads, particularly in regions where geopolitical instability may delay infrastructure development.

6.3.6 Reducing Carbon Footprints

Reducing the carbon footprint of transport systems is the prime objective of all decarbonization actions for climate change mitigation that were described in Chap. 3. Artificial Intelligence (AI) can emerge as a transformative tool in this quest, particularly through its capacity to optimise routing and logistics, thereby significantly reducing greenhouse gas (GHG) emissions and giving a tremendous boost to mitigation efforts and policies. AI-driven systems can streamline transportation networks at both macro and micro levels, ensuring that every journey, from long-haul freight to daily commutes, is as energy-efficient as possible in an era of increasing uncertainty and resource constraints (Machado et al., 2018). Dynamic routing optimization is essential for reducing emissions in both public and private transport operations. By analyzing real-time traffic patterns, road conditions, weather data, and fuel consumption rates, AI can recommend the most fuel-efficient routes for vehicles, minimizing time spent idling in traffic and reducing fuel use and so minimise GHG emissions in the atmosphere. For freight and logistics, this means fewer empty miles and more efficient delivery schedules, which not only cuts operational costs but also contributes to a substantial reduction in GHG emissions. In urban areas where traffic congestion is a persistent problem, AI can dynamically reroute vehicles to less congested roads, ensuring smoother traffic flow and decreasing the overall carbon footprint of transport systems.

The impact of AI extends beyond optimizing individual vehicle routes. It can also transform logistics networks at a systemic level when, by analysing data on vehicle loads, delivery schedules, warehouse locations, and fuel availability, AI can design comprehensive logistics systems that minimize energy use and carbon output (Rane et al., 2024). As an example, AI can suggest consolidating deliveries to reduce the number of trips or recommend shifting certain deliveries to off-peak hours when roads are less congested. This has a direct impact on reducing GHG emissions, particularly in the freight sector, where inefficient logistics account for a significant portion of global emissions. The role of AI in multimodal transportation—the integration and seamless coordinated operation of different modes of transport, such as rail, road, and sea—offers another avenue for reducing emissions from logistic systems (Makarova et al., 2023). By analyzing the carbon intensity of different transport modes in real time, AI can recommend optimal modal combinations that minimize the carbon footprint for a given journey. Another key area in the freight and logistics sector where AI can reduce carbon emissions is through predictive maintenance of the freight vehicle fleet (Hasan et al., 2024).

In the air and maritime logistics sectors, which are traditionally more difficult to decarbonize due to their heavy reliance on fossil fuels and the high cost of converting to clean fuels, AI can also provide valuable solutions. In air transport, it can optimise flight paths based on weather patterns, wind speeds, and air traffic, reducing fuel consumption and emissions. In maritime logistics, it can optimise shipping routes to take advantage of favourable ocean currents and minimize fuel use (Tsolakis et al., 2022). This is especially important in regions where geopolitical disputes over airspace or maritime boundaries complicate traditional air or maritime route planning. By offering alternative, fuel-efficient routes, AI helps reduce the carbon footprint of global trade, even in politically sensitive regions.

6.3.7 *Supporting Economic and Financial Sustainability*

The intersection of climate change, transportation infrastructure / operation, and economic/financial sustainability presents a complex and multifaceted landscape of challenges and opportunities. As the global community grapples with the urgent need for rapid adaptation and mitigation strategies, the economic and financial aspects of climate-resilient transportation infrastructure and operation have moved to the forefront of policy discussions and investment decisions. There is an intricate web of financial mechanisms and economic implications, where AI-powered innovative approaches can literally help shape the future of climate-adapted transportation systems by facilitating the economic and financial viability and sustainability of the relevant projects. In the following we explore some of these approaches.

6.3.7.1 Anticipatory Financing

Anticipatory financing is a forward-looking strategy powered by artificial intelligence (AI) that is revolutionising how we manage climate risks and allocating resources for relevant infrastructure upgrades (Huynh & Kiang, 2023). AI systems can provide new insights into future climate-related risks with a level of precision that was previously unattainable. They are therefore particularly suited for use in anticipatory financing applications and their role can be transformative. Examples of such applications:

1. *Risk assessment and prioritization*: AI algorithms can analyse complex datasets to identify which transportation assets are most vulnerable to climate impacts. This granular level of risk assessment allows authorities to prioritize investments in areas of highest risk, ensuring that limited resources are allocated where they can have the greatest impact. For instance, an AI system might identify a coastal highway that is particularly susceptible to sea-level rise and storm surges, flagging it for immediate attention and investment.
2. *Adaptation cost–benefit analysis*: By simulating various climate scenarios and their potential impacts on infrastructure, AI can help quantify the long-term benefits of different adaptation measures. This enables more informed decision-making about which projects to fund and how to structure financing. For example, an AI model might compare the costs of elevating a road against the potential damage and economic losses from future flooding, providing a clear picture of the costs and benefits for calculating the return on investment.
3. *Innovative financing mechanisms*: AI supports the development of novel financial instruments tailored to climate resilience and adaptation. *Climate resilience*

bonds, catastrophe bonds, and *parametric insurance products* are instruments enhanced by AI-driven risk assessments that trigger mechanisms based on predictive climate data. These instruments not only help to attract private investment into climate resilience projects but also provide a safety net for communities and industries at risk of climate disruption.
4. *Dynamic budget allocation*: Climate conditions are not static, and neither should financial planning be. AI systems can continuously update risks and recommend real-time adjustments to budget allocations, ensuring that financing remains responsive to changing needs. This dynamic approach allows for more flexible and adaptive financial strategies, crucial in the face of climate uncertainty.
5. *Cross-sector coordination*: Climate change impacts are not confined to transportation alone. AI can identify synergies between transportation infrastructure upgrades and other sectors such as energy, water management, and urban development. This holistic view facilitates integrated planning and can potentially unlock new sources of funding by demonstrating the wider benefits of resilience investments.

In the future, AI applications in anticipatory financing may extend beyond risk assessment and resource allocation, potentially including systems that automatically trigger pre-approved funding based on specific climate indicators. The development of this field represents a paradigm shift in infrastructure investment approaches to climate change (Cugurullo & Xu, 2024).

6.3.7.2 AI-Based Risk Assessment Models for Long-Term Economic Planning

The integration of AI into risk assessment processes has revolutionized long-term economic planning to include risk mitigation strategies for climate-adaptive transportation infrastructure (Haimes, 2002; Onwubuariri et al., 2024; Pirogova et al., 2022). Key aspects of AI-driven risk assessment include:

1. *Comprehensive risk data integration*: AI models can synthesize diverse data sources, including historical climate data, geological information, infrastructure condition reports, socio-economic indicators, and real-time sensor data. This holistic approach enables a more nuanced understanding of potential risks and their interconnections.
2. *Predictive modelling*: Advanced machine learning algorithms can forecast potential climate-related risks with increasing accuracy. These models can simulate various climate scenarios and their impacts on transportation infrastructure, allowing planners to anticipate and prepare for a range of possible futures.
3. *Dynamic risk mapping*: AI systems can generate and continuously update risk maps, highlighting vulnerable areas of transportation networks. These dynamic maps can enable an informed prioritization of adaptation efforts and resource allocation.

4. *Economic impact analysis*: By analyzing historical data and projecting future scenarios, AI models can estimate the potential economic impacts of climate-related infrastructure failures. This includes direct costs of damage and repair, as well as indirect costs such as disruptions to transport and supply chains and economic activity.
5. *Dynamic decision-making support*: AI-powered decision support systems can provide real-time recommendations for risk mitigation strategies, adapting to changing conditions and new data inputs.
6. *Uncertainty quantification*: Machine learning techniques can help quantify the uncertainties inherent in long-term climate projections, enabling more robust decision-making under conditions of uncertainty.

The potential of AI-based risk assessment models to enhance long-term economic planning and risk mitigation in climate-adaptive transportation infrastructure is immense. As AI technologies evolve, they are poised to play an increasingly crucial role in building resilient and sustainable transportation systems, offering powerful tools for navigating the complexities of climate change adaptation in infrastructure planning.

6.4 Challenges and Weaknesses

6.4.1 Key Challenges

While Artificial Intelligence (AI), according to the above, holds immense promises for enhancing the management and operation of climate-friendly mobility both as regards the adaptation as well as the operation of transport systems, this by no means is free of problems and challenges that must be carefully met. One can easily see that the role of AI in transforming transport systems is promising but at the same time fraught with challenges because its widespread adoption and efficacy are conditioned by several key obstacles and issues to monitor and safeguard. Below we refer to the main ones.

6.4.1.1 Data Quality and Non-bias

Data quality and availability, especially in developing countries, remains a major obstacle. Valid and good quality data are the sine-qua-non condition for the development and use of successful AI applications. AI systems are only as unbiased as the data they are trained on, and in the transport sector, this can lead to unintended distortions. In regions where access to transportation infrastructure is uneven, or where certain populations are marginalized, AI systems risk exacerbating existing inequalities. For instance, in urban areas that lack of comprehensive traffic data, AI-driven traffic management systems may prioritize wealthier districts or more developed regions, leaving poorer neighbourhoods underserved. This is particularly relevant in countries where geopolitical priorities may favour certain regions over others, leading to unequal distribution of resources and technological benefits.

Data privacy concerns must also be carefully managed and resolved to avoid using data that breach existing privacy and non-disclosure of private data legislation. The data sets used must not perpetuate existing biases or overlook factors not captured in historical data. In this sense the AI algorithms used must be transparent to allow for the necessary validation and control of these issues.

Key Challenge: How to ensure that the data used for the training and machine learning stage of AI development are as objective and free from biases and systematic errors, as possible.

6.4.1.2 Data Sovereignty

In relation to the previous need for securing data of good quality, one must not ignore the fact that such data may come from various sources and owners. Data sovereignty can become a significant barrier in AI applications development. The issue becomes bigger and more acute when the data sets used originate in different countries and jurisdictions. The cross-border exchange and use of data and information may be crucial for optimizing transportation networks that span multiple regions and countries.

In a world of diverging geopolitical priorities, nations are increasingly protective of their data, citing concerns over national security, economic competitiveness, and privacy. This fragmentation complicates efforts to create seamless, integrated AI-driven transport systems. Regional trade agreements may dictate localized data-sharing policies that inhibit global optimization, forcing AI systems to operate with incomplete datasets. Moreover, in politically sensitive regions, the fear of data breaches or misuse by foreign entities can lead to restrictive regulations that further hinder collaboration.

Key Challenge: How can the international community foster trust and cooperation in data sharing while respecting national interests, particularly in regions where geopolitical tensions are high?

6.4.1.3 Securing Trust, Transparency, and Equity

As technology evolves, and more data becomes available, issues of data trust and transparency need to be addressed through clear AI governance frameworks. Such governance frameworks must ensure that the data used for training AI and producing the AI-driven decisions are aligned with the true social and environmental conditions that exist in reality, rather than reflecting specific economic or political interests. Decisions made by AI systems are often opaque, making it difficult to assess whether these systems are aligning with broader social and environmental goals. Without robust mechanisms for algorithmic transparency, there is a risk of AI reinforcing systemic biases, rather than mitigating them.

Equity in the data to be used for AI, is also important especially in regions where national interests diverge from global climate goals. This requires a careful balance between technological innovation and political pragmatism, ensuring that AI serves as a tool for equity rather than a source of further division.

Key Challenge: How can AI systems be designed to ensure equitable access to transport solutions, particularly in regions with uneven infrastructure development or where certain populations may be politically marginalized?

6.4.1.4 Energy Consumption and the "AI Sustainability Paradox"

While AI is often touted as a solution for reducing the carbon footprint of transport systems, it is also very important to acknowledge the energy demands associated with AI technologies themselves. Training and operating AI systems, especially those that rely on massive datasets and complex machine learning models, can consume significant amounts of energy. This creates a paradox: AI can be used to significantly reduce emissions in transportation, but the computational infrastructure required to run these systems needs lots of energy to consume and thus could contribute to increased emissions. As regions transition to low-carbon energy sources, this issue has now become more pressing. In countries where energy security is tied to fossil fuels and geopolitical stability, the deployment of AI must be carefully balanced against the availability of renewable energy resources. Additionally, the carbon intensity of local energy grids can significantly impact the overall sustainability of AI-driven transport systems. The AI sustainability paradox is discussed in more length in the next Chap. 7.

Key Challenge: How can AI technologies be developed and deployed in ways that minimize their energy footprint, particularly in regions where energy infrastructure remains dependent on fossil fuels?

6.4.1.5 Public Trust and Resistance to AI Adoption

Another significant challenge is the issue of public trust. In many regions, there is growing scepticism towards the increasing role of AI in decision-making processes, particularly when it comes to critical infrastructure like transportation. Concerns over job displacement, privacy, and the potential for AI systems to make imperfect or harmful decisions have led to resistance in both developed and developing nations. In some cases, this resistance is further amplified by political actors who leverage public fear of AI to advance nationalistic or protectionist agendas.

Moreover, the perception that AI systems are controlled by large corporations or foreign entities can trigger political backlash in regions where economic sovereignty is a key concern. This makes it difficult for governments to implement AI solutions without facing public opposition, even when these technologies offer clear environmental and economic benefits.

Key Challenge: How can governments and private companies build public trust in AI systems, ensuring that these technologies are seen as tools for public good rather than threats to jobs, privacy, or national sovereignty?

6.4.1.6 Ethical AI and Algorithmic Accountability

The future of AI in transport operation and adaptation will hinge on the development of ethical AI frameworks that prioritize algorithmic transparency and accountability. As AI systems become more integrated into critical infrastructure, it will be essential to ensure that these technologies are aligned with broader social, environmental, and geopolitical goals. Governments, corporations, and civil society organisations must work together to establish guidelines that ensure AI systems are fair, transparent, and accountable. This will require ongoing investment in education, training, and public engagement, ensuring that AI is seen as a tool for collective progress rather than a source of division or exploitation. These issues are further explored in Chap. 9.

Key Challenge: How can governments work together to develop legal frameworks for ethical AI?

6.4.1.7 Regulatory Frameworks

The rapid advancement of AI technologies has outpaced the development of regulatory and ethical frameworks, leaving many governments struggling to establish guidelines that ensure AI is used responsibly. In the transport sector, this regulatory gap creates uncertainty for both public and private actors looking to invest in AI-driven solutions. Without clear international standards for data privacy, algorithmic accountability, and AI ethics, the risk of unintended consequences—such as environmental degradation, job losses, or social unrest—remains high. The challenge is more acute in regions where geopolitical priorities clash with global efforts to establish AI norms. For instance, countries prioritizing economic growth over climate action may be less inclined to adopt stringent regulations that limit AI's potential for short-term economic gains, even if those gains come at the expense of long-term sustainability.

At the time of writing this book, the most notable step towards the establishment of a regulatory and ethical framework for the use of AI was taken by the EU government (the EU Commission) through the publication of its *AI Act* (EC, 2024). This Act, which is now part of the EU's legal system, aims to secure trust, transparency and accountability for the development and use of AI systems by both private and public actors and ensure respect for the fundamental rights of the citizens while stimulating investment and innovation for AI in Europe. A more detailed presentation of the EU's

AI Act is given in Chap. 9 (section: *Safety, privacy, intellectual property, competition rules*).

Key Challenge: How can international organisations and governments work together to develop robust, flexible international regulatory frameworks for AI that accommodate the different geopolitical priorities while ensuring that AI is used ethically and responsibly?

6.4.2 The Need for International Cooperation

While geopolitical tensions may complicate data sharing and regulatory alignment, there are opportunities for regional collaborations and public–private partnerships that can drive innovation forward. Global AI standards will eventually be essential for ensuring that these systems are transparent, secure, and beneficial to all. The future success of AI-driven mobility applications will depend not only on technological breakthroughs but also on the ability of governments, businesses, and civil society to work together in a world of competing priorities. By working together to establish common frameworks, countries can ensure that AI technologies are used to solve global challenges, rather than exacerbating local political divisions. The future success of AI-driven transport adaptation will depend not only on technological breakthroughs but also on the ability of governments, businesses, and civil society to work together in a world of competing priorities. AI's potential to adapt infrastructure to the realities of a warming planet is not only a technical opportunity but a strategic imperative which should (and will at the end) be treated through common international efforts and cooperation.

The full promise of AI will be realised only through scalability and inclusivity. The future of AI-driven climate adaptation lies in its ability to be flexible enough to function across a variety of geopolitical settings and sensitive enough to address local needs while remaining aligned with global sustainability goals. So, while the potential of AI in climate-adaptive transportation is substantial, the persistence of internationally agreed technical, ethical, and geopolitical norms for AI, calls for deeper research and inquiry. The intersection between AI and climate science demands more sophisticated models that can account for the uncertainties inherent in climate projections. The ability to create self-learning AI systems that adapt to rapidly changing environmental conditions which will be crucial as the impacts of climate change accelerate is another area where international cooperation is necessary especially as regards the input data for machine learning.

The future of AI in climate-adaptive transport depends on collaboration across borders, sectors, and disciplines. Geopolitical fragmentation presents a significant barrier to the widespread implementation of AI solutions. To overcome this, there must be an emphasis on creating international frameworks for data sharing, regulatory harmonization, and ethical governance. This is particularly important in regions where data sovereignty and technological sovereignty are rising concerns. Without

AI diplomacy and multilateral cooperation, the potential for AI to transform transport systems will remain constrained by political and economic divisions.

Investment will also play a critical role. Governments, private enterprises, and international organisations must coordinate and prioritise funding for AI research, infrastructure development, and capacity building. This includes the creation of public–private partnerships that can accelerate the deployment of AI technologies while ensuring that public interests are safeguarded. Additionally, targeted investments in developing regions are necessary to bridge the global digital divide and ensure that AI-driven transport solutions benefit all nations, not just the technologically advanced. The development of multinational transport corridors such as the TEN-T in the European Union or those of the China's Belt and Road Initiative, could serve as testbeds for international cooperation for AI-driven solutions that prioritize both climate adaptation and geopolitical compatibility.

The ability of AI to navigate the complexities of climate adaptation, while simultaneously addressing the geopolitical realities of a divided world, positions it as a critical tool in both mitigating the effects of climate change and enhancing global cooperation. Success will depend on our collective ability to move beyond short-term national interests and embrace a shared vision of sustainability. By investing in AI-driven transport adaptation, we are not merely advancing technology, we are laying the foundation for a world that is more resilient, equitable, and prepared to face the challenges of the future.

References

Alqasi, M. A. Y., Alkelanie, Y. A. M., & Alnagrat, A. J. A. (2024). Intelligent infrastructure for urban transportation: The role of artificial intelligence in predictive maintenance. *Brilliance: Research of Artificial Intelligence, 4*(2), 625–637.

Bharadiya, J. (2023). Artificial intelligence in transportation systems: A critical review. *American Journal of Computing and Engineering, 6*(1), 34–45.

Chen, Z., Sivaparthipan, C. B., & Muthu, B. (2022). IoT-based smart and intelligent city energy optimization. *Sustainable Energy Technologies and Assessments, 49*, Article 101724. https://doi.org/10.1016/j.seta.2021.101724

Chhetri, P., Corcoran, J., Stimson, R., & Mahmood, A. (2020). Assessing the workforce adaptive capacity of seaports to climate change: An Australian perspective. *Maritime Policy & Management, 47*(7), 903–919. https://doi.org/10.1080/03088839.2020.1729433

Choksi, A., & Shah, M. (2024). Machine learning-based centralized dynamic clustering algorithm for energy-efficient routing in vehicular ad hoc networks. *Transactions on Emerging Telecommunications Technologies, 35*(1), Article e4914. https://doi.org/10.1002/ett.4914

Cirianni, F. M. M., Comi, A., & Quattrone, A. (2023). Mobility control centre and artificial intelligence for sustainable urban districts. *Information, 14*(10), 581. https://doi.org/10.3390/info14100581

Cugurullo, F., & Xu, Y. (2024). When AIs become oracles: Generative artificial intelligence, anticipatory urban governance, and the future of cities. *Policy and Society*. https://doi.org/10.1093/polsoc/puae025

Damaj, I., Al-Roubaiey, A., & Haidar, A. (2022). Intelligent transportation systems: A survey on modern hardware devices for the era of machine learning. *Journal of King Saud University-Computer and Information Sciences, 34*(8), 5921–5942. https://doi.org/10.1016/j.jksuci.2021.07.020

European Commission (EC). (2024). *The Artificial Intelligence Act: Regulation (EU) 2024/1689. Official Journal of the European Union,* 12 July 2024.

Haimes, Y. Y. (2002). A risk assessment methodology for critical transportation infrastructure (No. VTRC 02-CR5). *Virginia Transportation Research Council.*

Hasan, M. R., Islam, M. Z., Sumon, M. F. I., Osiujjaman, M., Debnath, P., & Pant, L. (2024). Integrating artificial intelligence and predictive analytics in supply chain management to minimize carbon footprint and enhance business growth in the USA. *Journal of Business and Management Studies, 6*(4), 195–212. https://doi.org/10.32996/jbms.2024.6.4.17

Haseli, G., Deveci, M., Isik, M., & Aydin, N. (2024). Providing climate change resilient land-use transport projects with green finance using Z-extended numbers based decision-making model. *Expert Systems with Applications, 243,* Article 122858. https://doi.org/10.1016/j.eswa.2023.122858

Huynh, B. Q., & Kiang, M. V. (2023). AI for anticipatory action: Moving beyond climate forecasting. *Proceedings of the AAAI Symposium Series, 2*(1).

Jackson, L., Beatty, M., & Mehta, P. (2019). Including vulnerable populations in the assessment of data from vulnerable populations. *Frontiers in Big Data, 2,* 19. https://doi.org/10.3389/fdata.2019.00019

Jain, H., Dhupper, R., Shrivastava, A., & Kumar, P. (2023). AI-enabled strategies for climate change adaptation: Protecting communities, infrastructure, and businesses from the impacts of climate change. *Computational Urban Science, 3*(1), 25. https://doi.org/10.1007/s43762-023-00100-2

Jha, M. K., & Dev, M. (2024). Impacts of climate change. In M. Azrour, J. Mabrouki, A. Alabdulatif, A. Guezzaz, & F. Amounas (Eds.), *Smart Internet of Things for environment and healthcare. Studies in computational intelligence*, vol. 1165. Springer. https://doi.org/10.1007/978-3-031-70102-3_10

Loukea, M., Bekiaris, E., Panou, M., Goger, T., & Adesiyun, A. (2016). Skills and competences development for the professionals of the future transportation sector—The SKILLFUL project. *Transportation Research Procedia, 14,* 1251–1260.

Machado, C. A. S., de Salles Hue, N. P. M., Berssaneti, F. T., & Quintanilha, J. A. (2018). An overview of shared mobility. *Sustainability, 10*(12), 4342. https://doi.org/10.3390/su10124342

Makarova, I., Serikkaliyeva, A., Gubacheva, L., Mukhametdinov, E., Buyvol, P., Barinov, A., & Mavlyautdinova, G. (2023). The role of multimodal transportation in ensuring sustainable territorial development: Review of risks and prospects. *Sustainability, 15*(7), 6309. https://doi.org/10.3390/su15076309

McMillan, L., & Varga, L. (2022). A review of the use of artificial intelligence methods in infrastructure systems. *Engineering Applications of Artificial Intelligence, 116,* Article 105472. https://doi.org/10.1016/j.engappai.2022.105472

Nikitas, A., Michalakopoulou, K., Njoya, E. T., & Karampatzakis, D. (2020). Artificial intelligence, transport and the smart city: Definitions and dimensions of a new mobility era. *Sustainability, 12*(7), 2789. https://doi.org/10.3390/su12072789

Ogungbire, A., & Mitra, S. K. (2024). Workforce forecasting for state transportation agencies: A machine learning approach. *International Journal of Transportation Science and Technology.* https://doi.org/10.1016/j.ijtst.2024.05.004

Olatunde, T. M., Okwandu, A. C., Akande, D. O., & Sikhakhane, Z. Q. (2024). Reviewing the role of artificial intelligence in energy efficiency optimization. *Engineering Science & Technology Journal, 5*(4), 1243–1256. https://doi.org/10.51594/estj.v5i4.1015

Onwubuariri, E. R., Adelakun, B. O., Olaiya, O. P., & Ziorklui, J. E. K. (2024). AI-driven risk assessment: Revolutionizing audit planning and execution. *Finance & Accounting Research Journal, 6*(6), 1069–1090. https://doi.org/10.51594/farj.v6i6.1236

Paramesha, M., Rane, N., & Rane, J. (2024). Generative artificial intelligence such as ChatGPT in transportation system: A comprehensive review. Available at SSRN 4869724.

Patil, D., Jadhav, S., Kulkarni, A., & Rao, P. (2024). Artificial intelligence and generative AI, such as ChatGPT, in transportation: Applications, technologies, challenges, and ethical considerations. *SSRN Electronic Journal*. https://doi.org/10.70593/978-81-981367-4-9_6

Pirogova, O., Plotnikov, V., & Uvarov, S. (2022). Risk-based approach in the assessment of infrastructure transport projects. *Transportation Research Procedia, 63*, 129–139. https://doi.org/10.1016/j.trpro.2022.05.015

Porter, G., & Dungey, C. (2021). Innovative field research methodologies for more inclusive transport planning: Review and prospect. *Advances in Transport Policy and Planning, 8*, 273–303. https://doi.org/10.1016/bs.atpp.2021.06.006

Qiao, Q., Zhao, F., Liu, Z., He, X., & Hao, H. (2019). Life cycle greenhouse gas emissions of electric vehicles in China: Combining the vehicle cycle and fuel cycle. *Energy, 177*, 222–233. https://doi.org/10.1016/j.energy.2019.04.080

Rane, N. L., Singh, R., Sharma, A., & Patel, R. (2024). Artificial intelligence, machine learning, and deep learning for enabling smart and sustainable cities and infrastructure. *Artificial Intelligence and Industry in Society, 5*, 2–25. https://doi.org/10.70593/978-81-981271-1-2_2

Singh, M., Arora, V., & Kulshreshta, K. (2024). AI and the environment: Innovative approaches to climate change. In *Maintaining a sustainable world in the nexus of environmental science and AI* (pp. 1–22). IGI Global. https://doi.org/10.4018/979-8-3693-6336-2.ch001

Sinha, K. C., Labi, S., & Agbelie, B. R. D. K. (2017). Transportation infrastructure asset management in the new millennium: Continuing issues, and emerging challenges and opportunities. *Transportmetrica a: Transport Science, 13*(7), 591–606. https://doi.org/10.1080/23249935.2017.1308977

Tsolakis, N., Zissis, D., Papaefthimiou, S., & Korfiatis, N. (2022). Towards AI-driven environmental sustainability: An application of automated logistics in container port terminals. *International Journal of Production Research, 60*(14), 4508–4528. https://doi.org/10.1080/00207543.2021.1976571

Yazdani, M., Mojtahedi, M., & Loosemore, M. (2020). Enhancing evacuation response to extreme weather disasters using public transportation systems: A novel simheuristic approach. *Journal of Computational Design and Engineering, 7*(2), 195–210. https://doi.org/10.1093/jcde/qwaa017

Yuan, T., da Rocha Neto, W., Rothenberg, C. E., Obraczka, K., Barakat, C., & Turletti, T. (2022). Machine learning for next-generation intelligent transportation systems: A survey. *Transactions on Emerging Telecommunications Technologies, 33*(4), Article e4427. https://doi.org/10.1002/ett.4427

Chapter 7
Data Needs and Utilsation

Abstract This chapter deals with the issues regarding the utilisation of AI in the management and analysis of big data for climate friendly mobility applications. After a brief but concise description of the modern collection methods for such data it goes through the main AI techniques for data processing to ensure data quality. The main advantage of using AI for this sort of quality assurance exercise is its ability to train algorithmic models so that they can recognize trends and hidden issues and extract useful information with very short processing times hundreds or even thousands of times faster than traditional data processing methods. The text explains how machine learning or deep learning algorithms and other AI based technology can automate complex data analysis tasks, excel at identifying complex patterns and relationships from big data, discover valuable information hidden in these data and increase its accuracy and reliability. Consequently, there is a concise reference to the many AI tools and software libraries that can be used for such big data analytics. A few indicative examples of existing applications of big data in the transport and mobility sector follow as an indication of the use of AI techniques for this purpose in practice. With the help of AI, big data gets a number of quality attributes which include Accuracy, Availability, Integrity, Consistency, Completeness, and Precision. All these are explained and discussed in length. Finally, this chapter duels on the issue of energy consumption that is necessary for AI-assisted big data analytics and calls this situation "a mixed blessing". This characterization is justified by the fact that whereas the use and utilisation of AI for big data analysis and handling has many advantages and can be a beneficial transformational force, it necessitates large quantities of electricity which if not generated by "clean" renewable sources it can generate more carbon emissions than saved from transport operations by using AI.

Keywords Big data · Transport data · Data collection methods · AI in data processing · Big data consolidation · Using big data · AI powered databases · Big data quality · Energy requirements · Weaknesses of e-data · Data carbon emissions

G. A. Giannopoulos and Y. Li, *Human Mobility, Artificial Intelligence and Climate Change*, Lecture Notes in Mobility, https://doi.org/10.1007/978-3-032-08171-1_7

7.1 Introduction

If AI is set to revolutionize the way we run our future transport and mobility systems and at the same time analyze the two-way impacts of transport on the environment and climate change, we need a multitude of data. Data that supports a comprehensive understanding of all aspects of human travel, from planning and executing journeys or cargo transport, to designing and developing the necessary infrastructure. This understanding not only relies heavily on the collection and analysis of copious amounts of data, but it must be done within tight timeframes and adequate quality assurances. When climate change is factored in, the demand for data grows exponentially and so does the need to do most of the analysis and utilization of this data in real-time (Chen et al., 2016). It is within this increasingly intricate and dynamic context that Artificial Intelligence (AI) has been recognised as having a pivotal role, enabling the handling and processing of vast amounts of data to extract meaningful information and insights (Zheng et al., 2013, 2014).

Mobility and climate data may originate from diverse sources, including sensors placed at mobile or stationary points, smart visual or radar-based devices, social media, weather stations, and other data repositories. The handling of all this data involves complex and highly specialized computing capabilities that refer to the data sources, the processing of the data to enhance and secure their reliability, evaluating their quality, and feed-in and produce the various applications, as well as face the various challenges that relate to this whole process. This is diagrammatically shown in Fig. 7.1. Through machine learning and deep learning algorithms, AI can process vast amounts of data in an intelligent way to uncover hidden patterns and correlations and provide accurate predictions and recommendations and formulate a great number of applications.

On the data handling side i.e., the collection and processing of the data to be reliably used for producing the various applications, AI is called to perform a unique and very specialized set of actions for handling mobility and climate change big data. These actions include (see also Fig. 7.1):

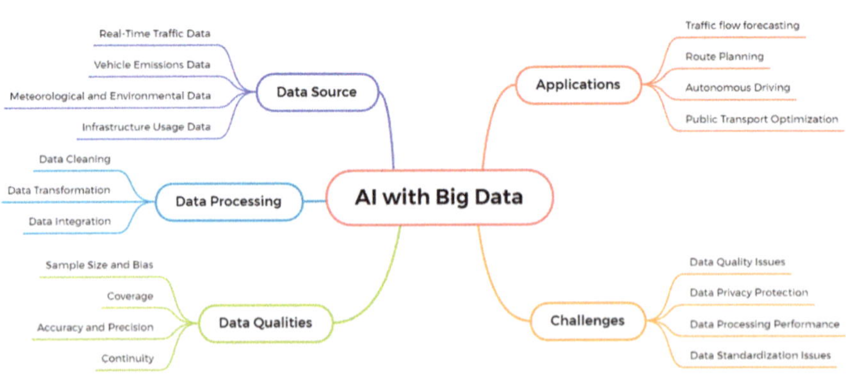

Fig. 7.1 The key elements of AI involvement with Big Data

- Selection and utilisation of data from various sources (traffic flows, vehicle emissions, infrastructure utilization, social media use, and many others).
- Integration and harmonization of the different data sets (data cleaning, interoperability through transformation, etc.).
- Qualitative assessment and evaluation of the data through assessing sample sizes, coverage, accuracy levels, and other qualitative characteristics (see later sections). Data quality and reliability must be ensured and this calls for making sure that all data sets are judged as representative, accurate, unbiased and comprehensive enough to be used. Data privacy and security issues must be prioritized as big data may contain substantial amounts of personal data and other sensitive information, warranting appropriate data protection policies and measures.
- Enabling collaborative use and data sharing among the various stakeholders through creating interoperability. This makes data security a crucial issue, but it also necessitates appropriate organizational steps such as creation of public–private partnerships and data-sharing frameworks, fostering innovation and enabling the development of comprehensive solutions.

In handling such big data, the involvement of AI results in specific outputs that include (indicatively):

- Forecasting short and longer-term future congestion (from travel demand and traffic flow data).
- Optimizing route planning (from traffic flow data).
- Formulating smarter and more efficient traffic management strategies for reducing energy consumption and minimizing emissions (various sources).
- Planning and maintaining intelligent and sustainable transport infrastructures (Data on infrastructure conditions and maintenance records as well as weather records).
- Simulating and optimizing smart city operations, especially operations enhancing energy efficiency and reducing carbon emissions (various sources).
- Optimizing power supply and demand matching (smart grid management) with full utilization of renewable energy sources (from national electric power distribution administrators).

As climate-friendly mobility evolves rapidly, data requirements surge accompanied by the complexities of their analysis and utilization. Large-scale data collection becomes paramount to understanding travel behaviour, managing traffic flows, reducing energy consumption, and minimizing environmental impacts. Timely and accurate collection and organization of such data is therefore essential in providing decision-makers with real-time insights and guidance into transportation demands, travel patterns, energy consumption, identification of potential areas for improvement, and formulation of more effective policies and measures to reduce carbon footprints.

This chapter deals with issues regarding the involvement and utilisation of AI in management and initial analysis of big data for climate friendly mobility applications and the needs for such data.

7.2 AI in Handling Big Data

7.2.1 *Types of Transport and Mobility Data*

The core characteristic and value of big data in the transport and mobility area is the extensive diversity and richness of the data that can be collected with both conventional and modern (digital) data collection methods (Nie et al., 2023). The following is an indicative list of the main types of transport and mobility data that are (or can be) collected in the field:

Travel demand: These data refer to the characteristics of travel demand and consist of several sub-types among which: the number of trips generated per unit of space or time, trips per land use type, trip distribution in space (origins and destinations), other trip characteristics such as travel purpose, mode used, etc., and so on. These are collected via conventional or modern (digital) ways and sensors (see below).

Real-time traffic: Real-time traffic data such as traffic volumes, speeds, densities, intersection movements, delays, travel times, and similar within any given transport network. These data are collected in real-time using GPS devices on the vehicles, road cameras, intelligent traffic signal installations and other, more conventional methods (traffic counts and other). Their use is mainly made by city managers to optimize traffic signals, reduce traffic congestion, and improve the overall efficiency of transportation systems.

Vehicle emissions: Vehicle emissions data is becoming increasingly important as awareness of environmental protection increases. These data are usually collected by emission monitoring devices on-board the vehicles and cover the emissions of various pollutants such as carbon dioxide (CO_2), nitrogen oxides (NO_x), and particulate matter (PM) under different operating conditions. Analysing these data can help develop stricter emission standards and promote the development of green and climate friendly zones as well as in measuring the impact of the various emission reduction policies in transportation.

Meteorological and environmental conditions: Meteorological and environmental conditions data include temperature, humidity, wind speed, precipitation, air quality, and more. These are acquired through weather stations, environmental monitoring stations, and satellite remote sensing. The comprehensive analysis of meteorological and environmental data is of great significance for early warnings optimising transport operations in all modes of transport but especially for maritime and air transport and in improving the living environment.

Infrastructure usage: Infrastructure usage data relate to the usage of transportation facilities such as roads, bridges, tunnels, etc. They include data on relevant traffic flows, accidents, structural elements of the infrastructure, and related weather conditions. They are usually collected by sensors and cameras and on-board the vehicle devices.

Public transport operation: Data in this field, include passenger flows per mode of public transport, adherence to schedules, capacity utilisation and similar. This data is collected through smart card systems, GPS devices, and control centres to optimize public transport operations.

Freight and logistics: This type of data covers volumes of cargo carried, cargo tracking data, supply chain processes, warehouse management data, types of freight vehicles used, and similar. These can be obtained via IoT devices, RFID systems, and logistics platforms.

Tourism and recreation activities: Tourism and recreation data are related to transportation in many ways as they encompass visitor movement patterns and crowd or vehicle movement levels especially near the various cultural or recreation areas.

Shared mobility operations: These refer to the operation of shared mobility services and aim at quantifying the demand as well as the operation of these services. Shared mobility data include ride-hailing, car-sharing, and micro-mobility usage data, number of trips made in this way, vehicle maintenance data, and similar.

Accessibility and equity in mobility services: These data refer to the travel behaviour of different demographic groups and aim at measuring the degree of accessibility of the different groups to public transport or other mobility services. They are collected and analysed through focussed interviews and surveys and public transport usage surveys.

7.2.2 Data Collection Methods

The conventional, still mostly used, data collection methods are both manual and automatic and utilise conventional types of technologies such as for example, stationary detectors for traffic data (e.g., for road transport: pneumatic tubes, piezoelectric sensors, passive or active magnetic, microwave, infrared, ultrasonic or acoustic detectors), radar or video image detectors and so on. Other "conventional" types of data collection that mostly refer to road transport, are Floating car data (FCD) which are time related (geo-localization and speed data are directly collected by moving vehicles equipped with the appropriate technical equipment), and probe vehicle data (PVD) which are collected by suitably equipped probe vehicles that act as individual probes. PVDs gathering data send status messages (called CAM—Cooperative Awareness Messages, standardised by the European Telecommunications Standards Institute—ETSI) to receiving stations in the infrastructure who forward them to a central station for analysis.

Conventional are also the traditional techniques of conducting Household Travel Surveys (HTS), intercept surveys (for data collected by survey agents on-board public transport vehicles or at stops or stations), stated choice or preference surveys, traffic and passenger counts, number plate surveys, and a few others of similar nature. The data collected by these traditional methods are used for the planning and operation of transport services and are relatively modest in volume. Their analysis is performed mainly off-line by use of traditional statistical tools and conventional information technology data management techniques. The result of these data collection methods is to get a static (i.e., at a given point in time) representation of the travel and trip making characteristics to better understand travel behaviour and embed it in the functional forms and parameters of transportation planning models.

The new generation of traffic and transport data collection methods use digital technologies based on the use made of modern mobile phones and other intelligent applications and sensors (Chen et al., 2016). Examples of such digital methods are, Bluetooth, Wi-Fi signals, public transport smart cards, smartphone apps, mobile network data and other vehicle telematics (Jiang et al., 2017). Broadly speaking there are five main categories of modern-day transport data collection sources as diagrammatically shown in Fig. 7.2 (ITF, 2021): Mobile phone data (generated by a global navigation satellite system microchip integrated in all smartphones), smart cards, use of apps and app aggregators (social media), GPS and Navigation, Bluetooth and WiFi devices as well as Automatic Number Plate Recognition—ANPR. The most notable of these modern-day data collection methods are the ones using mobile phone signals and other floating cellular data i.e. cellular data from CDMA (Code-division multiple access), GSM (Global System for Mobile communication), UMTS (Universal Mobile Telecommunications Service), GPRS (General Packet Radio Service) networks. Tracking mobile phones and using them as traffic probes to collect anonymous travel data and information is now routinely used for the calculation of traffic volumes, travel speeds, and congestion. There is no special equipment needed in the cars or other devices, but simply reliable filtering and data mining algorithms that are applied to obtain reliable traffic data from the sources mentioned above.

The data collected by modern-day methods of data collection are known as digital trace generated data because of their digital nature and sources. They are all passive and non-disruptive and take advantage of the digital traces of modern-day telecommunication devices. They are obtained in collaboration with the Mobile Network Operators (MNOs), i.e., the companies that provide wireless communications services to consumers, private companies and businesses operating around the world. MNOs generate some of the largest data sets that contain massive amounts of travel related data with sizes measured in terms of petabytes (PB—1000 terabytes) or

Fig. 7.2 Modern-day sources for data collection in the transport sector and indication of their basic use. *Source* ITF (2021)

exabytes (1000 PB) per day, or even bigger. This constitutes the modern-day digital trace generated big data i.e., vast collections of structured and unstructured data sets. Digital trace data are difficult to process using traditional data processing information technology tools (due to their volume and complexity) and this is where AI comes as the technology of choice to handle these data through its machine learning algorithms and other capabilities for data mining and analysis in real time as explained in more detail below.

7.2.3 *AI Techniques for Data Processing*

AI, besides being a "user" of big data, is also the main tool used for their processing and ensuring data quality as mentioned in subsequent sections. Using machine learning algorithms for data analysis and pattern recognition, AI trains algorithmic models so that they can recognise trends and hidden issues and extract useful information with very short processing times hundreds or even thousands of times faster than traditional data processing methods. Through machine learning and deep learning algorithms, AI technology can automate complex data analysis tasks, excel at identifying complex patterns and relationships from big data, discover valuable information hidden in these data and increase its accuracy and reliability (Djavadian & Chow, 2017).

By identifying effective patterns in massive traffic data, AI can detect problems of consistency and integrity in the collected data thus providing a basic focus in treating such data and facing the quality problems caused by their inherent weaknesses as mentioned in later sections (Ferrer-Cid et al., 2025). Machine and deep learning techniques are the tools of choice that can be used in developing the necessary AI algorithms (Vartziotis et al., 2024). There are already hundreds of AI tools and software libraries that can be used for big data analytics. Below, a few indicative examples:

- DeepSeek (https://www.deepseek.com/). This is a set of open-source AI large language models (LLMs) that can be used to analyse data using various AI and machine learning algorithms as well as natural language processing tools. The DeepSeek company was a small startup at the time of writing this book that became world famous when its open-source software challenged the quality of similar products of well-established giants in the field at a fraction of the cost. Their machine learning algorithms improve searches by analysing past queries and trends and make database integration with data streams from different sources easily possible.
- icCube (https://iccube.com). Developed by a Swiss-based analytics company, this is a Software as a Service (SaaS) product crafted for B2B development teams aiming to integrate advanced data analytics directly into their applications.
- Oracle Analytics (https://oracle.com/business-analytics/analytics-platform/). This is a platform designed for all data analytics user roles, integrating AI and

machine learning for productivity and smarter business decisions. It offers two main deployment options: Oracle Analytics Cloud (OAC), a cloud service, and Oracle Analytics Server (OAS), an on-premises solution.
- Deepnote (https://deepnote.com). Powered by GPT-4, this tool is an AI notebook for teams aimed at connecting data sets, exploring and analysing them with real-time collaboration and versioning.
- RapidMiner (https://altair.com/altair-rapidminer). This is a data analytics tool that streamlines data analysis processes. It features a UI and drag-and-drop framework that make it accessible to users with varying skill levels.
- Talend (https://www.talend.com/). A platform for data integration, monitoring, and administration. It enables users to handle and assess data across various big data channels such as Hadoop, Spark, and Hive.
- Knime (https://www.knime.com/). An open-source platform that offers a full suite of data science tools, including machine learning, predictive modelling, and ETL (Extract, Transform, Load) processes.
- Google Sheets (https://workspace.google.com/intl/en_in/products/sheets/). This is a spreadsheet application integrating machine learning features for data analysis and visualization.
- Tableau (https://www.tableau.com/). A data visualization tool that helps users create dynamic dashboards and visualizations without coding. It includes AI-powered features for quick data exploration.

Of a more elaborate nature, are the various AI software libraries that can be used in developing custom made AI data handling software using a variety of programming languages. Examples:

- Tensorflow (https://www.tensorflow.org/), the scalable and production-ready library for deep learning, developed by Google. It can be used with Python, Javascript, C++, and other languages.
- Keras (https://keras.io/), a high-level Application Programming Interface (API) for the Tensorflow platform that provides a highly productive interface for solving machine learning problems with a focus on deep learning.
- Pytorch (https://pytorch.org/), the flexible deep learning library used for natural language processing, originally developed by Meta AI and now part of the Linux foundation.
- Scikit-learn (https://scikit-learn.org/stable/), the open-source machine learning library developed originally for the Python programming language that includes material suitable for classical machine learning models and data processing.

The application of AI, LLMs and other tools, in big data processing and analysis does not mean that other more traditional tools are not used in AI-powered big data handling. Traditional techniques such as the following are also used regularly:

- Regression analysis for short term prediction of the values of continuous variables (e.g., traffic flows and vehicle speeds).
- Classification algorithms for dividing data into different categories (e.g., by vehicle type or road congestion level, etc.).

- Cluster analysis for discovering naturally connected groups in the data (e.g., traffic patterns in different regions, decision trees for decision-making), etc.

In short, big data processing—in general and in the transport and mobility field in particular—is increasingly using AI as the technology of choice. AI big data analytics is continuously being technically improved and expanded and will soon form the basis for all analytical tools and models in this area (Abirami et al., 2024).

7.2.4 Big Data Consolidation

Big data consolidation includes all the processes and AI-powered analyses for data cleaning, data transformation, and data integration and sharing. Data cleaning is the process by which the data are checked for logical as well as other errors such as duplicates, outliers, and similar. In the cleaning process, errors and redundant data are eliminated to ensure the accuracy and consistency of whole data set. Logical errors are those that are revealed when the collected data contain areas that do not make sense or disrupt the normally expected values (e.g. car occupancies of 10 or 20). They can also reveal missing data i.e., unexplained gaps. AI can be trained to easily detect such logical errors and quickly amend the integrity and validity of a data set by removing duplicates or identifying outliers or even restoring missing data with methods such as mean filling and interpolation.

Data transformation is the process by which the data are transformed from one format or structure into another format or structure suitable for their subsequent use and analysis. The data transformation process is also referred to as ETL (for Extract, Transform, Load). The extraction phase involves identifying and pulling data from their various source systems and moving it to a single repository. The Transform phase involves the transformation of the raw data into a target format suitable for the operational system to be used for its analysis or a data warehouse, a data lake or other repository. Before transformation the raw data must be "cleaned" if needed.

Data integration and sharing is necessary when data from different sources are to be used or if the current data being processed need to be effectively integrated to form a comprehensive and integrated whole with other data sets. In the process of integration, data are transformed so that all formats are unified to facilitate subsequent sharing and fusion. Data integration (or fusion) is the process by which multi-source data are associated and merged to form a complete and unified set. This process is completed by use of AI and the establishment of some standardisation of formats that can make data availability and utilization feasible across different domains (academic institutions, city or industrial departments and administrations). The establishment of uniform standards as regards the format and other specifications, can improve the interoperability and availability of big data and promote their efficient use (Djavadian & Chow, 2017).

Data sharing mechanisms include three major levels of actions: access management, sharing protocols, and security guarantees. Access management ensures that data is only used within authorized limits to prevent misuse and unauthorised disclosures. The data sharing protocols are agreements that specify the scope, methods and responsibilities of data sharing to ensure the interests of all parties involved. Finally, security guarantees are necessary to be put in place to protect data through encryption, anonymization and other technical means. Security is a crucial issue related to the overall quality and reliability of big data and for maximizing their utility for AI applications in climate-friendly mobility.

7.3 The Use of Big Data in Transportation

The big data that is collected and qualitatively handled (by AI) in the field of transportation, can be utilised for a variety of purposes. The users of such data are mainly the infrastructure owners (such as for example motorway companies), the transport/mobility service providers, and other third-party entities such as for example internet-service providers, research organisations, etc. The proposed services can be cloud-based, available for all users with access, or they can be implemented directly on the servers of an interested party to run locally. Some typical areas of big data utilisation to produce applications for the benefit of mobility system users, are the following:

- Travel time and congestion estimation: Reliable travel time estimation in real time is on the top of the list of the most sought-after applications. This goes together with the determination, in real-time, of the average speeds of movement along any road section and the determination whether the section is congested or not. Using digital traces from mobile phones, beacons, and tele pass gates, travellers are matched along consecutive positions and travel time distributions are estimated in real time. Used primarily on road networks the travel time estimation is based on elaborate algorithms that are fed with these digital traces to calculate travel times and average speeds of movement. Then, based on the estimated speeds, congestion maps can be drawn for all links in an urban or interurban road network, all in real time.
- Short term traffic forecasting. Extra modules in the AI software can employ algorithms for short term forecasts of the travel times and speeds based on various parameters such as the rate of increase of traffic flows and the spatio-temporal features of the specific road network and its traffic. There are many short-term traffic prediction algorithms that have been developed over the years, including autoregressive integrated moving averages (ARIMA), or Kalman filters, and more recently AI-powered dynamic models that include machine learning and algorithms for artificial neural networks, Bayesian networks, and the k-nearest neighbour technique, or deep learning models such as deep confidence network, long short-term memory models and others.

- Navigation services: Reliable and real-time navigation services, with re-routing capabilities in case of unexpected events or high congestion. These navigation services are provided based on travel time estimation, incidents detection and scheduled events. The routes are continuously updated to avoid congestion or unexpected incidents offering smoother travel conditions.
- Personalised information: Updated information regarding the on-going travel of specific travelers can be personalized and offered to them aiming at improving their travel experience. Such information is continuously calculated and updated and based on the personal preferences of these travelers, en-route information is provided including e.g., driving rest areas, touristic locations, local events, points of interest and so on.
- Traffic incidents and events detection: Traffic incidents such as accidents, stopped vehicles, and other incidents are automatically detected, and relevant info is given to all travelers in real time. Based on the location and the type of incident, safety warnings and traffic status information are provided to all traveling in the affected area.
- Fuel information. For all types of fuels, gas or petrol or electricity, fuel station locations as well as fuel prices can be provided along with the expected times for refuelling. Also, fuel consumption until each fuel station is calculated and made available to the drivers.

Various other services for the drivers or the travellers can be developed (Liu et al., 2014; Toole et al., 2015). All this information is made available directly to the various in-vehicle devices or broadcasted through radio channels or VMS.

For the infrastructure owners and/or managers other applications can be developed for improving their management and planning capabilities. A list of possible application services in this category is the following:

- Traffic control strategies: Advanced dynamic traffic management plans can be produced based on short-term traffic flow predictions as well as driver needs and infrastructure status. Short-term traffic prediction models are quite well established now (Vlahogianni et al., 2014) while of crucial importance are also the algorithms for identifying the travel patterns of the travelers (O-D of trips) to adapt better the existing traffic management plans.
- Traveler patterns identification: This is done through calculating the Origin and Destination (O-D) matrices that correspond to the current traffic flows on the network. Through these matrices the AI system can understand better the spatial and temporal distribution of the vehicles and their drivers and provide dedicated services (including billing for the use of road space when a policy of congestion charging is followed).
- Traffic models calibration: This has to do with the calibration of existing traffic models and increasing their reliability. As already mentioned, big data from digital trace sources are not usually employed for developing predictive or other types of models due to their inherent weaknesses (see later sections). However, this data can be used for calibrating existing traffic models for planning and decision-making issues.

- Roadworks and maintenance planning: Dynamic planning of road works and maintenance tasks can be performed and managed based on real-time (traffic and other) conditions analysed by AI and aimed at minimizing the impact on travel times and congestion.
- Network planning and design: By analysing big data on the traffic flows and other traffic characteristics, AI based algorithms can evaluate options and produce recommendations regarding new infrastructure planning or expansion / modification of existing ones. A major advantage here is the better ability of AI to simulate the travel patterns on the network before and after the implementation of the planned new elements and based on these to calculate quantified indicators for the evaluation of each option.
- Incidence detection and response: Efficient response to incidents using real-time traffic status data based on early detection and info communication, is made possible much more efficiently through use of AI-powered big data analytics. Any incidents can be detected and broadcast almost in real time to control centers and the rest of the traffic, increasing in this way significantly the safety levels.
- Weather conditions monitoring: Dynamic and real-time weather information provided by roadside as well as in-vehicle devices can be transmitted and analysed in real time. External weather data can be combined with data coming from the vehicles' CAN-bus for early detecting weather conditions (basically raining conditions) and inform all other users early almost proactively. AI can then propose driving reactions accordingly.

7.4 Real-Life Examples of AI-Powered Data Bases

A few indicative examples of existing applications of big data in the transport and mobility sector, will be presented here with a view to showcasing the current uptake of AI in handling data. In all cases mentioned below, AI is used for the collection, consolidation, and analysis of the relevant data sets. A comprehensive review of AI-powered big databases in the transport and mobility sector can be found in Lee et al., (2023) while current and potential future mobility big data architectures can be found in Abirami et al., (2024).

7.4.1 The Beijing Municipal Commission Data Base

The Beijing Municipal Commission of Transportation (BMCT) has installed a large number of ground-based magnetic sensors, video surveillance cameras, and microwave radars at key locations (such as major roads, intersections, and expressways) throughout the road network of Beijing. The aim is to collect digital trace data, as well as from other sources, to better understand traffic demand, travel patterns, energy consumption, and to improve traffic management efficiency. These devices monitor in real-time vehicle speeds, flow levels, and vehicle types. The BMCT huge traffic data collection system can update traffic flow data every few minutes and thus create a huge pool of data whose analysis is made in real time to ensure timeliness and accuracy. Based on this analysis, estimates and short-term predictions of traffic volumes and congestion are made and based on them the BMCT dynamically optimises traffic signals at major intersections. For example, at intersections with peak-hour heavy traffic and pronounced tidal patterns, smart traffic signal control systems are applied which automatically adjust signal light durations based on the traffic volumes in the respective directions. Through this approach, the capacity of the intersections is improved, the waiting times for vehicles and the energy consumption is reduced.

7.4.2 The Baidu ACE

The Baidu ACE Intelligent Transportation Engine is known as the first full-stack intelligent transportation solution in China that uses extensive digital trace big data which are used for many applications. The system provides personalised cloud-based mobility information to users and has many other uses including the operation of autonomous driving and intelligent transportation systems. ACE is the abbreviation for Autonomous driving, Connected road, and Efficient mobility. It basically consists of a comprehensive intelligent transportation AI-based platform that is used for big data collection and analysis to support the company's intelligent transportation vision. ACE is comparable to platforms such as Tescent's WeCity, Microsoft's CityNext, Amazon's AWS Wavelength, or IBM's Smart City.

The Baidu Inc is China's main internet service provider, known for its strong search engine platform. As such, it is interested in producing elaborate traffic information for its customers and uses for this purpose a massive data collection, storing and analysing system based on sensors and other devices to collect and store the relevant data. These include real-time location information, driving speeds, road conditions and other information based on sensors and include feedback from data generated by the users of the Baidu maps for navigation. The Baidu big data bases are constructed for specific urban areas with which the company has a contract to dynamically reflect the road traffic flow and congestion conditions. One of its first applications was the Baidu ACE Intelligent Transportation system for the public security bureau, traffic police detachment, of Baoding City (population 2 million). The system developed, the Baoding City AI Traffic Management Brain and AI Signal Optimization Control System, integrates video, magnetometer, radar, and internet multi-source traffic data, to realize a panoramic visualization of the traffic situation in that city that includes road conditions, signal control status, and other data on items of interest in real-time. The system also provides intersection adaptive traffic signal control and other dynamic traffic control functions.

7.4.3 The Uber Big Data Platform

The Uber platform creates and maintains big databases on the trips of its vehicles, the drivers, the locations visited, and each vehicle status. This data is continuously updated, tested and utilized by AI for foretelling the demand, supply, accurate location of drivers, and trip fares. The system monitors its data to predict spikes in demand and variations in driver availability. That information is used to set the proper pricing of rides and provide incentives to drivers, so the necessary number of vehicles are always available to keep up with demand. Data analytics form the basis of Uber's big data platform for calculating estimated times of arrival and other customer related information (satisfaction levels, driver evaluation, etc.). Big data analytics are used to get users safely to their destination and provide insights into routes, travel times and traffic conditions which can also be pulled from other data bases that provide a look at travel conditions and vehicle demands in real time.

7.4.4 The UPS Database

The UPS (United Parcel Service) is a global parcel delivery company that uses big data analytics to map the most efficient delivery routes for their vehicles. Their ORION system (On-Road Integrated Optimization and Navigation) is responsible for the analysis of big data that consist of traffic flows, congestion levels, weather patterns, as well as truck sensor data, to establish the most optimal routes for their vehicles to reach their destinations. ORION allows multiple drivers to access pre-planned multi-stop delivery routes on their personal handheld devices instantly. The system has led to substantial costs reduction and energy savings. A most interesting result of their AI use in handling their collected big data, was the discovery that by turning left against opposing traffic streams vehicles were wasting a lot of fuel. As a result, UPS introduced a 'no left turn' policy and designed their vehicle routing software to eliminate as many left-hand turns as possible (in countries with right-hand traffic). As a result, the company claims that at global level it uses 10 million gallons less fuel, emits 20,000 tonnes less CO_2 and delivers 350,000 more packages every year!

7.4.5 Boeing's Airline Health Management System

Aircraft manufacturer, Boeing, operates a big database with billions of measurements across their entire fleet that range from in-flight metrics to aircraft mechanical analytics. Every day, the system analyses these data and uses AI to predict potential failures, the servicing required, and other safety issues improving airplane safety both at Boeing factories and across the airline industry at large.

The same or similar systems are also used by individual airlines which apply AI on big data from Boeing or Airbus as well as from sensors on planes for not only predictive purposes for aircraft maintenance but also for the prediction of component failures. This makes it possible to preventively replace the faulted parts and reduce the time of downtime while enhancing safety.

7.5 Big Data Quality and Energy Consumption Issues

7.5.1 Quality of Digital Trace Generated Data

With the traditional methods of data collection, the focus was on sampling data quantity since the data collected was quite limited. Nowadays the amount of available data is enormous, so the focus has lifted to data quality aiming at providing reliable data to infrastructure managers as well as mobility service providers and users, most of them in real time. Quality must be checked since no data source is error-free and eliminating or compensating for errors is key to the quality of the results. In this respect, AI comes as a revolutionizing technology to give concrete and very beneficial results (ITF, 2021).

The quality of the data collected is characterised by the following attributes which are further detailed below:

- Accuracy
- Availability
- Integrity
- Consistency
- Completeness, and
- Precision.

Accuracy refers to the level of precision with which the identified, via the digital traces, trips and trip chains correspond to reality e.g., in terms of the exact location of their origin and destination or the actual travel time, etc.

Availability refers to the degree of continuity and the time duration of the commitment under which the Mobile Network Operators (MNOs) provide the digital trace data. The lack of continuity degrades the usefulness of the data, and this may happen in different ways. Some MNOs re-anonymise the data with some frequency, for example once a day or even once every couple of hours. This makes it impossible to track sufficient movements at each state of affairs and makes correcting for bias and expansion of the samples much more difficult. Smartphone app data seems to suffer less from this issue as the marketing IDs of the apps used are more persistent. However, people install and uninstall apps much more frequently than they change their mobile network operator, so the temporal series available tend to be shorter, which limits the possibilities for certain types of longitudinal analyses.

Integrity—lack of bias. Most digital trace data is sufficiently well spread over space and socio-economic groups to be in need to correct for bias, but still a check is necessary. This is achieved through appropriate sample expansion techniques by considering the socio-economic characteristics of the population in the area. The potential bias that may be present in data originating in smartphone apps (social media apps) is more difficult to know and to correct. App consolidators merge data from different apps and do not disclose them. So, AI can be used to detect if certain population groups are underrepresented or even absent from the sample and provide the necessary corrections.

Consistency. There are two dimensions to the consistency of geocoded data: time and space granularity. This is defined by how often a digital transaction identifies a trace (trip) and how accurate is the location obtained for that trace. MNO data is providing more and more traces as people intensify the use of mobile data. In terms of spatial precision, smartphone apps offer better geolocation than mobile network data (from MNOs). MNO data is also dependent on the size of the cells, and this can be quite small in dense urban areas and much bigger in rural underpopulated regions.

Completeness relates to the geographic coverage of the data. Both MNO and app sourced data can provide coverage for a whole country or region, but app data are better because they allow the tracking of the users in any country or region and therefore are not limited by country limits. However, it is often the case that the apps that are more popular in one country are less so in others and so the sample size may be heterogeneous depending on the target markets of the app aggregator. AI can be used to indicate whether the user is a visitor from abroad (international roaming) identifying the country of origin of the mobile phone contract while some MNOs also provide roaming-out registers, which allows the identification of international trips.

Precision relates to the case that the data refer to a part (sample) of the total population of the measured parameter so then the precision of the data refers to the accuracy with which the sample represents the whole population. Most MNOs generate very large data samples which are well representative of the total population. Their size is only limited by the market share of a given MNO i.e., telecommunication company. Similarly large, although smaller than those of the MNOs, are the sample sizes from smartphone apps. In most countries the sample sizes used represent between 15 and 40% of the population not just for one day but on 365 days a year. This is generally a much higher precision than the sample achieved with household travel or other conventional surveys.

Of particular interest are also two issues which can potentially be of crucial importance and need to be checked and safeguarded in parallel with the above-mentioned quality attributes. The first is the need for data privacy issues protection. As the scope of data collection expands, the issue of protecting personal privacy becomes more and more prominent. How to protect personal privacy while ensuring the value of data utilization has become an important issue in big data applications. Anonymization, encryption and access control are the usual means of safeguarding privacy issues in big data. The second issue is the data processing performance. When dealing with massive amounts of data, data processing performance becomes critical. Efficient data storage, retrieval and analysis methods are the basis for improving performance. The development and use of distributed computing, in-memory computing and big data platform technologies are usually techniques employed to improve data processing performance.

7.5.2 *Weaknesses of Digital Trace Generated Data*

The quality of big data in the transport and mobility field, which is expressed by the attributes mentioned previously, is also affected by some key weaknesses that digital trace generated data appear to have. These weaknesses relate to the nature of this data and the enormous volumes of digital traces that are collected and stored.

The first, relates to the representativeness of the data due to the lack of a controlled sampling framework. For example, some age or social groups might have a greater tendency to use a specific mobile network or smartphone app or connected device (in-vehicle). Some trip purposes and transport modes may attract a more intense use of navigation apps, and so on. These are issues that must be checked and taken care-off especially when using AI to handle these data.

Another weakness is that most big data on mobility is missing the socio-demographic attributes of the travelling individuals. This prevents researchers from drawing insights that are meaningful and statistically sound about causal relations that could lead to the development of predictive models. So, there are several transport researchers who, because of this issue, believe that digital trace big data alone are insufficient in developing travel prediction models or transport demand models (ITF, 2021).

Possible dependence on external interferences is another weakness that needs to be checked as some types of digital trace data suffer from outliers' interferences. For example, devices in shared e-scooters, send and receive data over the mobile communication network and are traced through that medium sometimes as part of an Internet of Things (IoT) connection. Contrary to mobile phones that are carried by people throughout the day, e-scooters stay stationary for varying lengths of time. However, their signal can be caught by nearby antennas and thus they are creating misleading pictures of their use ("noise") when left stationary.

Lack of homogeneity is another weakness. The collection of digital trace big data is not always repeatable in a way that offers homogeneity and comparability of the different data sets. This could be due to changes in the technology used, or in the market penetration of mobile devices, or in the market share of the various service providers. In this way, digital trace data collected over longer periods of time, even if collected in the same area, they may not be comparable sets. The same problem occurs when private sector companies change priorities and stop sharing data. Data homogeneity in the above sense must be ensured by appropriate action to eliminate any discrepancies before the different sets of data can be integrated in one big database. Public–private partnerships and data-sharing frameworks must always be alert to these issues when they collect and exchange data.

In conclusion, big data in transportation collected via digital trace networks or other means, need careful additional refinement and control vis-à-vis the above weaknesses, before they can be reliably used. This is, also, one of the main justifications and benefits of using AI i.e., that it can perform the necessary tasks of quality assessment and evaluation and produce at the end amble and reliable mobility related data on which to build the various mobility related applications. Of interest is to note that the weaknesses mentioned above are much easier to control and consider in traditional data collection survey methods so ultimately, both conventional data collection methods and those based on digital sources will remain valid and necessary as valuable sources of data collection on which to build the future AI-powered mobility (Scrocca et al., 2025). Big data with the help of AI will enable us to provide answers to many new questions in a timely manner but more conventional sources will often be necessary to complement and validate the information derived from such big data.

7.5.3 Big Data and Carbon Emissions—A Mixed Blessing!

There is no doubt that the use of AI to manage and analyse big data in transportation can be a transformational force in formulating policies and measures to enhance safety and reduce carbon emissions. By collecting and analysing big transportation data, transport planners and engineers can better understand traffic conditions in relation to travel demand, implement effective traffic management schemes, optimize signal timings, and reduce vehicle idle times. All this leads to lower carbon emissions. More specifically, the cases for reducing carbon emissions in the transport and mobility sector have been analytically presented in Chap. 3 (section *measures and policies for net-zero transport emissions*).

Transportation demand management. Through big data analysis, and AI deep learning techniques, a better understanding of travel demand and passenger behaviour can provide decision-making support for governments and transportation management departments. Travel guidance, along with incentives like preferential policies and rewards, can encourage green travel choices such as walking, cycling, and shared mobility, reducing reliance on motor vehicles and lowering carbon emissions.

Improved public transport operation. Based on travel demand forecasting and real time monitoring of traffic conditions, public transportation resources can be allocated more efficiently to provide convenient and efficient mobility services thus encouraging passengers to reduce private car usage and thereby lowering carbon emissions.

Intelligent transportation systems management and operation. Systems like traffic navigation and ride-sharing services can guide users to select energy-efficient routes and modes based on real-time traffic data. Utilizing big data and their analysis, AI technologies offer fully adaptive control of traffic signals by dynamically adjusting signal timings based on real-time traffic conditions with a view to reducing vehicle waiting times and fuel consumption.

Intelligent parking guidance and parking systems operation. Vehicle parking demands and parking supply availability are matched through AI application aimed at optimizing the utilisation of parking resources, avoiding unnecessary vehicle-circling in search of a parking space.

There are many other AI and big data applications that can be mentioned for climate friendly mobility that hold significant promise in optimizing the way that transport and mobility systems operate, reducing congestion and cutting down carbon emissions.

However, while the benefits of the use of AI for reducing emissions are substantial they come with a most notable cost in terms of the energy required to sustain the computational infrastructure that is necessary behind AI and big data analytics. The training of AI models requires enormous computational and thus energy resources. Training large-scale AI models can take weeks or even months, using vast amounts of electric energy in the process (Feng et al., 2023). The computational power needed to process this data is immense, involving thousands of GPUs or TPUs (Tensor Processing Units) operating in parallel in massive data centers. The high-performance computing machines used for these procedures consume energy at a staggering rate, far outstripping the energy usage of traditional computing tasks.

Unfortunately, it's not just the training phase that is energy intensive. AI models need to be constantly fine-tuned, updated, and deployed across networks, each of these processes contributing to increasing their overall energy and emissions footprint. Real-time traffic optimization, for example, involves continuous data collection from multiple sources, including sensors, GPS systems, and cameras and this constant influx of data requires not only real-time analysis but also efficient data storage and retrieval systems, further compounding these systems' energy needs. The data centers, which form the "homes" of AI and big data processing, consume large amounts of electricity to power their servers, cooling systems, and network infrastructure. Cooling systems themselves are very energy-intensive, particularly in regions where the climate is warm. The larger the data center, the greater the cooling demands, which in turn leads to higher energy consumption and potentially high carbon emissions.

Data centers are often strategically located in regions where electricity is abundant and cheap, but their energy sources may not always be renewable. As a significant portion of the energy used in global data centers comes from fossil fuels, all the above considerations lead to the conclusion that the use of AI and big data (in Transportation as well as in other sectors) has an overall negative impact in increasing the carbon emissions at least for as long as the energy generation is not made 100% clean i.e., from renewables. This is the "AI paradox", in that the very application of the new technologies of AI and big data that are expected to significantly reduce overall carbon emissions through their climate friendly applications, are themselves potential high emitters. Addressing this paradox is crucial if the use of AI and big data in the transport and mobility field can be declared to have a net beneficial effect.

References

Abirami, S., Pethuraj, M., Uthayakumar, M., & Chitra, P. (2024). A systematic survey on big data and artificial intelligence algorithms for intelligent transportation systems. *Case Studies on Transport Policy, 17*, September 2024. https://doi.org/10.1016/j.cstp.2024.101247

Chen, C., Ma J., Susilo Y., Liu Y., & Wang M. (2016). The promises of big data and small data for travel behavior (aka human mobility) analysis. *Transportation Research Part C: Emerging Technologies, 68*, 285–299.

Djavadian, S., & Chow, J. Y. (2017). Agent-based day-to-day adjustment process to evaluate dynamic flexible transport service policies. *Transportmetrica B: Transport Dynamics, 5*(3), 281–300.

Feng, W., Yao, C., Zhang, Z., Zheng, Y., & Huang, S. (2023). Exploring the role of energy Big Data in promoting green development practices in the digital economy era. In *2023 Panda forum on power and energy (PandaFPE)* (pp. 1723–1730).

Ferrer-Cid, P., Barcelo-Ordinas, J. M., & Garcia-Vidal, J. (2025). A review of graph-powered data quality applications for IoT monitoring sensor networks. *Journal of Network and Computer Applications, 236*, Article 104116.

ITF. (2021). Big Data for travel demand modelling: summary and conclusions. In *International Transport Forum, Roundtable Report No. 186, author Luis Willumsen, Nommon solutions and technologies*. OECD Publishing.

Jiang S., Ferreira J., & González M. C. (2017). Activity-based human mobility patterns inferred from mobile phone data: A case study of Singapore. *IEEE Transactions on Big Data, 3*(2), 208–219.

Lee D., Camacho D., & Jung J. J. (2023). Smart mobility with Big Data: Approaches, applications, and challenges. *Applied Sciences, 13*(12). https://doi.org/10.3390/app13127244

Liu L., Andris C., Biderman A., Ratti C., & Batty M. (2014). Revealing taxi driver strategies in a congested urban road network using big data. *International Journal of Geographic Information Science, 28*(1), 1–20.

Nie, L., Wang, X., Zhao, Q., Shang, Z., Feng, L., & Li, G. (2023). Digital Twin for transportation Big Data: A reinforcement learning-based network traffic prediction approach. *IEEE Transactions on Intelligent Transportation Systems*, 1–11.

Scrocca, M., Grassi, M., Comerio, M., Carriero, V. A., Delgado Dias, T., Vieira Da Silva, A., & Celino, I. (2025). Intelligent urban traffic management via semantic interoperability across multiple heterogeneous mobility data sources. In *Lecture notes in computer science, The Semantic Web—ISWC 2024* (pp. 218–235).

Toole, J. L., Colak, S., Sturt, B., Alexander, L. P., Evsukoff, A., & González, M. C. (2015). The path most travelled: Travel demand estimation using big data resources. *Transportation Research Part C: Emerging Technologies, 58*, 162–177.

Vartziotis, T., Dellatolas, I., Dasoulas, G., Schmidt, M., Schneider, F., Hoffmann, T., Kotsopoulos, S. D., & Keckeisen, M. (2024). Learn to code sustainably: An empirical study on LLM-based green code generation. ArXiv, abs/2403.03344.

Vlahogianni, E. I., Karlaftis, M. G., & Golias, J. C. (2014). Short-term traffic forecasting: Where we are and where we're going. *Transportation Research Part C: Emerging Technologies, 43*, 3–19.

Zheng, Y., Capra, L., Wolfson, O., & Yang, H. (2014). Urban computing: Concepts, methodologies, and applications. *ACM Transactions on Intelligent Systems and Technology, 5*(3), 1–55.

Zheng, Y., Liu, F., & Hsieh, H. P. (2013). U-Air: When urban air quality inference meets big data. In *Proceedings of the 19th ACM SIGKDD international conference on knowledge discovery and data mining* (pp. 1436–1444). ACM.

Chapter 8
AI in Adapting the Human Travel Behaviour

Abstract This chapter identifies some of the basic characteristics of human travel behaviour and investigates how AI can influence it towards meeting the strategic goals of sustainable and climate friendly mobility. It stresses the view that human travel behaviour and transportation system operation are two interlinked elements that interact in many and complex ways. Travel behaviour has been a major field of study in the Transportation science for many decades, and this chapter summarises many of the findings so far and explores ways in which AI can influence human behaviour to adapt and "match" the available transportation system capacity and make it more aware of the climate change and its impacts on transport and mobility. The chapter refers to the various Transportation Demand Management (TDM) policies (e.g., incentives, information dissemination campaigns, and other) and how AI can support their formulation and implementation. AI is seen as the catalyst for a whole new series of TDM models and policies so its potential application areas in this field are many and these are presented and discussed in detail. To support the view that when TDM policies are properly designed and consistently applied, they can significantly reduce vehicle travel even without the use of AI, the chapter refers to a recent report by the Australian Victoria Transport Institute where a series of data are presented showing the results of specific TDM policies that have been applied in the past in several urban areas. As TDM policies tend to have synergistic effects, their results are more pronounced if implemented as an integrated program that includes a combination of measures of different but complimentary nature. With the wider use of AI in TDM, it is reasonable to expect that the impacts will be higher, more widespread, and varied. Of particular interest is the list of application areas of AI in this field which are given in this chapter. These applications are aimed at influencing travel demand while at the same time optimising the transport system's performance and operation.

Keywords Travel behaviour · Transportation demand management · TDM · Modelling human behaviour · TDM results · TDM policies · AI in TDM · TDM and climate change · AI application areas · AI techniques for TDM

G. A. Giannopoulos and Y. Li, *Human Mobility, Artificial Intelligence and Climate Change*, Lecture Notes in Mobility, https://doi.org/10.1007/978-3-032-08171-1_8

8.1 Introduction

Speaking of human travel behaviour, Daniel McFadden[1] in his landmark paper "The behavioural science of transportation" (McFadden, 2007), stated: "*One can think of transportation as a technological behemoth bedevilled by human behaviour. Transportation research contributes technological and management innovations that drive this beast forward and can also offer insights into the limits that human actors and institutions can impose on implementation of an efficient transportation system. Transportation is affected by human behaviour through its consumers (drivers, riders, vehicle buyers, and shippers); through its managers and workers; and through the policymakers and voters who determine transportation infrastructure and policy*". McFadden was referring to the crucial role of human behaviour in shaping the operation of modern transportation systems besides the other technological and management influences. Today, human behaviour is becoming even more critical in driving the "beast" forward because there are more dimensions to consider one of them being the need for mitigation and adaptation to the looming climate crisis.

Horowitz, two decades before McFadden, had studied the behavioural mechanisms underlying the day-to-day evolution of traffic in modern road networks and formulated his famous *stochastic equilibrium theory.* According to that theory, the demand for travel in a given network increases until matching the supply of transport capacity following a stochastic process with the congestion that is created, when demand exceeds supply, as the main regulating mechanism (Horowitz, 1984). Since the demand for travel is the result of decisions made by each individual prospective traveller, the equilibrium theory demonstrates the impact upon the traffic conditions of human behaviour. Understanding human travel behaviour and its main drivers, is therefore of paramount importance.

Human behaviour is a complex process that researchers have traditionally tried to understand using the laws of economic, behavioural and other related sciences. Scientific research in these areas is providing valuable insights regarding also the human travel behaviour. Humans are consumers of transportation services and the driving force behind the search for efficiency of operation and sustainability in modern transportation systems. AI technologies are providing new ways and means not only for understanding human travel behaviour and its functional mechanisms but also in providing the policy makers with the tools of affecting this behaviour. They enable the collection and analysis of large amounts of data that are then used as input for AI model-training and building that fully replicates human travel behaviour. They can also create the necessary documentation for the public information and other campaigns that form part of influencing this behaviour.

[1] Daniel L. McFadden is an American econometrician who is the father of the discrete choice models in transportation that were successfully used for many decades in modal split research. He shared the 2000 Nobel Memorial Prize in Economic Sciences with James Heckman. McFadden's share of the prize was "for his development of theory and methods for analysing discrete choice".

The term *Travel demand management*—TDM, refers to all actions, incentives, information dissemination campaigns, and other policies that aim to shape the desired travel behaviour of people. Reducing the harmful emissions from transport activities and contributing to the success of the various transport decarbonisation measures is now a major objective of TDM. The scope of TDM is quite wide. It includes not only the travellers or prospective travellers, but also other stakeholders that are involved in the transport and mobility system function such as the vehicle drivers and other people working in the transport sector, the vehicle buyers, the various operators of transport and mobility services, and the public at large. As the new technologies in transportation take effect in the coming years, and as environmental and climate change related restrictions are becoming more and more strict, adapting human behaviour takes a prominent position in the spectrum of available policy choices.

In this chapter we are first clarifying some of the basic characteristics of human travel behaviour and then we investigate how AI can affect travel demand and how it can support TDM policies that aim at a sustainable and climate friendly mobility.

8.2 Understanding Human Travel Behaviour

8.2.1 The Modelling of Travel Behaviour

The use of AI in managing travel behaviour, besides understanding the mechanisms that motivate such behaviour, also calls for the ability to represent it in the form of a model through the construction of reliable algorithms and logical interrelations suitable for mathematical representation or simulation. Traditionally, human behaviour has been modelled based on the view of individuals as consumers of transportation services. Such view has been pursued at three levels of consideration:

a. The *aggregate level* in which the individuals involved are considered in terms of large groups having (more or less) homogeneous characteristics.
b. The *disaggregate level* in which the individuals are examined as separate "units" having rational behaviour i.e., making choices that maximise their net benefit from a trip, and
c. The more complex *cognitive level* that examines the individuals in groups based on their behavioural characteristics that derive from brain science, cognitive psychology and to some extent sociology.

At the *aggregate level*, human travel behaviour in choosing trip destinations, transport modes, or the routes on the respective networks, has been modelled by reference to well-known physical laws such as the gravity law of classical physics (Newton), or the hydrodynamic or thermodynamic laws, with differing degrees of success (Liao et al., 2019; Vanni & Lambert, 2024). At the *disaggregate level*, the discrete individual choices for trip making have been depicted through models normally used in economics or biology. The pioneering work of Domencich and McFadden, published in 1975, is the most characteristic work at this level (Domencich & McFadden, 1975). This work has created the family of *discrete choice models* replicating urban trip generation, destination choice, and mode choice. These are based on the premise that individuals obtain a utility from activities that require travel and therefore make travel choices to maximise that utility (Manski, 2001). The fact that different individuals have different individual behaviours, is captured by making that "utility" random and then use observed discrete travel choices to estimate its distribution and its dependence on variables. This type of analysis has been implemented for the last 50 years or so and one of its most well-known results is the *multinomial logit* or *nested logit* models (Börsch-Supan, 1987).

The third level of consideration i.e., the so-called *cognitive level* links human mobility behaviour to the findings of brain science and cognitive psychology (Cartenì et al., 2022). Brain science has now established how our brains are wired, how we parse and process information, and how we experience things like pain, pleasure, change, choice-making, and so on. It has for example been shown that, typically, human behaviour seems to mistrust anything new and avert change be it in a simple decision like changing a driving lane or larger issues like changing jobs or residences. Travellers make their choices influenced by the same cognitive asymmetries that make us feel losses looming larger than gains and future events heavily discounted relative to the present. The status quo is always preferable, and the fear of change affects human choices and decisions in many instances of trip making. Cognitive psychology results have been used to forecast human behaviour in congested traffic situations (Schumann et al., 2023), or the outcome of complex interactions between the various involved stakeholders in cases of implementation of new transportation projects. Also, in cases where there are multi-agent decision-making processes involving different types of decisions and uncertainty levels (Cartenì et al., 2022). Cognitive model applications in transportation have used static, semi-dynamic, or dynamic parameters for representing driver/traveller behaviour focussing on the ability of drivers and travellers to drive or travel as passengers, in a specific environment and under specific circumstances (Giannopoulos et al., 2006; Schumann et al., 2023).

To the above basics, there is a multitude of variant theories that have been developed, over the years contradicting, or complementing the original ones. Two examples are very indicative. In 1979, Kahneman and Tversky developed and published a critique of the utility theory of McFadden as a descriptive model of decision making in modal choice and developed an alternative model called *Prospect Theory* (Kahneman & Tversky, 1979). The Prospect Theory posits that humans make their choices based on risk i.e., that decisions are taken based on the probability of potential gains or losses and the risks of doing so are assessed using heuristics. The second example, is the so called *contrarian route choice behaviour* theory (Meneguzzer, 2023). According to this, among the many facets of travellers' behaviour that may significantly affect the dynamic evolution, and the stability properties of a traffic system is the so-called *contrarian behaviour*. This is defined as the tendency to act "contrary to the crowd" and is shown that it may have significant impacts on key properties of the traffic system such as the stability of network equilibria, or various system performance metrics (e.g. average travel time). Contrarian behaviour is a kind of strategic behaviour that can be framed within the wider context of cognitive hierarchy theory. The emergence of contrarian behaviour has been addressed by studies relating to various fields, most notably investment strategies in financial markets (Park & Herding, 2012) or social psychology and opinion dynamics (Gambaro & Crokidakis, 2017).

All the above, constitute an exciting field of transport research which, due to its gravity and importance, has been suggested that it could become a distinct new field of study within the Transportation field with the name "*Mobilistics*" (Gunnarsson, 2000).

8.2.2 Transportation Demand Management

Transportation Demand Management (TDM) is a holistic approach to strategically influencing travel behaviour aimed at reducing congestion and enhancing overall efficiency in the transport system. With the realisation of the climate change phenomenon and the urgent need to reduce and eventually zero the transport related emissions, TDM takes on an added importance and gravity in influencing human behaviour towards choosing environmentally friendly and reduced or no-carbon emitting travel and transport activities. The basic notions upon which TDM has been formulated in the past, are the following:

A. *An individual, when deciding to make a specific trip from an origin A to a destination B, determines the characteristics of this trip to maximise his/her expected utility from this trip.* In taking that decision, the prospective traveller, normally goes through a rational decision-making stage in which he/she evaluates alternatives and identifies consequences in a short-term perspective. This means that more direct factors such as, for example, time savings have a higher ranking than long-term consequences such as risks for a serious injury. According to this thesis the decision to make a trip from A to B is based on an evaluation of the (known) utility for the trip at the point of origin A plus the (estimated or expected) utility at destination point B minus the (estimated or expected) sacrifice or "travel cost" in terms of time, money, physical effort, discomfort, risks etc., for the movement from A to B by the selected mode of transport. This thesis is one of the first theoretical bases for expressing travel behaviour and forming related models (Ben-Akiva & Lerman, 1985; Ben-Akiva et al., 2002).
B. *Mobility is part of the daily activity patterns of an individual and is determined by them.* According to this, trip making is the result of the daily patterns of activities which all individuals develop and maintain as part of their daily routines. The daily activity pattern consists of a series of activities such as going to work, shopping, visiting friends, etc. These, generate the need for movement—on foot or by vehicle—to gain physical access to specific destinations. Some of

these activities are bound or forced in space and time, e.g. going to work and school or to an appointed meeting, while others are freer and more open to individual choice, e.g. shopping. Based on this premises, the so-called activity-based mobility models have been developed (Bhat & Koppelmann, 2003; Gunnarsson, 2000).

C. *Transport demand is generated by dissatisfaction with one's own space at a certain moment, event, or situation* so an individual makes a set of decisions to gain access to a specific activity because of dissatisfaction with his/her current situation. Decisions are made to physically move (or not move) or use a web-based communication or a telephone conversation and when they are made other decisions follow-up regarding the destination, the time, the transport mode, the route to follow and so on. All these decisions are linked together in a chain of choices, some made before the trip is started, while some can be changed during an ongoing trip (Sun, 2023).

D. *Transport decisions are the result of a complex interaction between many factors that can be represented by discrete variables* such as, for example, choice of mode 1 or 2, choice of destination 1 or 2 or 3, car ownership 1 or 2, and so on. Complex considerations and constraints are considered for the decision to make a trip such as, for example, if the trip is to satisfy a regular and repetitive need or if the trip purpose is to buy a seldom-purchased product, or if there are other travellers that travel with the original one, and so on (Mwale et al., 2022; Qin et al., 2022).

The traditional studies for the formulation of TDM measures and policies were first focused on adopting or developing a theory on how the behaviour of individuals in travel choice situations can best be influenced and then formulating the most appropriate actions to affect and manage this behaviour to achieve specific goals. Through such studies we have increased our understanding on how individuals react to regulation, pricing and information and have also identified the basic influencing factors of travel behaviour (Hu et al., 2024; Le & Teng, 2023; Mwale et al., 2022; iMotions (2022); US DoT, 2021). A summary of the main such influencing factors is the following:

- Perceived cost of the proposed trip.
- Expected travel time.
- Availability of parking at destination.
- Perceived safety of the mode to be used.
- Level of privacy during the travel.
- Level of convenience.
- Walking distance to bus stops or public transit stations.
- Habitual selections.
- Nice scenery.
- Avoiding of transfers.

AI is expected to be the catalyst for a whole new series of travel demand management models and policies formulation through simulating human travel behaviour and applying machine learning approaches. The AI-powered application areas as well as the technologies and techniques used are presented in the sections after next.

8.2.3 TDM Experience and Results

Typical results from the application of (traditional) Transportation Demand Management policies in various cities and countries around the world, over the last decade, is shown in Table 8.1. The contents of this Table are compiled from data given in a recent report by the Australian *Victoria Transport Institute* (Litman & Pan, 2024). They are provided here in a summary way for the purpose of giving indicative orders of magnitude of TDM related impacts (before AI is used widely for TDM purposes) to serve as examples. The interested reader is encouraged to look for more details in the reference sites provided in the last column of Table 8.1, or elsewhere in the bibliography.

As regards the effects of typical TDM individual policies, we can formulate Table 8.2 also derived from data of the survey by the Victoria Transport Institute (Litman & Pan, 2024). As shown in Table 8.2 vehicle travel reduction policies when properly designed and consistently applied (even without the use of AI) can significantly reduce vehicle travel. The impacts of the various TDM policies can vary considerably depending on their design, application procedure, and local conditions. TDM policies tend to have synergistic effects, i.e. if implemented as an integrated program that includes a combination of policies and measures of different but complimentary nature their results are more pronounced. With the wider use of AI in TDM, it is reasonable to expect that the impacts will be higher than the ones shown in Table 8.2. They will also be more widespread and varied.

8.2.4 Travel Behaviour and Climate Change

Human travel behaviour and transportation system operation are two interlinked elements that interact in many and complex ways. Human behaviour, of travellers in deciding, choosing, selecting, and acting during a trip of their own, or of senders and receivers of goods, or of people working in the transport sector, is linked to the operation of the transport system in a reciprocal way. Human behaviour and decisions affect the operation of the system and the level of service it offers but at the same time the quality and level of service achieved affects human travel behaviour and decisions. Travel behaviour has been a major field of study in the Transportation science for many decades. The aim of the research work has been to better understand this behaviour and find ways of influencing it to adapt and "match" the available transportation system capacity. When climate change and the need to control and eventually eliminate carbon emissions became evident, understanding and managing human travel behaviour took on an added weight as it is increasingly evident that

Table 8.1 Examples of implemented TDM schemes or TDM and other measures, and a sample of the results

A/A	Name of scheme	Area/Country	Short description	Summary of results	Reference
1	Paris sustainable transport program	Paris, France	Improved public transit services, extensive bike-sharing and electric car-sharing, reduced city centre parking supply, reduced traffic speeds in the centre, New bus and bike-lanes. Policies to improve travel options in automobile dependent areas	Reduced vehicle traffic by 60% in the centre and 35% in suburbs. The intended TDM policies to increase fuel tax, sparked the yellow vest protests that led to the abandonment of these policies	https://bit.ly/3CTwKoo and https://slate.com/business/2023/03/paris-car-ban-bikes-cycling-history-france.html
2	Brussels, *Good Move Plan*	Brussels, Belgium	The city's TDM is the *Good Move Plan* that started being implemented in 2022. The main objective was to limit traffic in the city centre and increase the use of alternative modes of travel (cycling, public transport)	By 2024, total vehicle traffic declined 19%, bicycling increased 23% during the morning rush and 13% in the evening	https://www.brussels.be/what-good-move

(continued)

Table 8.1 (continued)

A/A	Name of scheme	Area/Country	Short description	Summary of results	Reference
3	Jak Lingko program	Jakarta, Indonesia	The Jak Lingko program improved urban rail and bus services with integrated physical connections and fares. Road space was removed from cars and given to walking, cycling and transit facilities. It also implemented more participatory planning. Jakarta won for this, the 2021 Sustainable Transport Award	Reduced car trips, increased walking, and cycling trips. The city saw a tenfold (1000%) increase in cyclists in the central area, and a 340% increase in walking, overall	https://itdp.org/wp-content/uploads/2021/11/MOBILIZE-Jakarta-Case-Study-11.16.21_pages.pdf
4	FHWA's Non-motorised Transport Pilot Program	US: Columbia, Missouri; Marin County, Calif.; Minneapolis area, Minnesota; and Sheboygan County, Wisconsin	Spent a total of about $100 per capita to improve active travel conditions and increase walking and bicycling trips	Increased walking trips by an average of 19% and bicycling trips by 62%. This growth continued after the program ended, with 3.7% and 10.5% average annual growth rates for walking and bicycling respectively	https://www.fhwa.dot.gov/environment/bicycle_pedestrian/ntpp/2014_report/hep14035.pdf

(continued)

Table 8.1 (continued)

A/A	Name of scheme	Area/Country	Short description	Summary of results	Reference
5	Minneapolis-St. Paul TDM Program	Minneapolis-St. Paul, US	Nine typical office buildings introduced TDM programs for the people working there	The 9 office buildings generated, on average, 34% to 37% less traffic and need 17% to 24% fewer on-site parking spaces than the Institute of Transportation Engineers' predicted rates	https://www.mikeontraffic.com/wp-content/uploads/2014/01/TDM-Plan-Effectiveness-Study.pdf
6	Suburban TDM Program	US, Fairfax County	Thirteen residential and commercial developments that implemented TDM plans, were studied	The TDM policies resulted, on average, in 63% fewer trips generated as compared to those predicted by ITE's trip generation models	https://www.wellsandassociates.com/research/property-trip-forecasting-fairfax-county/
7	Portland Mode Shift program	Portland, Oregon, US	Integrated TDM and Smart Growth policies	Per capita vehicle travel declined to nearly 30% lower than the U.S. average	https://www.oregonmetro.gov/transportation-system-monitoring-daily-vehicle-miles-travel
8	Capital Regional Vehicle Travel Reduction program	Greater Victoria, British Columbia, Canada	Combination of TDM with extensive bikeway networks construction (70% in Victoria and 44% in Downtown)	Total automobile trips declined 13%, since 2022. Walking and bicycling trips increased	

(continued)

Table 8.1 (continued)

A/A	Name of scheme	Area/Country	Short description	Summary of results	Reference
9	Boulder, mode shifts	Boulder, Colorado, US	Boulder, Colorado invested about $1000 per capita in non-auto modes over two decades, i.e., about $50 annual per capita. Also, TDM policies implemented	Non-auto mode shares increased to about a third of all trips, and single occupant vehicle (SOV) shares declined about 17%	https://doi.org/10.1016/j.tranpol.2014.09.012
10	Strategic TDM plan	Vancouver, Canada	Implementing TDM and improving pedestrian, cycling, and PT facilities, since 2014 In 2021 the Strategic TDM Plan started with new ambitious vehicle travel reduction targets	Between 2014 and 2019 automobile mode shares declined 12% of total trips and 11% for commute trips. Transit shares held steady and walking and bicycling increased significantly. Also, the kilometres travelled per vehicle were reduced (16%) and per capita (12%)	https://vancouver.ca/files/cov/transportation-demand-management-action-plan.pdf

(continued)

Table 8.1 (continued)

A/A	Name of scheme	Area/Country	Short description	Summary of results	Reference
11	School and campus TDM programs	Many schools, colleges and universities, in the US	For each campus TDM programs were implemented. These typically included incentives such as public transit service improvements, highly discounted transit service for all students (the campus purchases them wholesale), more accessible campus design, bike- and car-sharing, and efficient parking pricing	These schemes have been found to reduce car trips to campus by 20–50%	Litman and Pan (2024—Sect. 8.2)
12	Capital regional district (CRD)	Greater Victoria, British Columbia, Canada	Public transit service improvements, promotion of cycling and walking,	Total car trips declined 13%, between 2017 and 2022, despite 9% population growth. Average daily car trips per capita declined 20%, from 1.52 in 2017 to 1.22 in 2022. Cycling grew in areas that expanded bikeway networks (70% in Victoria and 44% in Downtown)	https://www.vtpi.org/CRDts.pdf

Source Litman and Pan (2024)

for achieving net-zero levels by 2050 (i.e., cutting carbon emissions to zero), technological and technical measures alone will not be enough. The consent and active support and cooperation of the people in their daily lives is equally important and necessary.

Understanding the human travel behaviour and taking measures to change and adapt it to overall transport and environmental objectives, is a basic element of a

Table 8.2 Examples of individual TDM policies and typical results

TDM policy	Description	Typical impacts
Efficient parking pricing and management	Charge cost-recovery parking fees with rates that vary by demand. Cash out and unbundle parking. Eliminate parking mandates	5–15% reduction in vehicle ownership and 10–30% reduction in effected vehicle trips
High quality public transport	Frequent, fast, convenient, comfortable transit services. Amenities such as free wifi, and improved payment systems	Service improvements increase public transport travel by 20–50%, and reduce auto travel by 5–15%, and sometimes more
Smart Growth, New Urbanism, Transit- oriented development	Develop compact, mixed-use neighbourhoods around high quality public transport lines	Residents tend to walk, bike and use public transit 20–100% more, and drive 20–60% fewer annual miles
Promotion of micro mobility modes (walking, bicycling, e-bikes and variants)	Improve walking and bicycling conditions and encourage use of these modes. Create more compact, walkable neighbourhoods	Infrastructure improvements increase active and micro mobility mode travel by 50–100% and reduce driving by 5–15%. Compact, walkable communities reduce driving by 20–40%
Commute, school and campus transport management programs	Improve non-auto travel options and encourage their use with financial incentives (parking pricing and cash out)	Programs that only use persuasion reduce driving 5–15%, those that provide financial incentives reduce auto trips 10–30%
Roadway redesigns to favour sustainable modes	Improve sidewalks, add bike- and bus lanes, and reduce traffic speeds. Apply complete streets policies	Non-auto travel typically increases 20–100%, and auto travel declines 10–30% Reducing traffic speeds reduces VMT
Efficient road pricing	Motorists pay cost-recovery tolls on urban highways and fees to enter city centres	10–30% reduction in affected road traffic volumes
Distance-based pricing	Vehicle insurance and registration fees are prorated by average annual mileage	Up to 15% if total insurance premiums and registration fees are prorated
Vehicle sharing	Provide car- and bikesharing services in urban neighbourhoods	12–15 private cars replaced by each shared car
Freight transport management	Require or encourage shippers to use efficient vehicles and logistics	Can reduce freight vehicle travel and emissions 10–30%
Limited traffic zone	Limit vehicle trips to central city areas	10–20% reduction in city-centre cars

(continued)

Table 8.2 (continued)

TDM policy	Description	Typical impacts
Personalised travel planning	Residents encouraged to use non-auto modes. Transit fare discounts	6–12% reduction in car use by participants
Sustainable mobility apps	Mobile apps provide user information, payments and rewards for reduced driving	73% of app users reduce their vehicle travel

Source Litman and Pan (2024)
Notes The stated impacts are shown as given in the source reference. The method for their measurement or calculation is not mentioned in the source reference and thus they are given here with reservation and as indicative orders of magnitude

climate sensitive transport and environmental planning in its pursuit to contribute to climate change mitigation and adaptation actions. Behavioural changes may affect many and diverse aspects of transport related activities from the choice of public transport, cycling or walking instead of driving, to actions like turning down car heating or going on holiday nearer to home. If such behavioural changes do not occur, then reducing transport related emissions to the required net-zero level by 2050 would require ramping-up all other decarbonisation measures and policies (shown in Chap. 3) to much higher levels than anticipated now, thus making them practically unattainable.

The International Energy Agency in a 2021 report, in which they put forward their net-zero 2050 scenarios, estimated that for the period 2021–2050 behavioural changes can reduce energy demand by an average of 4% all around (for all sectors) on a yearly basis. This percentage is higher in the richer and more developed regions where energy lifestyles are more pronounced, and lower to lesser developed regions (IEA, 2021). In the road transport sector, the reductions to energy consumption from travel behavioural changes are shown in Fig. 8.1 (for the period 2020–2050). As shown in that Figure, the largest reductions (and thus behavioural changes) are expected in the 2030–2040 period with the maximum amount being – 11.8% for the year 2035. This amount is shared as follows: – 3.6% from phasing out internal combustion engine vehicles, – 0.5% from teleworking from home 3 days/week, – 3.8% from ridesharing in urban travel (the largest reduction), – 2.4% from setting a maximum speed limit 100 kms/hr on the motorways, and -1,5% from eco-driving (IEA, 2021). Teleworking may indeed effect a larger percentage reduction if implemented systematically as a central TDM policy Bonifacio (2024).

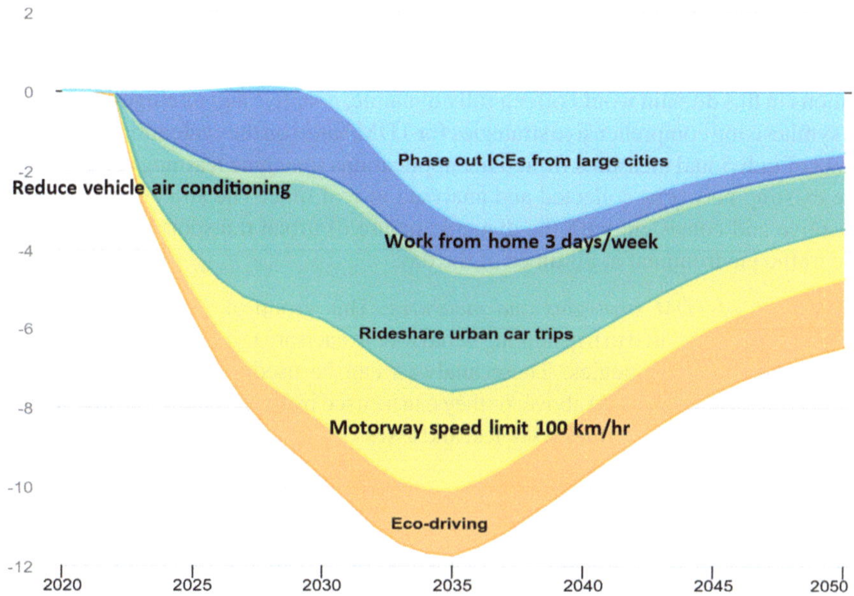

Fig. 8.1 Estimated % reduction in road transport energy consumption from travel behavioural changes, as a percent of the total energy consumption in road transport, in the net-zero scenario of the International Energy Agency for the period 2020–2050. *Source* IEA (2021)

8.3 How AI Can Transform Travel Demand Management

8.3.1 Application Areas for a Climate-Friendly, AI-Assisted, TDM

The potential of AI in travel demand management (TDM) and the adoption of a climate friendly human mobility behaviour, should be a major goal of future AI mobility applications. In doing so, AI will be utilising its ability to analyse enormous amounts of traffic, weather, and other data that are now becoming available daily in many urban areas. There are now clear governmental policies and specific guidelines issued for the collection and use of such big data in the EU, US, and China (EC, 2023; Mingtao Shi & Lechner, 2019; US DoT, 2020; Wu, 2024). By utilising this capability, AI-powered chatbots can assist travellers in resolving queries, providing information about travel destinations, offering real time support in many aspects of a traveller's journey and generally assisting travellers in making travel related decisions that conform to climate friendly mobility. At the same time the mobility service-providers, i.e. the "supply" side, can be helped in delivering climate friendly services to the users while optimising their travel experience.

Indicative areas for AI applications in the development of climate-friendly TDM policies, are:

(A) For the authorities:

Formulation of climate friendly Transport Demand Management strategies. AI applications in this domain would offer a fully dynamic, adaptive and intelligent approach in synthesising comprehensive strategies for TDM based on the evaluation and assessment of individual measures from other applications (machine learning) or data from the existing networks collected and analysed in real time. The strategies should be adaptive and consistent with the dynamic nature of urban transport conditions and the traffic environment at all times of the day.

Evaluation of TDM strategies and measures. This group of AI applications will analyse big data related to trip making behaviour in each area with a view to evaluating the applied TDM strategies. These analyses can be used in formulating a TDM strategy, as described in (1) above, or they can be used by traffic control centres and/or researchers and/or policymakers to test and assess specific trip or traffic management strategies.

Assistance in local government *monitoring and control of urban and interurban transport system operation.* The need to have a dynamic and accurate picture of the overall operation of the transport system within the jurisdiction of a regional or local authority can form a major area of application of AI. Many AI-applications can be formed here, aimed at informing authorities on problems and issues that exist in their quest for creating a more seamless, efficient, and sustainable transport and mobility system that is:

Safety advise provision (by evaluating and communicating potential safety hazards—continuously monitoring traffic conditions—detecting erratic driving behaviour—identifying potential collision scenarios, etc.).

Advice on accessibility issues (by assessing selected routes for individuals with mobility challenges and ensuring that everyone can use and participate in the transportation system services), and

Climate friendly decision making (by offering guidance for policies reducing the need for travel, making better use of non-polluting modes of transport, and improving the overall traffic efficiency minimising idle times and unnecessary fuel consumption).

(B) For the travellers:

Personalised Travel Planning. This group of applications is aimed at the individual prospective travellers to provide travel advice with the primary goal of making their trips with low or no carbon footprint while, at the same time, satisfying the personal preferences of the individual traveller. AI interfaces, integrated into the smartphones or smart home devices, would provide travellers with personalised recommendations based on their travel history, preferences, and the real-time conditions that exist on the networks regarding all modes of transportation. The advice should also take account and notify users of accidents and other disruptive events as they happen (in real time), excessive delays on parts of the network, and so on. There are many individual applications that can be developed within this framework. Their results will offer capabilities for creating a personalised itinerary by recommending destinations, routes to follow, combination of modes to use, booking links where necessary, maps and so on and all of them harmonised with the personal preferences and past experiences of the specific traveller.

Examples of applications within the "personalised travel planning" item, include:

- *Route optimization for minimum energy consumption*: AI algorithms can streamline travel behaviour towards choices that minimise fuel consumption and emissions in tandem with the traffic management system options that exist or can be created dynamically to match changes in demand.
- *Calculating sustainability scores and rewards.* Users could receive a sustainability score based on their mobility choices, incentivizing greener decisions with rewards such as discounts for public transport journeys or access to premium features in a select number of mobility services.
- *Calculating and making available to users, mobility indicators and other traffic related data analytics.* "Users" in this respect are not only the prospective travellers but also other mobility related stakeholders such as traffic control centres, infrastructure owners, automated vehicles and their passengers, and so on.
- *Offering chatbot and virtual assistant services to answer specific questions in real*-time. Potential or actual travellers (during their trip) can get answers to all sorts of questions about the execution of a trip or other related issues and obtain valuable insights for decision-making.

Provision of real time advice and information during the trip. This would include alternative route suggestions in cases of sudden incidents on the network, extreme weather or other unexpected events. These applications should also help travellers in adapting their travel or driving behaviour to the traffic conditions as they develop dynamically in time.

Calculating carbon footprints and carbon offsetting. AI-powered mobility platforms can calculate the carbon footprint of individual trips and at the same time calculate and suggest potential offsetting actions for the produced carbon by the trip. It can for example, offer the possibility to automatically invest in carbon offset projects such as reforestation or renewable energy initiatives, or reduced fares in urban public transport or intercity trips, and so on. This application is particularly suitable for large companies where mobility decisions by the management can affect a great number of daily trips by the employees or visitors. In this way, sustainable mobility is made an integral part of the company's policy towards its employees.

(C) For the mobility service providers:

Accurate traffic predictions. AI can be utilised in making accurate traffic predictions both short-term and longer term. This AI application is particularly important for urban road network operators in connection with the planning and implementation of the respective TDM strategies. AI algorithms can analyse historical and current traffic data, weather patterns, traffic congestion patterns, etc., and calculate the impacts of any traffic situation in terms of resulting emissions, peak travel times, potential traffic bottlenecks, and other related parameters. By using these as input, AI related algorithms can produce reliable short-term predictions for urban traffic conditions and assist traffic control centres to proactively implement measures to mitigate congestion and improve traffic flows.

Optimisation of public transportation schedules and networks. These apps will analyse passenger data, public transport usage patterns, public transportation schedules, public transportation O-D flows and other data to ensure that buses and trains are deployed efficiently, reducing waiting times and improving the overall transit experience. The public transport controllers can adapt their schedules and plans to match the anticipated conditions and by consequence relay back all information on the changes made to the users.

Demand driven traffic management. This application area is the most common as the management of traffic in any network requires a multitude of complex tasks and data analyses for optimising performance in road traffic signals coordination, minimising delays and reducing travel times, with tangible climate change mitigation impacts as well as increasing the resilience of the system. By utilising inputs from many individual travellers, and vast data from many other sources, AI can estimate the emerging magnitude of the demand for transport and traffic services at any point in time and inform traffic control centres to adjust their capacity and traffic flow management policies. It can provide all sorts of predictive analytics and suggest ways to mitigate safety and congestion risks. AI can also be used to adjust the traffic signal cycles to match, in real-time, the changing traffic conditions reducing vehicle delays and improving traffic flow conditions. Most importantly, AI can also improve the safety conditions in emergency situations such as road closures, road accidents, etc., by alarming and then facilitating the approach of emergency vehicles like police, ambulance, or fire brigade vehicles.

Assistance in promoting the planning, operation, and management of shared mobility services. Efficient shared mobility services are perhaps the key to the future sustainable mobility system. Such services are to be promoted in relation to TDM policies and AI is particularly suited for applications that promote the use of shared services such as car or bike-sharing, car-pooling, ride-sourcing etc.

Assessment of safety levels. In all traffic and transport systems operating in the various transport mode domains, maintaining a high safety level is of paramount importance. AI by analysing real-time data from different sources can be used to identify safety risks and assess the safety level of any operation. Consequently, it can suggest measures and steps in averting these risks.

Monitoring and reporting applications. AI will be used for monitoring and reporting applications in relation to video, radar, or other means of monitoring traffic conditions along specific road, or rail networks as well as in ports and airports. By use of the appropriate algorithms and learning routines it can identify all sorts of incidents and automatically inform the corresponding authorities as well as the users.

Facilitating the introduction of automated (autonomous) vehicles into mixed traffic situations. This is an application area that is related to the support of connected vehicles as well as control centres in dealing with the mixed traffic situations i.e., in which automated vehicles circulate on the same road space with human driven vehicles. Indicative examples of AI assisted "mixed traffic" applications include recognizing and assessing vehicle priorities in non-signalised intersections, recognizing and assessing vehicle priorities at signalised intersections, assessing safety risks at specific situations, recognizing abnormal driving by human-driven vehicles (cars and motorcycles), assessing the impact of weather conditions on traffic safety, relaying information and two-way communication between human-driven and automatic vehicles as well as between them and control centres. In car connectivity and digital services are perceived as highly desirable by potential users (McKinsey, 2023).

Assistance to tourists. AI-powered chatbots and virtual assistants in many languages can provide tourists with real-time information about local sustainability initiatives and eco-friendly options for their travel. They can provide language assistance and translations, or information on places of interest and/or places of need (hospitals, pharmacies, etc.). In areas of high tourist flows this application of AI is already used and is of great value (Gao et al., 2023; Kim et al., 2024).

8.3.2 Technologies and Techniques

To perform the roles and develop the applications mentioned above and effectively provide a reliable basis for accurately modelling the human travel behaviour, AI can utilise several novel technologies and techniques to profit from the findings of scientific research not only in computer science but in disciplines like cognitive psychology and behavioural economics. In the traditional transport demand models, travel behaviour was assumed to be rational, consistent, perfectly informed and always following the laws and prescriptions of traditional economic theory, i.e. maximizing the economic utility of a trip by trading off costs and benefits. In AI related applications, travel decisions are treated as deviating from the predictions of the traditional economic theory. Some of these deviations are systematic and consistent, while others are irrational and largely unpredictable made in a random way. Evidence on systematic deviations from rational behaviour have emerged from studies on financial, consumer, or health behaviour (Ariely, 2010; Thaler & Sunstein, 2009) as well as from studies on travel behaviour) (Avineri & Chorus, 2010; Van de Kaa, 2008). Utilising the research on these deviations and focused on cognitive biases and the cognitive bounded rationality of decision makers, AI can offer a novel approach by incorporating systematic deviations of travel choice from the rational models of utility maximization and in addition irrational random deviations.

Incorporating the irrational deviations of human travel behaviour, can be done by utilising the *fuzzy sets theory* to construct a fuzzy-based model of travel choice

(Avineri, 2009). This is a more realistic approach because imprecision and vagueness are inherent to the traveller's cognitive model of behaviour and choice. Also, because the information available to the traveller about the constraints and parameters of the intended trip is vague and not precisely measurable. The fuzzy sets theory combined with AI and big data analytics can provide a promising basis to model the biases, flaws and limitations in judging, inferencing, learning, and problem-solving of individuals in their travel behaviour.

Another appropriate AI related technology is that of *neural networks*. This is one of the oldest AI technologies that has already been applied to a wide range of transport problems (Ishak & Trifiro, 2007; Lin et al., 2019; Qin et al., 2022; Wu et al., 2022). There is a variety of different types of neural networks that have been developed, with different focus and/or specialisations but their basic structure is as follows. A neural network consists of layers of nodes, or artificial neurons, and an output layer. Each node connects to others and has its own associated weight and threshold. If the output of any individual node is above the specified threshold value, that node is activated, sending data to the next layer of the network. Otherwise, no data is passed along to the next layer of the network. Combining neural network-based approaches with fuzzy rule-based ones produces the so-called *Adaptive Neuro Fuzzy Inference System*—ANFIS approaches, in which a fuzzy inference system can be represented with a neural network structure. In principle, the fuzzy rule-based neural network approach enables adjustment of rules to improve the overall performance of the model in replicating human travel behaviour (Lee et al., 2021; Liu et al., 2017; Ünal & Başçiftçi, 2022). A well-structured and documented presentation of the capabilities, characteristics, and structural details of neural fuzzy inference machine deep learning systems can be found in Kala (2024), Lv et al., (2020).

Another use of AI as a tool for TDM policy formulation, is its potential to be combined with technologies such as *Blockchain* and *Internet-of-Things* (IoT) to advance applications for tradable travel credit schemes. These are normally conducted according to a virtual smart contract between the traveller and the transport service provider or any other relevant stakeholder (Qin et al., 2022). The blockchain and Internet-of-things (IoT) interaction with AI can be a key to developing applications related to shared mobility systems and increasing the automobile occupancy by shared rides. Already several ride hailing and shared mobility operators are using such combination of AI with blockchain and IoT technologies (e.g., Uber™, Lyft™, Via™, UberPool™, Shared Lyft™, Ruqi Travel™, Xiangdao Chuxing™, Caocao™ and Didi™). In China there are more than 350 shared mobility companies most of them using intelligent apps through AI in which users are enabled to make better and more climate friendly decisions incentivised by use of these technologies (Su et al., 2022).

Finally, the construction of *multi-agent simulations* in which "agents" representing individual travellers, households, or other decision-making entities, are maintained as artificial individual entities with individual attributes and individual states. Then they are simulated to make individual decisions based on these attributes and states. Two agents, when submitted to the same situation, can make different decisions. For a review of agent-based approaches, see Bernhardt (2007), and Huang et al. (2022). There are many new theories and techniques that are continuously being developed in relation to AI technologies that can result in intelligent software systems that can perform tasks such as learning, problem-solving, decision-making, and mimicking human intelligence. By doing so, AI appears as the ideal tool for modelling travel behaviour and providing an opportunity for testing the potential impacts of travel demand management or other transport related policies.

References

Ariely, D. (2010). *Predictably irrational: The hidden forces that shape our decisions* (384 pages). Harper-Collins. ISBN: 978-0061353246.

Avineri E. (2009). The fuzzy meaning of reference-based perceptions in travel choice modelling. In *Transportation Research Board* (TRB)*, 88th annual meeting, Washington D.C., 11–15 January 2009. TRID data base Accession Number: 01125546, Paper Numbers:09-0803.*

Avineri, E., & Chorus, C. (2010). Editorial: Recent developments in prospect theory-based travel behaviour research. *European Journal of Transport and Infrastructure Research, 10*(4), 293–298.

Ben-Akiva, M., & Lerman, S. R. (1985). *Discrete choice analysis: Theory and application to travel demand*. MIT Press.

Ben-Akiva, M., McFadden, D., Train, K., Walker, J., Bhat, C., Bierlaire, M., Bolduc, D., Boersch-Supan, A., Brownstone, D., Bunch, D. S., Daly, A., De Palma, A., Gopinath, D., Karlstrom, A., & Munizaga, M. A. (2002). Hybrid choice models: Progress and challenges. *Marketing Letters, 13*(3), 163–175. https://doi.org/10.1023/A:1020254301302

Bernhardt, K. L. S. (2007). Agent-based modelling in transportation. Artificial intelligence in transportation: information for application. In *Transportation research circular* E-C113 (pp. 72–80). Artificial Intelligence and Advanced Computing Applications Committee, TRB, National Research Council.

Bhat, C., & Koppelmann, F. S. (2003). Activity-based modelling of travel demand. In *Handbook of transportation science* (pp. 39–65). https://doi.org/10.1007/0-306-48058-1_3

Bonifacio R. (2024). Teleworking: A comprehensive guide to remote work success. *Shiftbase*, 9 April 2024. https://www.shiftbase.com/glosary/teleworking

Börsch-Supan, A. (1987). The nested multinomial logit model. In A. Börsch-Supan (Eds.), *Econometric analysis of discrete choice. Lecture notes in economics and mathematical systems* (vol. 296). Springer. https://doi.org/10.1007/978-3-642-45633-6_4

Cartenì A., Marzano V., Henke Il., & Cascetta E. (2022). A cognitive and participative decision-making model for transportation planning under different uncertainty levels. *Transport Policy, 116*, 386–398. https://doi.org/10.1016/j.tranpol.2021.12.013

Domencich, T., & McFadden, D. L. (1975). *Urban travel demand: A behavioral analysis*. North-Holland Publishing Co., 1975. Reprinted 1996.

EC. (2023). Creation of a common European mobility data space. *European Commission, Communication no. COM (2023) no. 751 final. Brussels 29–11–2023.*

Gambaro, J. P., & Crokidakis, N. (2017). The influence of contrarians in the dynamics of opinion formation. *Physica A: Statistical Mechanics and Its Applications, 2017*(486), 465–472.

Gao, Z., Cheah, J.-H., Lim, X. J., & Ng, S. I. (2023). Can travel apps improve tourists' intentions? Investigating the drivers of Chinese gen Y users' experience? *Journal of Vacation Marketing, 30*(4). https://doi.org/10.1177/13567667231152938

Giannopoulos, G. A., Panou, M., & Bekiaris, E. (2006). Driver cognitive engineering and challenges for the future. In *Transport Research Arena—Proceedings TRA Conference, 12–15 June 2006, Gothenburg, Sweden.*

Gunnarsson, S. O. (2000). Studies in travel behaviour and mobility management need a special scientific discipline: "Mobilistics". *IATSS Research, 24*(1), 69–75. https://doi.org/10.1016/S0386-1112(14)60019-4

Horowitz, J. L. (1984). The stability of stochastic equilibrium in a two-link transportation network. *Transportation Research B, 1984*(18), 13–28.

Hu, S., Xiong, C., Ji Eric, Y., Wu, X., Liu, K., & Schonfeld, P. (2024). Understanding factors influencing user engagement in incentive-based travel demand management program. *Transportation Research Part A: Policy and Practice, 186*, August 2024. https://doi.org/10.1016/j.tra.2024.104145

Huang, J., Cui, Y., Zhang, L., Tong, W., Shi, Y., & Liu, Z. (2022). An overview of agent-based models for transport simulation and analysis. *Journal of Advanced Transportation, 2022*, Article ID 1252534, 27 February 2022. https://doi.org/10.1155/2022/1252534

IEA. (2021). Do we need to change our behaviour to reach net zero by 2050? *International Energy Agency*, report, Paris 29 October 2021, Licence: CC BY 4.0. https://www.iea.org/articles/do-we-need-to-change-our-behaviour-to-reach-net-zero-by-2050

iMotions. (2022). Human behavior research: The complete guide. *iMotions A/S*, Bryn Farnsworth, July 28, 2022. https://imotions.com/blog/learning/research-fundamentals/human-behavior/

Ishak, S., & Trifiro, F. (2007). Neural networks: Artificial intelligence in transportation information for application. In *Transportation Research Circular* E-C113 (pp. 17–30), Artificial Intelligence and Advanced Computing Applications Committee, TRB, National Research Council. ISSN 0097-8515. https://onlinepubs.trb.org/onlinepubs/circulars/ec113.pdf#page=23

Kahneman, D., & Tversky, A. (1979). Prospect theory: An analysis of decision under risk. *Econometrica, 47*(2), 263–292. www.jstor.org

Kala, R. (2024). Chapter 12—An introduction to machine learning and deep learning. In *Autonomous mobile robots: Planning, navigation, and applications in modelling* (pp. 569–625). Elsevier. ISBN: 978-0-443-18908-1. https://doi.org/10.1016/C2022-0-00636-X.

Kim, M. J., Hall, C. M., & Chung, N. (2024). The influence of AI and smart apps on tourist public transport use: Applying mixed methods. *Information Technology & Tourism, 26*, 1–24. https://doi.org/10.1007/s40558-023-00272-x

Le, J., & Teng, J. (2023). Understanding influencing factors of travel mode choice in urban-suburban travel: A case study in Shanghai. *Urban Rail Transit, 9*, 127–146. https://doi.org/10.1007/s40864-023-00190-5

Lee K., Eo M., Jung E., Yoon Y., Rhee W. (2021). Short-term traffic prediction with deep neural networks: a survey. *IEEE Access*, 9, 5 April 2021, pp. 54739–54756. https://doi.org/10.1109/ACCESS.2021.3071174. Electronic ISSN: 2169–3536.

Liao, Y., Yeh, S., & Jeuken, G. S. (2019). From individual to collective behaviours: exploring population heterogeneity of human mobility based on social media data. *EPJ Data Science, 8*(34). https://doi.org/10.1140/epjds/s13688-019-0212-x

Lin Z., Feng J., Lu Z., Li Y., Jin D. (2019). DeepSTN+: Context-aware spatial-temporal neural network for crowd flow prediction in metropolis. *Proceedings of AAAI Conference on Artificial Intelligence*, Vol 33, No. 1, (17 July 2019), pp. 1020–1027. https://doi.org/10.1609/aaai.v33i01.33011020.

Litman, T., & Pan, M. (2024). TDM success stories: Examples of effective transportation demand management policies and programs, and keys to their success. Victoria Transport Institute. https://vtpi.org/tdmss.pdf

Liu, W., Wang, Z., Liu, X., Zeng, N., Liu, Y., & Alsaadi F. E. (2017). A survey of deep neural network architectures and their applications. *Neurocomputing, 234*, 11–26. https://doi.org/10.1016/j.neucom.2016.12.038

Lv, M., Hong, Z., Chen, L., Chen, T., Zhu, T., & Ji, S. (2020). Temporal multi-graph convolutional network for traffic flow prediction. *IEEE Transactions on Intelligent Transportation Systems, 22*(6), 3337–3348.

Manski C. F. (2001). Daniel McFadden and the econometric analysis of discrete choice. *The Scandinavian Journal of Economics, 103*(2), 217–229. https://www.jstor.org/stable/3440992

McFadden, D. (2007). The behavioural science of transportation. *Transport Policy, 14*(4), 269–274. https://doi.org/10.1016/j.tranpol.2007.07.001

McKinsey. (2023). How do consumers perceive in-car connectivity and digital services? *McKinsey Centre for Future Mobility* (MCFM), report by Alexander Matthey, Felix Rupalla, Tobias Schneiderbauer, and Dasha Zuyeva, September 2023. https://www.mckinsey.com/industries/automotive-and-assembly/our-insights/how-do-consumers-perceive-in-car-connectivity-and-digital-services

Meneguzzer, C. (2023). Modelling the combined effect of travelers' contrarian behaviour, learning and inertia on the day-to-day dynamics of route choice. *Applied Sciences, 13*(5). https://doi.org/10.3390/app13053294

Mingtao Shi, C., & Lechner, M. (2019). Big-Data in China: An overview. In *Industrie 4.0 management* (Vol. 35, ed. 2, pp. 61–65).

Mwale M., Luke R., Pisa N. (2022). Factors that affect travel behaviour in developing cities: A methodological review. *Transportation Research Interdisciplinary Perspectives*, Vol 16, December 2022. https://doi.org/10.1016/j.trip.2022.100683

Park, A., Herding, S. D. (2012). Contrarianism, and delay in financial market trading. *European Economic Review, 56*, 1020–1037.

Qin, X., Ke, J., Wang, X., Tang, Y., & Yang, H. (2022). Demand management for smart transportation: A review. *Multimodal Transportation,1*(4). https://doi.org/10.1016/j.multra.2022.100038

Schumann, J. F., Srinivasan, Ar. R., Kober, J., Markkula, G., & Zgonnikov, Ar. (2023). Using models based on cognitive theory to predict human behavior in traffic: A case study. In *Proceedings of the IEEE 26th international conference on intelligent transportation systems, ITSC 2023, pp 5870–5875.* ISBN 979-8-3503-9946-2.

Su, S., Liu, A., & Ma, J. (2022). Status and opportunities of shared mobility systems in China. *Volvo Research and Educational Foundation—VREF*, May 2022.

Sun, J. (2023). How is AI reshaping the travel experience? *World Economic Forum*, Trip.com Group, Dec 19, 2023.

Thaler, R., & Sunstein, C. R. (2009). Nudge: Improving decisions about health, wealth and happiness. Yale University Press. ISBN: 978-0143115267.

Ünal, H. T., & Başçiftçi, F. (2022). Evolutionary design of neural network architectures: A review of three decades of research. *Artificial Intelligence Review, 55*, 1723–1802. https://doi.org/10.1007/s10462-021-10049-5

US DoT. (2020). Enhancing active transportation and demand management with advanced and emerging technologies and data sources. US Department of Transportation, Office of Operations, report no. FHWA-HOP-19-010, March 2020.

US DoT. (2021). Analysis of travel choices and scenarios for sharing rides: Final report. US Department of Transportation, Office of Operations, report no. FHWA-HOP-21-011, March 2021.

Van de Kaa, E. J. (2008). *Extended prospect theory: findings on choice behaviour from economics and the behavioural sciences and their relevance for travel behaviour* [Doctoral Thesis, Issue Number: T2008/11. TRAIL Thesis Series, the Netherlands, 446 pages]. ISBN: 9789055841059.

Vanni, F., & Lambert, D. (2024). On an aggregated estimate for human mobility regularities through movement trends and population density. *Entropy, 26*(5), 398. https://doi.org/10.3390/e26050398

Wu, G. (2024). Development status and future trend of China's Big Data. *Applied and Computational Engineering, 114*(1), 102–106, December 2024. https://doi.org/10.54254/2755-2721/2024.18218

Wu, H., Aguilar, V., & Montgomery, J. (2022). Applications of machine learning to foundation design for transportation structures. *Transportation Research Record*, September 20, 2022. https://doi.org/10.1177/03611981221122779

Chapter 9
Ethical and Societal Implications

Abstract The ethical implications of AI arise with respect to the moral values of humans and human societies. They are examined in this chapter under the three headings: *Ethical implications related to the traveller, i.e.* the decision making about travel (trip making) or transport (freight movement), *Ethical implications related to the operation of the transport system, i.e., of the measures related to* maximising efficiency and safety of the system and *Ethical implications related to the vehicle, which refer to the use of the info and data about the driver of the vehicle. As regards the* societal implications these relate to ways and means to safeguard that AI applications abide with commonly accepted social rules, conditions, and norms. The elements in each category of implications (ethical or societal) are discussed under names *such as diversity, non-discrimination, fairness, privacy and security of users*, and several others and then the various potential applications are classified under the following three categories: (a) Privacy, accountability, and security during the trip making. (b) Equitable use and accessibility of transport infrastructures and services, and (c) Maximum adherence to societal values, goals and objectives. As the need for ensuring equitable access to sustainable transportation is becoming a central issue of adaptation of the transport system in the face of climate change, this chapter further elaborates on the ability of Artificial Intelligence to support underserved and vulnerable communities in accessing climate-resilient transport services, to identify and support the mobility of vulnerable populations, to reduce mobility disparities, and to support inclusive transportation planning. At the end, the ethical implications of four specific mobility technologies are examined in more detail. These technologies are automated mobility, shared mobility, public transport, and freight transport and logistics.

Keywords AI ethical implications · AI societal implications · Ethical AI programming · Societal requirements · Vulnerable populations · Inclusive transport planning · Legislation · Ethical issues in automated mobility · Ethical issues in shared mobility · Ethical issues in public transport · Ethical issues in freight transport

G. A. Giannopoulos and Y. Li, *Human Mobility, Artificial Intelligence and Climate Change*, Lecture Notes in Mobility, https://doi.org/10.1007/978-3-032-08171-1_9

9.1 Introduction

The rapid development and increasing use of AI technologies in all fields of human activities raises many ethical as well as wider, societal, safety and security implications. The significance of knowing these implications is rooted primarily in the issues that must be safeguarded when individuals give, knowingly or unknowingly, personal information to AI systems which potentially could then exploit or misuse it. The input of personal data in AI mobility applications is somewhat inevitable because if the AI application is to produce personalised travel recommendations or other info relevant to the individual user and data adapted to a specific situation, it will have to utilise information of a personal nature for the person concerned. Such information may include age, social and cultural background, car ownership, income level and other economic data, preferred destinations, usual trip purposes, and other. Ensuring and guaranteeing that this information will not be released to third parties or in any way misused, is of paramount importance as it directly affects the social acceptance of the technology and thus its future development and growth.

As the phrase "Artificial Intelligence" implies, it is all about trying to replicate human intelligence by a computer software that can learn and develop logic like a human. This process takes place (for the time being) within the strict confines of the prescribed software instructions and impeded functional algorithms under which an AI software program operates. The current know-how and capabilities of AI computer programming tools do not exclude the possibility that in the not-too-distant future the current confines within which AI operates may be abolished and AI is enabled to develop its own logic and its own ways of operation that practically have no ethical or other boundaries! This situation will create a whole new playground for AI which may well be out of human control with many and uncharted implications, mostly unwelcome. This is not something that we anticipate happening so in this book when we talk about AI's ethical and societal implications, we refer to the currently allowed AI capabilities in which there is no "autonomy" in the AI logic.

It is taken for granted that the AI technologies that we will be dealing with in the future will use "intelligence" within boundaries and that valid and accurate input data will be available. Without valid input data biases and other erroneous outcomes will arise and then besides the wrong answers we will also get ethically questionable results. A typical example of this, is found in many of the answers of current generative AI apps (Gen-AI) known as *chatbots* (i.e. the non-scripted, conversational "robotic" responses to human text and voice input). These answers are derived mostly from internet searches and the blogs that are "floating" in the web whose validity the AI software is not yet trained to detect. This situation, results in chatbot texts whose validity and accuracy are often questioned on account of biases, lack of fairness, lack of accuracy and transparency, and other issues.

The societal impacts and implications of AI refer to its compliance with current societal values and moral precepts. The analysis and confrontation of ethical and societal impacts is instrumental in fostering public trust to AI, mitigating potential risks to society and reducing liabilities. The dimension of climate change and the need to materialise a green transition and a decarbonised way of life, brings added implications that make the application and use of socially and ethically responsible AI applications an urgent task for governments and public administrations. Such task consists of safeguarding the application of ethical and societal principles in the development, implementation, and evaluation of AI technologies and also the implementation of ethical oversight procedures, auditing processes, and continuous evaluation of such AI systems to ensure their compliance with the ethical and well-established societal principles and standards. In fact, ethical values and accepted societal norms should be embedded within the AI development technologies to provide the necessary guiding frames and limits to what these technologies can achieve.

This chapter addresses all these issues from the point of their relevance to climate-friendly transport and mobility. It also discusses the various secondary but relevant issues that these implications create for the traveller, the vehicle, and the system's operation in general.

9.2 Ethical Implications

The ethical implications of AI arise when assessed in terms of a moral point of view i.e. according to the moral values of humans and human societies. In mobility related applications, this covers the full spectrum of potential applications from the management of the demand for travel (i.e., related to the traveller), to the operation of the transport system and management of the resulting traffic flows, and the operation of the vehicles (automated or not). It's under these three headings that we examine the ethical implications of mobility related AI applications in the following.

9.2.1 Ethical Implications Related to the Traveller

For the traveller, AI provides support to decision making about travel (trip making) or transport (freight movement) choices. This support can take the form of detailed answers to traveller's inquiries, or of specific proposals for their trip making or (freight) transport choices, not only before the actual travel or transport activity but also during it. The AI tools used for this interaction, usually take the form of AI *bots* i.e., advanced software applications that answer questions or inquiries automatically, by performing tasks of data analysis (autonomously) and shaping their answers through machine learning and natural language processing capabilities. They understand user inputs, improve responses over time, and provide more personalised interactions. *Bots* can serve unique purposes with versatility and adaptability in different domains. It is expected that in the future, AI bots in combination with other technologies like blockchain and internet of things will augment human capabilities for assessing alternative travel plans and assist travellers in making efficient and climate friendly trip choices. Already, in the realm of mobility planning and travel demand management, AI bots have become indispensable for providing personalised travel experiences to users.

The other AI technology likely to be used in this respect is that of *virtual assistants*. These are designed to assist with specific tasks, propose travel plans, answer specific questions and generally stand by an individual to always assist. By analysing user preferences and behaviour, virtual assistants can tailor travel recommendations and streamline information about the trip and its environmental and climate change implications to the specific traveller. Virtual assistants in the, not too distant, future can also take the form of *humanoid robots* that can perform all the tasks of a virtual assistant and many more enabling them to interact in the same way as humans. They are expected to gradually form indispensable assistants to humans in a similar way that personal computers are now.

Both bots and virtual assistants create ethical concerns and have ethical implications that can be summarised as follows:

(a) *Safeguarding the privacy of the individual user*. All trip planning AI bots gather information not only from network or traffic control centres and other external sources but also from the traveller. The traveller's personal data that are requested, may range from facial recognition data to age, gender, preferences, habits, and similar as well as financial data (income, expenditures, car ownership, etc.). In such situations there is a potential for invasion of privacy and the line between security and personal surveillance blurs. In some applications it is even foreseen that AI will have access to bank accounts to draw income data and/or pay for some aspects of a trip (e.g., for booking and ticket payment, or as compensation for the negative environmental impacts of the trip). These capabilities raise important privacy concerns and for this reason, several countries have introduced legislation to impose data anonymity and implement measures to check and maintain such condition.
(b) *Transparency*. Transparency in this sense is about giving the traveller, and all other users of AI mobility apps, the possibility of knowing how the AI system makes decisions and how it uses their data. This requirement raises the particularly difficult technical problem to reveal the "hidden layers" of the AI algorithm, where all the processing happens. Even if this becomes possible in a practical way, the deep learning processes of AI algorithms will be difficult to understand by a non-expert, even if they are plainly laid out by the AI's creator. Since none of this information is currently disclosed, it is quite likely that users will use AI without the possibility to have context for the output they get and thus the transparency of future mobility applications is likely to be low. This has been noted also by one of the major players in AI mobility apps production (Microsoft, 2023).
(c) *Confidentiality*. A breach of confidentiality occurs when AI bots, trained and used for consulting a specific traveller, use this information to prepare advice for others too and in doing so release all or part of the personal details of a given individual to others. A case of breach of confidentiality is also when posting a seemingly innocent social media comment that could potentially allow others

to identify the specific traveller, or when posting an unredacted or partially redacted part of a traveller's record of trip making to show (for publicity) how much progress an AI user has made using the particular AI services. There is, therefore, a clear ethical obligation to keep confidential any personal information of those participating in AI-assisted travel advice and management services.

(d) *Handling practices for user data.* This refers to establishing and following practices within the AI developer company, or in national legislature, that ensure the proper handling of user data and that users are fully informed and have consented to how their data is used. Non-authorised use of user data should be considered a crime, and this is already treated as such by the legislature in many countries (e.g. in the EU's AI act).

(e) *Biases and inaccuracies.* Users should be aware that the answers of AI mobility apps may be influenced by two kinds of bias. One that may be due to the data used, and one that may result from the AI software creator. The first, is due to systematic errors or influences that may be inserted into the data sets that are used. These are normally searched and retracted by the AI software itself as already described in Chap. 7. The second, are due to biases that may exist in the beliefs and experiences of the owners and/or the developers of the AI software. These types of biases are generally difficult to detect and correct and it seems inevitable that they will come through in the results.

(f) *Difficulty in recognising fake or misleading information.* AI, in its current form, can fall into the trap of providing misleading or outright wrong information because some of its answers are based on searches in the web which include various blogs or other text that can be misleading, or biased, or outright fake and erroneous. As machines cannot fully distinguish between fake and accurate or inaccurate information, users will always need to do occasional fact-checking when using AI…

9.2.2 Ethical Implications Related to the Operation of the Transport System

AI applications for the operation of the transport system aim to maximise efficiency and safety as well as user information and comfort for all users of the transport system in a collective way. Areas of prime application of AI, include:

- Regulating traffic flows to become smoother and more adapted to the capacity of the infrastructure they use.
- Predict and quickly relieve traffic congestion.
- Minimise travel times.
- Making the travelling time of drivers and passengers more enjoyable and productive.
- Facilitating parking in dense urban areas.
- Making more comfortable and appealing car sharing and ridesharing use.

There are several ethical implications involved with the above applications of AI:

1. *Privacy issues*. These are like those mentioned in the previous section (about the rights of individual travellers) but they differ in that they refer to the collective data that are collected from various points in the network. Again, the concern here is that personal data may be compromised with potential misuse and abuse of personalised information that could be extracted from such collective data. The concerns are that although the vehicles as part of the general traffic and are the main item surveyed, driver or passenger data may also be derived compromising privacy issues. Data anonymization can help mitigate privacy concerns in these cases, but the effectiveness and extent of such anonymization practices are often questioned and are under continuous scrutiny (Bezrukov, 2024).
2. *Safety and security*. Safety of operation is always the key concern and primary ethical obligation for any traffic operation and control system and that includes applications of AI which must be designed to achieve practically fail-safe conditions. This means that the risk of malfunctions due to software bugs or hacking must be eliminated, or minimised to near zero levels, so that AI operates constantly at top level condition with the risks of failure minimised. The security aspect has to do with averting nefarious acts taking place in, and involving, all items of the transport and mobility system. Such nefarious acts could involve automated vehicles being used as part of a terrorist attack or if criminals use these vehicles to get away from the scene of a crime or to transport illegal substances, etc. Another major security concern is that hackers could exploit vulnerabilities in the traffic management and control system to cause accidents or major disruption to traffic and city life. Cyberattacks are a major security risk especially when autonomous vehicles will be circulating widely in the network together with human driven vehicles. Such security issues should be addressed and solved adequately at an early stage.
3. *Openness and explainability*. Building public trust requires that AI used in *the operation of the transport system* operates and makes decisions in an open and fully explainable way. The lack of such openness in the algorithms and data handling practices will eventually lead to public mistrust and unwillingness to follow the suggested routes or other actions.
4. *Non-bias of the AI-algorithms*. Existence of systematic biases in the AI algorithms could lead to discriminatory practices in navigation or traffic regulation which in turn could severely affect transport system operation. If not properly designed, AI systems might exhibit biases in managing traffic like e.g., favouring certain areas over others or certain users over others.
5. *Adhering to legal and regulatory frameworks*. Adhering to existing legal and regulatory frameworks is an obvious obligation of the AI systems for transport system operation. Governments should establish comprehensive legal frameworks that govern the use and safety standards of such AI applications.
6. *Equitable access to transport services*. All citizens should have equitable access to transport system operations and AI applications should respect and facilitate this principle. Special attention should be taken for equitable access to public

transport (mass transit) systems ensuring that AI-driven systems are accessible to all, including marginalized and low-income communities as well as to people with disabilities.

7. *Minimising AI-related labour issues.* As AI optimises the operation of transport systems it will inevitably cause some redundancies in the labour force that was previously operating these systems. This is particularly evident in AI applications in the public transport and the freight and logistics sectors. The AI impact on job displacement is an issue of strong ethical nature and should be attended carefully e.g., combined with proactive workforce development strategies (more on this issue in later section).

9.2.3 Ethical Implications of AI Related to the Vehicle

For human driven vehicles, AI apps can be classified in three main categories: *Advanced driver assistance systems* (ADAS), *diagnostics for the mechanical parts* of the vehicle, and *vehicle maintenance planning and management*. Most of the vehicle-related ethical implications refer to the first category (ADAS) and include issues of privacy, accountability, safety, and confidentiality. *Privacy* issues in ADAS arise from the records and data that the system collects and stores about the driver's mental state and condition during the drive. As the system monitors and controls the ability of the driver to drive safely it collects data about its drowsiness, irregular use of vehicle controls, and general way of driving. These data can be considered of a private nature and normally come under the privacy protection laws and safeguards in many countries. *Accountability* issues arise when the data collected about the driving condition and state of health or mental functionality of a given driver, are used for purposes other than the safety of the driver and the passengers during the specific drive. Examples of such other uses are when these data are given to third parties for research or training or other purposes. Of similar nature are the other two ethical considerations of *safety* and *confidentiality*. In all cases of ADAS related AI the vehicle manufacturer or owner should ensure that the AI apps used are subject to monitoring and periodic reviews as regards their fulfilling the above ethical considerations.

For automated vehicles (fully or partially) the ethical considerations that apply are many more and much more complex. They are considered in more detail in a separate section later in this chapter, but a brief account is also due here. The operation of an automated vehicle relies entirely on data collection and analysis inside and outside the vehicle and on decisions made by the AI software at all stages of the trip. Legal boundaries already apply in several countries concerning these data and their use and apply to specific items such as, for example, the Vehicle Identification Number (VIN). The issues of when and whether these data should be considered as "personal" remain vague and largely open. For example, the *European Data Protection Board* (EDPB), published in 2021 an opinion in which it claims that the Vehicle Identification Number (VINs) consists personal data. However, the

European Court of Justice in a judgement of 9 November 2023 judged that a VIN is to be considered as personal data only where the holder has additional data (or knowledge) to link it to a natural person.

As regards the in-vehicle kept data, certain protection rules also apply. Regulations in some countries require the vehicle manufacturers to explain to vehicle owners and users, at an early stage, what data is processed in the vehicle and how it is stored. For example, in April 2024, the French data protection authority (CNIL) issued practical guidelines regarding "personal" vehicle related data processing activities during the development stages of AI systems for automated vehicles (Schuler & Huck, 2024).

As regards the safety related provisions by AI applications in automated as well as human driven vehicles, these are aimed mainly at safeguarding against uncontrolled vehicle behaviour and in channelling the driver's behaviour towards more safe driving (in addition to any ADAS that may apply). The automated vehicles (beyond level 3 automation) should be able to function in fail-safe mode, hand over control to the user if necessary, and stop safely without endangering passengers or other road users. The vehicle should also be able to inform users of the type of failure and the functions concerned. For a comprehensive review of the European Union's approach to such safety safeguards and approvals for automated vehicles, see (Galassi et al., 2023). Several more ethical and societal implications of AI related applications in automated vehicles apply and these are considered in a later section dedicated to automated mobility.

9.2.4 Ethical Issues in AI Programming

AI programming for mobility applications is an area where ethical considerations and issues also apply. The need for *ethical AI design* results from the need to provide a common approach and impartiality to all AI applications to provide an equal basis for operation of these applications and avoid distortions which may threaten system operation, safety, and other harmful results. For example, in the case of an unregulated market for automated vehicles, some AI programmers may choose to give emphasis to passenger safety as opposed to pedestrians or persons in other vehicles. If this is the case, the buyers/users of automated cars will eventually seek those cars that are programmed to minimize the risks to passengers even if that significantly increases the risk to other persons. This could result in a catastrophic safety outcome in which the number of people killed or injured by automated cars would be higher than if cars were programmed to follow common risk-minimization strategies that are impartial between those inside the car and those outside it.

9.3 Societal Implications

9.3.1 Basic Societal Requirements

As AI empowered applications increasingly rule human mobility, human societies will look for ways and means to safeguard that such applications abide with commonly accepted social rules, conditions, and norms. *Diversity, non-discrimination and fairness,* and *ensuring the privacy and security of users*, may be listed as the top societal requirements for AI applications in all sectors including the mobility and transport one. *Diversity* is required to ensure coverage of the multitude of situations and use cases that will develop in practice to offer comprehensive and universal services. *Non-discrimination* refers to the risk of AI systems to inadvertently discriminate against certain social groups or areas, especially if the data used to train these systems is biased. Machine learning is based on a continues flow of relevant data on which to learn and if these data have accuracy gaps or biases in favour or against certain items (e.g. geographic areas or groups of residents) the end results will also be discriminatory and biased. Besides data biases there are, as already pointed out previously, the personal biases of the AI code developers. Non-discrimination is also referred to as *fairness.* Both are issues requiring regular audits and application of bias-mitigation techniques. For example, when using an AI tool to filter potential candidates for interviews for employment in a mobility related organisation, information should be sought on whether the AI tool that is used was trained with data that do not cause discrimination based on gender, age, race, skin colour, social or ethnic origin, nationality, sexual orientation, or any other characteristics who should not be discriminated according to applicable laws.

The potential for AI to unintentionally promote unfair bias or discrimination, due to improper programming or training, has been proven to cause prediction errors in AI powered travel demand forecasting tools (Zhang et al., 2024a, 2024b). "Fairness" for the purpose of travel demand forecasting, is about ensuring that the prediction accuracy is equal for advantaged and disadvantaged population groups. Consequently, a novel methodology has been developed and presented in Zhang et al., (2024a, 2024b) for fairness-aware, highly accurate travel demand AI forecasting models. Using such "unbiased" models for decision-making can make AI an appropriate tool for forming transportation policies that account for social inequalities.

Finally, the *privacy and security of the users,* is a highly appreciated social condition for the use of AI in modern societies. The issues of privacy and security have already been mentioned previously regarding the traveller, the vehicle, and the operation of the transport system. From the society's point of view, addressing privacy and security concerns is a general condition that has been secured by national legislation in many countries and requires formulation and instalment of robust data protection policies, transparent operations, and continuous monitoring for biases and vulnerabilities. Over and above these safeguards, data anonymization—by implementing measures to anonymize user data—can help mitigate privacy concerns, but the effectiveness and extent of such anonymization practices are often questioned and are under continuous scrutiny (Bezrukov, 2024).

9.3.2 Social Equity and Accessibility Issues

Of equal importance is the alignment of AI outputs with basic values in human societies such as respect for equality and non-discrimination, respect for human life, diversity and freedom of choice, safety and security, and so on. Development of a superintelligent AI which will surpass human intelligence and set its own values by which to act, is always a feared aspect of AI. This aspect, with the currently developed AI technologies seems unrealistic but future AI technological developments and capabilities may well make it realistic. The establishment of rules and safeguards to prevent AI applications to evolve in ways that would divert from human control is, therefore, necessary. They must secure that AI will always abide by human values and controls.

In the case of mobility related applications of AI, the main socio-ethical issues that are involved, relate to the need for:

a. Privacy, accountability, and security during the trip making.
b. Equitable use and accessibility of transport infrastructures and services, and
c. Maximum adherence to societal values, goals and objectives.

In the societal impacts of AI, one should not fail to include the potential labour implications, i.e., the potential loss of jobs due to the higher productivity that can be reached using AI. The issue of reduced employment is discussed later in this chapter in respect to AI applications in specific sectors. However, as a general comment at this stage, suffice it to say that this fear has always been the case when a new technology was introduced in the past. In all such cases, the initial fears have proven exaggerated because the labour market usually absorbed the job losses since the new technology invariably created new job opportunities. In the transport sector this was the case, for example, with the introduction of the jumbo jets in the late '70s, the use of computers and web-based applications in the '80s, the advent of automation in logistics and so on. As it was stated in the latest World Economic Forum's (WEF) jobs report, the largest job creation and destruction effects come from environmental (including climate change) and economic trends (WEF, 2023). As stated in that report, technological advancement through increased adoption of new and frontier technologies and increased digital access are expected to both create jobs as well as offset jobs (job displacement). The net job creation effect places these two trends of technological advancement and digital access in the 6th and 8th place respectively of employment influencers according to the 2023 job WEF survey (WEF, 2023).

In the face of accelerating climate change, the need for ensuring equitable access to sustainable transportation is becoming a critical issue of adaptation. The ability of Artificial Intelligence (AI) to support underserved and vulnerable communities in accessing climate-resilient transport, constitutes not only a technological innovation but also a moral imperative. As geopolitical priorities diverge and global cooperation on climate action becomes more fragmented, AI offers a means to transcend these barriers, ensuring that no community is left behind, by:

a. Identifying and supporting the mobility of vulnerable populations;
b. Supporting inclusive transportation planning; and
c. Reducing mobility disparities.

These issues are examined in more detail below.

9.3.3 Identifying and Supporting Vulnerable Populations

9.3.3.1 Identifying Vulnerable Populations with Precision

By integrating data on income levels, geographic isolation, demographic trends, and existing transportation infrastructure, AI can pinpoint "transport deserts" i.e., areas where residents lack access to sustainable, reliable, and affordable transportation (Jackson et al., 2019). This data-driven identification becomes particularly important in regions where governments may be preoccupied with short-term economic gains or geopolitical competition, often sidelining the needs of the most vulnerable. AI can also consistently highlight the socio-economic and environmental benefits of investing in underserved communities, offering a pragmatic argument for action even in complex political landscapes. For example, in regions where climate adaptation is deprioritized due to competing economic or security concerns, AI can provide clear evidence that investing in climate-friendly transport for marginalized populations leads to long-term economic benefits and resilience.

9.3.3.2 Predicting Future Needs in a Changing Climate

AI's predictive analytics can allow for a more detailed anticipation of future needs. As climate change accelerates, migration patterns, demographic shifts, and urbanization trends are likely to change dramatically. AI can model these shifts, enabling transportation planners to design systems that will meet the evolving needs of vulnerable populations well into the future. In regions at high risk of climate-induced displacement, for instance, AI can predict where populations are likely to relocate and what their transportation needs will be. This foresight allows for the proactive development of infrastructure that not only addresses current disparities but also anticipates future challenges (Porter & Dungey, 2021). AI-driven systems can diagnose disparities in mobility access across different socioeconomic groups and geographies and in doing so, it can help stabilize regions that might otherwise face increased social and economic strain due to inadequate planning.

9.3.3.3 Tailoring Solutions to Specific Communities

AI also excels at tailoring transportation solutions to the unique needs of different communities. By analysing localized data, it can recommend transport interventions that are not only sustainable but also culturally and demographically appropriate. For instance, AI can suggest on-demand electric shuttles for elderly populations in rural areas, while recommending shared e-bikes or micro-mobility options for younger, urban populations. Local governments, NGOs, and private sector actors can leverage AI to implement targeted projects that are both feasible and effective, even in the absence of broader political will.

9.3.3.4 Breaking Down Barriers with Multilingual and Non-accessible Systems

AI's Natural Language Processing (NLP) capabilities open new horizons for inclusivity, particularly in multilingual or highly diverse regions. By providing real-time translation of transport information, AI ensures that non-native speakers, refugees, migrants, or tourists, can access critical mobility services. Moreover, AI-powered navigation systems can enhance accessibility for people with disabilities, offering real-time information about the suitability of transport options for individuals with mobility, auditory, or visual impairments.

9.3.3.5 Dynamic Resource Allocation to Meet Real-Time Demands

AI's ability to monitor transportation systems in real time allows for dynamic resource allocation that can respond to the shifting needs of underserved populations. For example, it can analyse real-time data to identify where additional buses or ride-sharing vehicles are needed most, ensuring that resources are deployed efficiently and equitably. In regions where financial and infrastructure resources are limited, this dynamic allocation becomes even more critical. AI can also help prioritize investments in areas where they will have the greatest impact, ensuring that even in resource-constrained environments, the most vulnerable populations receive the support they need.

9.3.4 Support Inclusive Transportation Planning

There are several ways in which AI can support and enable an effective and inclusive transportation system planning (Cirianni et al., 2023; Nikitas et al., 2020). These are summarised below:

- AI-assisted predictive modelling enhances the planning process by allowing planners to anticipate future mobility needs based on evolving demographic trends, urbanization, and climate migration patterns. AI-driven demand forecasting provides a detailed understanding of community transport use by analysing real-time and historical travel data. It can, in this way, detect patterns in public transport usage, revealing gaps in service that disproportionately affect certain populations.
- AI's role in route optimization is also very important as it can design routes that better serve marginalized communities without sacrificing the operational viability of the overall system. This capability is particularly valuable in regions where climate change is altering daily travel patterns, forcing greater reliance on public transport systems.
- Dynamic pricing models driven by AI, can assist in public transport fares planning based on real-time economic conditions based on income data analysis, local economic trends, and household spending patterns. It can in this way assist planners to create fare structures that are both fair and sustainable.
- Lastly, AI-powered platforms can be increasingly useful in participatory transport planning to enable communities to have a greater say in the planning process.

Overall, by analysing and incorporating feedback from diverse stakeholders, AI can help create a more inclusive and responsive transport system planning.

9.3.5 Support in Reducing Mobility Disparities

AI is proving to be a powerful tool in bridging the mobility gaps that exist between urban and rural areas, as well as between affluent and low-income communities. These disparities are often stark, with rural areas and economically disadvantaged groups facing limited access to reliable, affordable, and climate-friendly transportation. By utilizing AI's capacity for real-time data analysis and predictive modelling, adaptive transport systems can be developed that cater to the specific needs of these underserved populations. They can also optimise last-mile connectivity, ensuring that residents have reliable access to integrated transport services particularly in areas where single-mode solutions are insufficient to meet the needs of low-income or geographically isolated populations (Bharadiya, 2023).

One of AI's most impactful contributions in addressing mobility disparities is in infrastructure prioritization (Nikitas et al., 2020). By analysing socioeconomic data, travel patterns, and geographic constraints, AI can identify critical areas where infrastructure investments will have the greatest impact on reducing mobility gaps and help ensure that investments are made strategically, maximizing their impact on the most vulnerable populations. Fare optimization is another area where AI can help bridge the gap between rich and poor communities. As already mentioned earlier, AI can design dynamic fare structures that adjust based on the economic capabilities of the users (lower fares for low-income riders).

9.4 Observing Ethical and Societal Issues by AI Developers

9.4.1 Ethical Issues in AI Development

Ethical and societal implications must be observed at the stage of developing the AI mobility applications, so they are impeded in them and the way they function. The issues to observe:

9.4.1.1 Accountability

All those responsible for developing, selling or running AI applications should be accountable for adherence to existing laws and guidelines, proper training of their staff and users, on AI use and utilisation and having an overall responsible use. Establishing clear lines of responsibility is critical because AI-in-mobility systems will make decisions that impact human lives. The case of accountability applies to the very end of the line, i.e. to the need for Al algorithm developers to find a way to cap the AI's learning capabilities so as not to go beyond the level of human control.

9.4.1.2 Responsibility

Responsibility in AI development, entails several actions on behalf of AI developers (and sellers) that would include the following:

- Ensuring that the development and use of an AI app is fully aligned with the existing ethical and legal rules.
- Requiring staff and vendors to check that ethical and legal rules are observed before acquiring and deploying new AI solutions and services.
- Be vigilant in inquiring where the data that are used for deep learning are acquired from, and what are the terms of their use and application.
- Ensuring that there are controls to prevent or reduce potential harms from AI, particularly where human empathy and judgement is needed.
- Ensuring that AI-controlled systems are subject to monitoring and periodic reviews.
- Producing a simplified description, capable of being understood by the public, as to how the AI makes its decisions.
- Considering how AI actions are recorded and kept in accordance with the company's record retention policy.

9.4.1.3 Safety Assurance

Safety concerns over the use of AI-powered applications are an obvious requirement technically as well as ethically. The ethical side refers to the responsibility in case of accidents and to minimising the fallibility of the systems (i.e., the risk of malfunction due to software bugs or hacking). Fallibility of AI can be a huge problem not only when companies are trying to run mobility operations, but also in running them for business models, or making projections and relevant analytics, and base on them future investments. The problem here is that if fallibility is high the users might not realize the faults until it is too late.

AI app developers and owners should develop and apply strong checks and testing practices to avoid unintended results that create risks of harm and install processes to mitigate these harms as necessary. Also, to ensure that the AI apps they produce are well designed, developed, and tested with appropriate training data. In doing these checks and testing all relevant stakeholders should be involved.

9.4.1.4 Addressing Labour Concerns

As already mentioned, one of the big questions surrounding the arrival of the AI technology, is whether it will result in net job losses. The net effect on jobs will depend on several factors but basically the following four (WEF, 2023):

1. The readiness of the individuals who must change jobs, to retrain and redirect their activity to the new or other technology sectors.
2. The rules imposed by the employers of the new technologies—in this case of AI—for the new jobs that will inevitably be produced by the new technology. For example, the terms of employment they offer, the type of skills they require, the openness in employing old-technology skilled workers, and so on.
3. The higher education system that will accept old-technology skilled workers with a view to training them in the new-technology skilled workforce.
4. The government and the extent to which it is ready to fund retraining and relocation actions.

All previous experience regarding labour losses when a new technology was introduced, shows that the net job losses from a new technology are in the long term small and may even be positive. However, the labour implications of AI in the transport and mobility sector have not yet been fully researched and analysed. For other sectors such as the industrial sector, there are studies which describe both a rather significant impact (Shen & Zhang, 2024) as well as the opposite (Albanesi et al., 2023; Economist, 2024). This shows the complexity of the issues involved. The anxiety and reactions for job losses from AI application in the transport and mobility sector, are here already and are seen e.g. in the opposition to automated taxis ("robo-taxis") from taxi drivers and in other similar cases.

9.4.1.5 Sustainability and Climate Change Mitigation Concerns

Developing "green" AI algorithms i.e., that result in overall energy-efficiency and sustainability should be a universal goal in any mobility related AI application. It is here that the use of AI can have substantial positive results because of its ability to analyse and learn from vast amounts of data and apply complex algorithms for energy use and emissions reduction at system's level.

Relevant to the sustainability, there is also the issue of "internal sustainability" of AI. This means that any AI app production and maintenance activity must observe the sustainability of AI-related computing power and the need to use renewable energy sources to reduce the overall carbon footprint of such AI related computing power consumption.

9.4.2 Confidential Information, Competition Rules

The AI owners in transport and mobility applications should ensure that the confidential information and intellectual property embodied in their AI products are held in strict confidence. "Confidential" information in this case includes all information related to trade secrets, copyrights, contents of databases, patents, software, algorithms, and key data (e.g., those used to train and test the AI). Such confidential information (including any key outputs) should be protected and only used in accordance with applicable laws and/or contractual obligations.

Competition rules must also be respected as companies using AI to gain market insights and trends will have a competitive advantage over those that don't. While this is part of how a free market and competition works, wrong practices could lead to monopolies, collusion, and other shady and avoidable practices that will affect final adoption of the new technology. Above all, is the issue of safety and public trust for AI. Building public confidence in AI systems is crucial.

9.4.3 The Urgent Need for Legislative Initiatives

Designing AI systems for transport operations with ethical considerations in mind, is essential. AI is a breakthrough technology that inevitably introduces complex ethical implications. These are only now beginning to be handled in a systematic and thorough way by the legislation of an increasing number of countries. The regulatory and policy landscape for AI is an emerging issue in many jurisdictions worldwide. To this extent it is useful to refer here to the most notable, legislative actions—of international importance—that are taking place:

A. Various data protection laws are being introduced to cope with the ethical and practical implementation issues regarding the collection and storing of big data that are involved in the AI applications examined here. The EU's *General Data Protection Regulation*—GDPR (of May 25th, 2018), or the *California Consumer Privacy Act*—CCPA (of June 28th, 2018), or the *Chinese Personal Information Protection Law* (Copp et al., 2023; NPCPRC, 2021), are examples of recent relevant legislation. These regulations dictate how user data should be handled and protected, in AI applications including those for the transport system.

B. One of the more sophisticated pieces of legislature dedicated to the development and use of AI applications setting ethical as well as operational rules, was the 2024 *Artificial Intelligence Act* of the EU (EC, 2024). This is part of the EU's *Digital Strategy* policy, launched in 2015, as it was complemented by the *Digital Europe* Programme (DIGITAL). This is the first comprehensive regulation on AI by a major regulator globally. It consists of 12 main titles, each of which contains a set of detailed Articles, and 13 Annexes that provide supplementary information alongside the Regulation. Some of the most important and pertinent main titles (chapters) are: Prohibited AI practices (Chapter II)/High-risk AI systems (Chapter III)/Transparency obligations for providers and deployers (Chapter IV)/ Measures in support of innovation (Chapter VI)/AI Governance (Chapter VII) Codes of Conduct and Guidelines (Chapter X)/Penalties (Chapter XII). The Act distinguishes the applications of AI in the following risk categories:

   a. Applications and systems that create an *unacceptable risk*. These are generally prohibited (an example of this category is the various social scoring systems and manipulative AI).
   b. *High-risk applications.* To these the Act gives most of its weight as they are fully regulated. For example, AI assisted CV-scanning that ranks job applicants, is considered a high-risk AI application and is subject to specific legal requirements.
   c. *Limited-risk applications* which are defined as applications not explicitly banned or listed as high-risk. These are subject to lighter transparency obligations which dictate that developers and deployers must ensure that end-users are aware that they are interacting with AI.
   d. *Minimal-risk applications* such as AI enabled video games and spam filters and which are largely left unregulated.

The EU's *AI act* aims to ensure respect of the fundamental rights of the citizens and also to stimulate investment and innovation on artificial intelligence in Europe. It is fully in line with the EU's *General Data Protection Regulation* (GDPR) legislation, mentioned earlier, so when AI makes automated decisions which have a legal or privacy breach effect on individuals, the processing will require consent as per article 22 of the GDPR unless the issue is covered by an exception.

C. The regulation of AI in China is mainly governed by the State Council of the People's Republic of China. The first legislative action in this respect was the July 8, 2017, plan named "*A Next Generation Artificial Intelligence Development Plan*" (State Council document No. 35). In this, the central committee of the Chinese communist party urges the governing bodies of China to promote the development of AI up to 2030.
D. In Germany, the Federal Ministry for Economic Affairs and Energy published, on November 2020, the "*German Standardization Roadmap for Artificial Intelligence*" (NRM KI) and presented it to the public at the Digital Summit of the

Federal Government of Germany. The first edition of the roadmap was a 200-page long document written by 300 experts. The second edition of the NRM KI was published during the German government's Digital Summit on December 9, 2022.

E. In the UK, on March 2023, the government released its White Paper named "*A pro-innovation approach to AI regulation*". This white paper presents general AI principles but leaves significant flexibility to existing regulators on how they adapt these principles to specific areas such as transport or financial markets.
F. In Israel, on December 2023, the Ministry of Innovation and the Ministry of Justice, published a joint AI regulation and ethics policy paper outlining several AI ethical principles and a set of recommendations including opting for sector-based regulation.
G. Similar initiatives are noted in several other countries which according to a Stanford university' index of 2023 numbered 37 countries (https://hai.stanford.edu/ai-index).

The need for socially accepted clear and robust rules, regulations, and legal safeguards regarding the use of AI is becoming widely accepted. In February 2025, the *Artificial Intelligence Action Summit* was held in Paris, France, co-chaired by French President Emmanuel Macron and Indian Prime Minister Narendra Modi. The AI Action Summit was attended by 58 countries all represented at the level of heads of state or prime ministers and including the EU, Canada, China, USA, UK, France and India. The purpose was to federate an ecosystem of stakeholders, all mobilised to strengthen the safety and security of AI. They signed[1] a joint declaration, the *Statement on Inclusive and Sustainable Artificial Intelligence for People and the Planet*, calling for policies "ensuring that AI is open, inclusive, transparent, ethical, safe, secure and trustworthy" and setting six main priorities:

- Promoting AI accessibility to reduce digital divides.
- Ensuring that AI is open, inclusive, transparent, ethical, safe, secure and trustworthy, considering international frameworks for all.
- Making AI innovation to thrive by enabling conditions for its development and avoiding market concentration driving industrial recovery and development.
- Encouraging AI deployment that positively shapes the future of work and labour markets and delivers opportunity for sustainable growth.
- Making AI sustainable for people and the planet.
- Reinforcing international cooperation to promote coordination in international governance.

Many other initiatives to promote a safe, open and secure AI are taking place around the world every year. As a result, there is a growing network of 'AI Safety Institutes' (introduced at the Bletchley Park Summit in UK, November 2023), and

[1] The USA and the UK did not sign this declaration the first, citing concerns about "excessive regulation stifling innovation in the AI sector" and the second, because the declaration "didn't provide enough practical clarity on global governance and didn't sufficiently address harder questions around national security".

many similar entities all dedicated to advancing the science, evaluation, and regulation of AI model development and building common best practices for testing advanced AI systems. The main elements reflected in such initiatives and regulatory frameworks, include:

a. Full protection of data privacy and establishment of proper surveillance and checking mechanisms to ensure against potential misuse and abuse of personal data.
b. Protection against security risks, i.e. to reduce or eliminate the risks of unauthorised access to data and key control mechanisms from hackers and similar types of criminal acts (cybersecurity).
c. Establishment of procedures that involve informed consent by the users, i.e. that the users are fully informed and have consented to how their data is used.
d. Introduce fairness and non-discrimination practices, to reduce the potential for biases in both the AI software and the data used, and avoid discrimination as regards the various geographic or personal characteristics of the users.
e. Ensure that large companies offering AI-empowered products or services establish internal Codes of Conduct regarding the development and commercialisation of AI products.
f. Other provisions such as arrangements on intellectual property, safety of operations, employment safeguards, developer liabilities and so on.

Continuous monitoring and updating of compliance of AI developers to the existing regulations and legislation is important.

9.5 Ethical Implications of Four Specific Mobility Technologies

The future trajectory of AI in the transportation and mobility sector is poised to be both transformative and challenging. Ethical and societal considerations and sustainable practices need to be at its core. Below we examine the more specific and focused ethical and societal implications of AI on the four technological application areas that appear as the most advanced currently, i.e. those of automated mobility, shared mobility, public transport operation, and freight transport and logistics.

9.5.1 Automated Mobility

The need for, and importance of, incorporating and integrating socio-ethical considerations into AI systems for automated (autonomous) vehicles (AVs) is all too evident. The ethical dilemmas and implications in the case of automated mobility are many and some of them quite grave to the extent that they threaten the final implementation of the new technology. The issues that have already been mentioned before, i.e.

those of *transparency and explainability*, *privacy and confidentiality*, *labour* reduction, *security*, *unbiased* and *non-discriminatory* software, etc., are valid here too, but in the case of automated mobility there are two more critical issues to be considered and satisfied. These are:

a. The moral decisions that will have to be made by the AI algorithms in the course of a trip by an automated vehicle, and
b. Defining responsibilities in case of accidents.

There is a lot of research and a huge debate going on about these two areas of AI application in automated mobility (Copp et al., 2023; Krügel & Uhl, 2024; Zhai et al., 2024; Zhang et al., 2024a, 2024b). So far, there is not yet any commonly (and more important, internationally) accepted position. The main issues are the following:

Defining moral guidelines for AI algorithms in AVs: This is about incorporating into the AI algorithms and software used for the operation of AVs, sensitive ethical decisions that will have to be made in unavoidable emergency situations like an imminent crash. For example, if a crash with a pedestrian or animal on the street is unavoidable what sort of reaction (manoeuvre) should the AV take? Should it harm the pedestrian or animal, or should it crash the vehicle at the side endangering in this way its passengers? The moral dilemma to solve here is to decide under what circumstances and conditions should an AV prioritize the safety of its passengers against that of pedestrians or passengers in other vehicles. There are several theories and approaches that are being researched to solve this dilemma (Gros et al., 2024). The two major ones are the *utilitarian* in which the moral calculation evaluates actions based on their consequences favouring outcomes that maximise happiness or "greatest good" for the greatest number of people, and disfavouring those that cause harm to the greater number of people. The other is the *deontological* one which argues against violating moral rules. Another distinction is between an *egalitarian* approach which advocates for equal consideration of all lives, and an *elitist* approach which prioritises lives with more value or potential. Interestingly, when people face choices, their preferences seem to tilt towards treating all lives equally, indicating a preference for fairness and equality over elitism in life-and-death decisions.

Responsibility in case of accidents: Determining responsibility and thus liability in accidents involving AVs is another complex and sensitive issue. When there is an accident, the question is to whom the consequences should fall: the manufacturer, the software developer, or the passenger? The simple straight forward answer to this question is that when the AV or its software is at fault the responsibility falls on the vehicle manufacturer or the AI developer whichever is the case. When the AV is not at fault, the responsibility should fall at the third party involved. However, the application of these principles in practice (if we assume they are accepted), is not easy. It will require the recording and safely storing of a continuous full record of all travel parameters as well as a clear account of the vehicle's functionality like the "black box" recorders in airplanes. This record should then be used by investigators to decide what went wrong and where they should put the blame. The issues relevant to responsibility in accidents of automated vehicles are still being researched and investigated and it will take a lot longer before they are fully resolved and settled.

The above two critical issues need to be settled in a universal and wholistic way before self-driving cars become a reality everywhere. The difficulty of finding universal and wholistic answers to the critical moral issues mentioned above, is demonstrated amply by a comprehensive study and survey that was conducted in a 5-year research project called the *Moral Machine,* performed by the MIT's Media Lab in the period 2016–2020 (Awad et al., 2018). This project collected the answers given by individuals all over the world on moral questions related to automated mobility and the decisions to be made by machine intelligence in self-driving cars. The project team generated moral dilemmas and was asking human participants to say what they think would be "moral" for a driverless car to choose. For example, a question asked was what would be the lesser of two evils for a driverless car to do, killing two passengers or five pedestrians?

In this way, the project gathered millions of human perspectives on several moral dilemmas, involving AVs, and tried to formulate generally accepted moral guidelines classifying the results according to respondents' demographics, cross-cultural ethical variations and cultural traits. The results of this exercise were not conclusive. They revealed diverse and complex viewpoints of humans around the world as to what AI should choose in moral dilemma situations. This shows that it will be extremely difficult to develop globally accepted, socially compatible principles for AV ethics.

Irrespective of the difficulties, the development of a comprehensive regulatory framework to govern the development of AI in automated transportation is a crucial issue for the further development and application of the new technology. Universal standards are necessary, but the legal frames will need to be settled on a country by country and region basis. At the time of writing this book the debate related to the moral issues in automated mobility is raging and there are numerous relevant publications (EC, 2020; Gros et al., 2024; Li & Miao, 2023; Microsoft, 2023; Papadimitriou et al., 2022; Singer & Yip, 2023; Zhao et al., 2023; Ziyan & Shiguo, 2021).

9.5.2 Shared Mobility

The advent of AI in shared mobility services, including ridesharing and bike-sharing programs, is a significant leap in operational efficiency and user experience. However, AI algorithms would have to rely on as many personal details as possible to improve service efficiency and match service to the personal needs and requirements of each user. This matching process is a key component. It involves checking the mutual compatibility of those participating in the shared mobility vehicles so that they have similar cultural and habitual characteristics so as to enhance efficiency and user satisfaction and convenience.

The ethical issues related to the handling and protection of user data in shared mobility are like the ones mentioned already previously concerning data handling in general. In shared mobility, however, more than in any other case it will be imperative to ensure that users are fully informed and have consented to how their data is used. This involves clear communication and information to the user about data practices and the signing of user agreements that contain full descriptions of how the personal data are used and the steps to be taken to ensure their integrity and confidentiality.

A possible variant to increase trust and confidence would be providing users with control over their data. This would entail giving them the ability to view, edit, or delete their personal information from the system. At the time of writing, there are regulations applicable in certain regions and/or countries that dictate how user data should be handled and protected, and these are adhering to various regional data protection laws, such as the GDPR in the EU or the CCPA in California. Continuous monitoring and regular updates regarding data handling practices in shared transport systems and apps is recommended.

9.5.3 Public Transport Operations

AI in Public Transport operations is a particularly challenging and effective application. It is expected to revolutionise how Public Transport services are managed and utilized to achieve the much-anticipated modal transfer from the private car to mass transit services. As in other application areas, use of AI in this domain, inevitably, raises ethical concerns related to surveillance and data privacy for the users of Public Transport systems. One of the most important and attractive AI applications for the public transport users, is the personalisation of the recommendations and advise given to them by the AI assistant i.e., the adjustment of the recommendations to the personal preferences, choices in the past, and expressed views of the user. Providing personalised information for the public transport system requires, besides the personal data, also information of a private nature such as voice or face recognition, details for the fare payment, data on disabilities or limited mobility, and so on. Developing robust privacy protection mechanisms in all these issues will be essential.

Another expected AI application for public transport is to dynamically adjust bus or train schedules based on real-time data about the expected demand. To achieve this, accurate short-term predictions will need to be made continuously by the AI algorithms based on real-time data from the users and their expressed travel preferences. Analyses will be needed about how the demand for public transport services changes over time in each area. This type of public transport smart routing and scheduling is raising the same ethical issues about data integrity and confidentiality that have been mentioned earlier.

Finally, AI applications in public transport operations must ensure equitable and all-inclusive access to the public transport system and services. This means ensuring that AI-driven public transport systems are accessible to all, including marginalized and low-income persons and communities. The inclusion of equity and accessibility provisions along with the provisions for public transport service quality in the comprehensive regulatory frameworks that have been suggested earlier, would also be a necessary measure.

9.5.4 Freight Transport and Logistics

The ethical issues of AI applications in freight transport and logistics are mainly focused on the ethical implications due to labour impacts both because these services are labour intensive, especially on the truck driver's side, but also because automated trucks are expected to be on the roads much earlier than for the rest of the automated vehicles. In the logistics sector, because of its widespread supply and delivery networks, AI seems an ideal application area so companies are expected to apply AI empowered solutions massively to use their resources more efficiently and reduce costs including labour costs.

All this, points to an urgency to examine any anticipated labour reductions proactively to avert potential negative reactions. The most affected areas will be those labour changes due to driverless trucks and the automated AI empowered loading/unloading as well as picking and other activities in warehouses. Currently there is a shortage of truck drivers in Europe and other regions' markets so operating companies are struggling to hire enough drivers to meet demand. This does not mean that the advent of automation in trucks will not raise labour issues, but it does mean that these problems will not be as severe as otherwise. Again, as it happened in other cases, the potential job losses will likely be gradually compensated by new jobs that will be created due to the expansion of business to new areas and services and due to the development of new job opportunities in areas that did not exist before. The establishment of a cyclical economy will help in overcoming the labour problems in this domain as the AI-empowered restructuring of the supply chains in a cyclical economy is likely to produce more jobs.

In concluding, we can say that the future of AI in transportation is not just about technological achievements and state-of-the-art hardware and software. It is also about satisfying the many ethical concerns and implications that arise from its use. Ethical considerations in conjunction with the sustainable practices necessary for the climate change mitigation dimension, will run into the very fabric of all mobility technological innovation. The focus must be put on creating systems that are not only efficient and convenient, but also ethically acceptable, safe, secure, and environmentally responsible. As we move forward, the collaboration between technologists, ethicists, policymakers, and the public will be pivotal in shaping an AI-enhanced future mobility in alignment with societal values and global sustainability goals.

References

Albanesi, S., Dias da Silva, A., Jimeno, F., Lamo, A., & Wabitsch, A. (2023). Reports of AI ending human labour may be greatly exaggerated. *European Central Bank Research Bulletin No., 113*, 28 November 2023. https://www.ecb.europa.eu/press/research-publications

Awad, E., Dsouza, S., Kim, R., Schulz, J., Henrich, J., Shariff, A., Bonnefon, J.-F., & Rahwan, I. (2018). The moral machine experiment. *Nature, 563*, 59–64. https://doi.org/10.1038/s41586-018-0637-6

Bezrukov, Ol. (2024). Navigating the future: Ethical considerations in AI-driven transportation. *Techstack Company Blog*, 16 January 2024, https://tech-stack.com/blog/ai-in-transportation/

Bharadiya, J. (2023). Artificial intelligence in transportation systems a critical review. *American Journal of Computing and Engineering, 6*(1), 34–45. https://doi.org/10.47672/ajce.1487

Cirianni, F. M. M., Comi, A., & Quattrone, A. (2023). Mobility control centre and artificial intelligence for sustainable urban districts. *Information, 14*(10). https://doi.org/10.3390/info14100581.

Copp, C. J., Cabell, J. J., & Kemmelmeier, M. (2023). Plenty of blame to go around: Attributions of responsibility in a fatal autonomous vehicle accident. *Current Psychology, 42*, 6752–6767. https://doi.org/10.1007/s12144-021-01956-5

EC. (2020). Ethics of connected and automated vehicles: Recommendations on road safety, privacy, fairness, explainability and responsibility. *European Commission, Directorate-General for Research and Innovation*, independent experts' recommendation, Publications Office. https://doi.org/10.2777/035239

EC. (2024). The artificial intelligence act. Regulation (EU) 2024/1689, *Official Journal of the European Union*, 12 July 2024.

Economist. (2024). Your job is (probably) safe from artificial intelligence: Why predictions of an imminent economic revolution are overstated. *The Economist*, May 7th, 2023.

Galassi, M.-C., Lagrange, A., Ciuffo, B., Suarez, B. R., Vass, S., Mattas, K., Donà, R., & Sollima, C. (2023). Safety approval of automated vehicles in the EU: Moving beyond highway applications. *Transportation Research Procedia, 72*, 4396–4403. https://doi.org/10.1016/j.trpro.2023.11.328

Gros, C., Kester, L., Martens, M., & Werkhoven, P. (2024). Addressing ethical challenges in automated vehicles: Bridging the gap with hybrid AI and augmented utilitarianism. *AI and Ethics*. https://doi.org/10.1007/s43681-024-00592-6

Jackson, L., Kuhlman, C., Jackson, F., & Keolu Fox, P. (2019). Including vulnerable populations in the assessment of data from vulnerable populations. *Frontiers in Big Data,* 2. Sec. Data Mining and Management. https://doi.org/10.3389/fdata.2019.00019

Krügel, S., & Uhl, M. (2024). The risk ethics of autonomous vehicles: an empirical approach. *Scientific Reports, 14*, 960. https://doi.org/10.1038/s41598-024-51313-2

Li, X.-W., & Miao, H.-Z. (2023). How to incorporate autonomous vehicles into the carbon neutrality framework of China: Legal and policy perspectives. *Sustainability, 15*(7). https://doi.org/10.3390/su15075671

Microsoft. (2023). The ethical implications of artificial intelligence. *Microsoft blog*, August 11, 2023. https://www.microsoft.com/en-us/microsoft-365-life-hacks/writing/ethical-implications-of-artificial-intelligence. Accessed February 2025.

Nikitas, A., Michalakopoulou, K., Tchouamou Njoya, E., & Karampatzakis, D. (2020). Artificial intelligence, transport and the smart city: Definitions and dimensions of a new mobility era. *Sustainability, 12*(7). https://doi.org/10.3390/su12072789

NPCPRC. (2021). Personal information protection law of the People's Republic of China. In *The National People's Congress of the People's Republic of China (NPCPRC), 30th meeting of the standing committee of the thirteenth National People's Congress, August 20, 2021.* http://en.npc.gov.cn.cdurl.cn/2021-12/29/c_694559.htm

Papadimitriou, El., Farah, H., van de Kaa, G., Santoni de Sio, F., Hagenzieker, M., & van Gelder, P. (2022). Towards common ethical and safe 'behaviour' standards for automated vehicles. *Accident Analysis & Prevention, 174*. https://doi.org/10.1016/j.aap.2022.106724

Porter, G., & Cl, D. (2021). Innovative field research methodologies for more inclusive transport planning: Review and prospect. *Advances in Transport Policy and Planning, 8*, 273–303. https://doi.org/10.1016/bs.atpp.2021.06.006

Schuler, M., & Huck, L. (2024). CNIL guidelines on AI: challenges for autonomous vehicle AI system providers. *TaylorWessing*, 3 June 2024. In: https://www.taylorwessing.com/en/interface/2024/onwards-and-upwards/cnil-guidelines-on-ai. Accessed February 2025. CNIL: The French data protection authority.

Shen, Y., & Zhang, X. (2024). The impact of artificial intelligence on employment: the role of virtual agglomeration. *Humanities & Social Sciences Communications, 11*(122). https://doi.org/10.1057/s41599-024-02647-9

Singer, P., & Yip Fai, T. (2023). Self-driving cars and AI ethics. *Project Syndicate,* Peter Singer and Tse Yip Fai, Sep 8, 2023. https://www.project-syndicate.org/commentary/san-francisco-self-driving-cars-ethical-implications-by-peter-singer-and-tse-yip-fai-2023-09. Accessed February 2025.

WEF. (2023). Future of Jobs report 2023. *World Economic Forum,* insight report, May 2023. ISBN-13: 978-2-940631-96*4.

Zhai, S., Wang, L., & Liu, P. (2024). Not in control, but liable? Attributing human responsibility for fully automated vehicle accidents. *Engineering, 33*, 121–132. https://doi.org/10.1016/j.eng.2023.10.008

Zhang, X., Ke, Q., & Zhao, X. (2024). Travel demand forecasting: A fair AI approach. *IEEE Transactions on Intelligent Transportation Systems*, 1–17. https://doi.org/10.1109/TITS.2024.3395061

Zhang, Q., Wallbridge, C. D., Jones, D. M., & Morgan, P. L. (2024b). Public perception of autonomous vehicle capability determines judgment of blame and trust in road traffic accidents. *Transportation Research Part A: Policy and Practice, 179.* https://doi.org/10.1016/j.tra.2023.103887

Zhao, H., Meng, M., Li, X., Yu, Q., Li, L., & Galland, S. (2023). The development of automated driving in China: A comparison to Germany regarding the government policies, laws, regulations, and industries. *Transportation Letters*, 1–9. https://doi.org/10.1080/19427867.2023.2238467

Ziyan, C., & Shiguo L. (2021). China's self-driving car legislation study. *Computer Law & Security Review, 41.* https://doi.org/10.1016/j.clsr.2021.105555

Chapter 10
Mobility 2050 and Beyond

Abstract This chapter encapsulates and utilises the main findings of all previous chapters in this book. It is doing so in the form of a "futures thinking" exercise in which the authors, based on the findings of the previous chapters as well as on the current trends and estimates of the future course of some key influencing factors, outline the likely mobility landscape in 2050 and beyond. The influencing factors that are considered are, political/geopolitical, economic, social, technological, legal/regulatory, and environmental/climate change related. Each of these categories of factors is examined in detail and estimates are made as to the expected trends and developments until 2050 and beyond. During this examination there are six *conditionalities* that are formulated i.e., basic assumptions of fundamental importance expressed as a sine qua non condition without which the described future cannot exist. There are also various other analyses made as regards the influencing factors such as, for example, the developing new international trade and competition environment, the national debt situation, and the economic outlook of four regions that are examined in more detail i.e., Asia, Europe, North America, and Africa. The future prospects and development potential are then examined for seven basic technological innovations that are likely to dominate the transport and mobility sector in the future, namely: Battery electric vehicles, Vehicles using hydrogen (in direct combustion or in fuel cells), Cooperative connected and automated vehicles, Shared automated mobility, Flying vehicles (vertical take-off and landing), Artificial Intelligence (software and apps), and Advanced freight transport and urban distribution technologies. After presenting the basic functional and technical characteristics of each of these technologies, the text discusses their technology readiness levels (TRLs) currently and expected by 2050+, as well as their estimated penetration levels by 2050. Similar analyses are made for the environmental and climate change factors including the degree of adaptation for each of the four regions, the energy related issues with the estimated renewable energy production per region, the legal and regulatory factors, and so on. The chapter ends with a 30-page description of the authors' vision of future mobility for urban and interurban/rural areas. This is presented in terms of three types of urban area, the estimated penetration levels of new technologies (including AI) in each of the four geographic regions considered, the estimated degree of adaptation of each region to climate change, and other related factors. The description of

G. A. Giannopoulos and Y. Li, *Human Mobility, Artificial Intelligence and Climate Change*, Lecture Notes in Mobility, https://doi.org/10.1007/978-3-032-08171-1_10

the 2050 + future mobility, is given in the form of two Tables describing the future mobility scene in each of the three types of urban areas in terms of the following ten attributes: (1) Types of networks and transport modes and V2X infrastructures (2) Degree of adaptation to climate change (3) Shared Mobility offered as a service (4) Accessibility, convenience and affordability of the urban mobility system (5) Personalised travel advice (6) Smart travel demand management (7) Cooperative connected and automated mobility (8) Smart traffic management (9) Car ownership and car use levels (10) Freight transport and urban delivery services.

Keywords Mobility 2050 · Influencing factors · Drivers of change · Conditionalities · Geopolitical factors · Economic outlook Asia · Economic outlook EU · Economic outlook N America · Economic outlook Africa · Battery electric vehicles · Hydrogen vehicles · Shared mobility · CCAV · Flying vehicles · Urban air mobility · Freight transport technologies · Technology readiness · Technology penetration · 2050 environment · 2050 adaptation · 2050 clean energy · Legal issues · Urban mobility vision · 2050 vision · 2050 urban mobility scenes · Interurban and rural areas

10.1 Introduction—Aim and Methodological Outline

In the words of the French author Marcel Proust "the real voyage of discovery consists not in seeking new landscapes but in having new eyes".[1] This chapter is about a voyage of discovery into the future of human mobility with the eyes of these authors based on the insights gained from the combined consideration of the emerging new transport and mobility technologies, the systematic and ubiquitous use of Artificial Intelligence (AI) and the impacts of climate change. The "voyage" will be our *futures thinking* exercise in which we consider the current trends and indications as well as some key influencing factors to make estimates for the mobility landscape in 2050 and beyond. "Futures thinking" is a flexible approach with which one can advocate future situations that are adaptive and responsive to the uncertainties of a fluid environment without having to develop comprehensive forecasting models or to resort to scenarios building (Canina et al., 2022), (WBCSD, 2014). This is our view of the future of mobility and subjective though it may be, it is considered useful because it is in effect a summary of the findings and conclusions of all the preceding chapters.

To try and make forecasts of the future or develop scenarios, which are the two alternative methods of future telling, would be unrealistic in the present circumstances. The use of forecasting models would entail collection and analysis of solid data with a considerable degree of sophistication, formulating mathematical models and then using them to project values and specific outcomes for the future. This is

[1] In his work "Remembrance of Things Past" (or *In Search of Lost Time*), volume 5, *The Prisoner* (1923) translated into English by C. K. Moncrief.

not within the scope of this book. In the case of scenarios, the future uncertainty is handled by developing hypotheses about the future of several key influencing factors ("drivers of change") and then describing the resulting most probable state(s) of outcome for the sector under study. These "scenario-like" descriptions give a "postulated sequence or development of events that will result in response to a possible future development of the influencing factors presented in the form of a real-life situation with a narrative of a person living in that scenario situation" (van Notten, 2005). Scenario building in the case of this book would not add anything new to what is already present in the literature (Aalto & Kuosa, 2024; ARUP, 2019; Cilliers et al., 2011; Cordova-Pozo & Rouwette, 2023; JRC, 2024; Kuhle & Günther, 2024; UK, 2023, 2024), etc.

By embarking in a *futures thinking exercise* and a subjective projection of the future of mobility at global level, we aim to provide the readers with a summary of the results and discussions in the previous chapters and project further a message of despair of the scientific community for the relative inactivity and lack of effective proactive action at political level to address the looming climate crisis and its impacts on transport and mobility. The methodology followed, is shown diagrammatically in Fig. 10.1. The target year of 2050 is simply a reference point and selecting another year e.g., 2055 or 2060 or even later, would not have made much difference to the described futures. This is why, in Fig. 10.1, the expected future is shown as HUMAN MOBILITY 2050+. As shown in the same Figure, there are four broad geographic regions that are considered: Asia/S E Asia–Europe–North America–Africa. There are also three types of urban areas differentiated according to their level of development (named as Level 1, 2, and 3), and reference to the degree of "adaptation" to climate change i.e. the degree of resilience that will have been achieved as regards both infrastructure and services.

The *influencing factors* or *drivers of change* that are considered are discussed and debated as to their future state of development (by 2050 and beyond), based on current (2023–2025) data and trends as derived from existing studies and reports and gaps covered by these author's estimates. The influencing factors considered, follow the PESTLE classification i.e. they are:

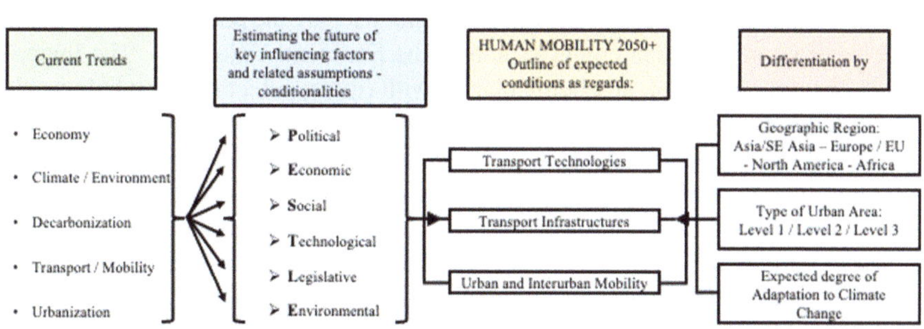

Fig. 10.1 Diagrammatic outline of the futures thinking exercise used in this chapter

1. **P**olitical/geopolitical,
2. **E**conomic,
3. **S**ocial,
4. **T**echnological,
5. **L**egal—Regulatory,
6. **E**nvironmental—climate change.

Within each of the PESTLE categories there are several "situations" or "dimensions" that are considered which refer to the geographical region of reference, the type of urban area of reference (according to our 3-level classification used), the expected potential impacts of climate change and the degree of adaptation. There is also a number of eight "conditionalities" that are put forward i.e., basic assumptions that are considered of fundamental importance and a sine qua non condition without which the described *futures* cannot exist. The development of AI and its relevant technologies is expected to proceed unabated to the maximum of its potential in the short to medium term (i.e., in the next 10 to 15 years) reaching a point of saturation and technical maturity before the target year i.e., 2040 or thereabouts.

10.2 Key Influencing Factors and Conditionalities

10.2.1 Political/Geopolitical Factors

Political and geopolitical factors include critical issues like peace, international cooperation and war conflicts. These issues have local, regional and world repercussions. International trade, economic development, financial means and priorities, social and political stability and many other critical issues of human societies depend on geopolitical stability and peaceful cooperation. Since 2020, the world has encountered a series of impactful geopolitical events at regional and world level that amply demonstrate the importance of these issues. The war conflict in Europe that started in February 2022, with the war in Ukraine, was the first such event in modern times that shook the continent changing almost instantaneously the political and economic policies of most European countries and of the European Union, too. The impacts were economic, political, and humanitarian with ripples affecting the whole world. On the economic side, the conflict has caused major economic disruptions primarily by surging energy prices upwards due to Europe's heavy reliance on Russian oil. Inflation rates have risen, and supply chains have been disrupted causing serious economic problems in all European economies. As it is mentioned in more detail in Chap. 3, many national governments and the EU retracted or delayed previously taken decisions about decarbonisation policies like for example many in the agricultural sector. On the political side, new security policies were forged and new priorities for spending have been established. On the humanitarian side, many millions of new refugees moved to EU member countries from Ukraine causing strains in social and economic systems and priorities. So, the impacts of this geopolitical event have had serious repercussions and will have for many years, among them in the transport and climate mitigation policies in all countries of Europe.

About 18 months after the beginning of the Ukraine war, a second war conflict flared up in the Middle East, initiated this time by the attack on Israel of October 7th, 2023. This war quickly escalated to a wider conflict in the Middle East affecting in a direct way the economies of the countries involved but also, indirectly, many other countries in the rest of the world. There are many more geopolitical events of similar nature that flare up all over the world from time to time with similar repercussions. So, the so-called geopolitical factors are both omnipresent and having serious consequences that can delay or even derail the progress of economic and social development in the areas affected. Having said this, it must be noted that after the first months or years of "shock" from such events the affected economies sooner or later recuperate and previously held policy targets may get rescheduled, with reconstruction activities planned and implemented. So—given time—new economic and political realities get established and earlier developmental trends can get restored. For economies that are not directly involved in a geopolitical conflict, the impacts are quicker to be absorbed as they normally involve just some delays in implementing planned projects for new infrastructures or technologies due to reallocation of priorities and resources. For directly involved economies, it may take many years or even decades to restore and recuperate.

The risk of a larger, global, war is also an eventuality but with the current nuclear arsenals stockpiled by many countries and the Armageddon that will result from their use, such a conflict is unlikely to happen. It is an eventuality, nevertheless, which we must exclude from our futures thinking as follows:

Conditionality 1: *The global destruction and risk of total annihilation of all human progress to be caused by a global nuclear war will prove a strong deterrent that will exclude such an event from happening in the twenty-first century.*

In contrast to the assumption that humanity will avoid a world war happening in this century, there is an almost certainty that local or regional war conflicts and other hot spot geopolitical crises will happen with the same frequency that they have been happening in the past. Many such potential hot spots are developing already, and it is quite likely that they will "erupt" one way or another in the next 15–20 years (ad minimum) in regions like the, Middle East, sub-Sahara, Central Africa, Aegean Sea/ Eastern Mediterranean, South China Sea. There will also be, most likely, other regions too that may experience major local geopolitical conflicts and the effects from such conflicts will be global i.e. affecting most other regions. Facing the impacts of such events will take the attention of the governments away from the developing climate crisis and the need for implementing rigorous decarbonisation policies including offering clean and sustainable mobility to all. Thus, our second basic hypothesis/ conditionality is the following:

Conditionality 2: *By the target year of 2050+, several regions of the world will experience a major "local" political or geopolitical disruption which will result in serious delays or even reversals in the process of implementing new mobility technologies and planned climate mitigation and adaptation policies.*

What this conditionality means in practice is that quite likely a good part of the 25 years between 2025 and 2050 (or 35, until 2060) will be consumed by delays caused by *"local"* political or geopolitical events and so the new mobility technologies and services that are being developed or anticipated now, as well as the fight to combat climate change, may not reach the level of maturity that could otherwise normally be achieved in this span of time. A good example of the kind of the "local political or geopolitical disruptions" that are foreseen in this conditionality, is the complete reversal of the US climate policies by the new US government under the 2^n Trump administration that took office in January of 2025. This development will most likely derail and delay for a good part of a decade, the process of climate change mitigation within this major polluting country and quite likely, globally.

However, it is anticipated that at some stage, the evolving climate crisis and its adverse impacts and consequences on infrastructures and human societies, will act as a "wake-up" call and will make the facing of these impacts a top priority for national governments and international organisations around the world. A plausible hypothesis is that this reversal of priorities will happen at the time horizon of the mid-2040s to mid-2050s (i.e. quite late enough) first for the governments of the most advanced economies and later for all the others. By that time, a new wave of international cooperation will develop to combat the many and varied climate crises around the world giving emphasis on implementation of the necessary actions. This expected change of policy priorities over time is diagrammatically depicted in Fig. 10.2 and is also expressed as conditionality 3, below:

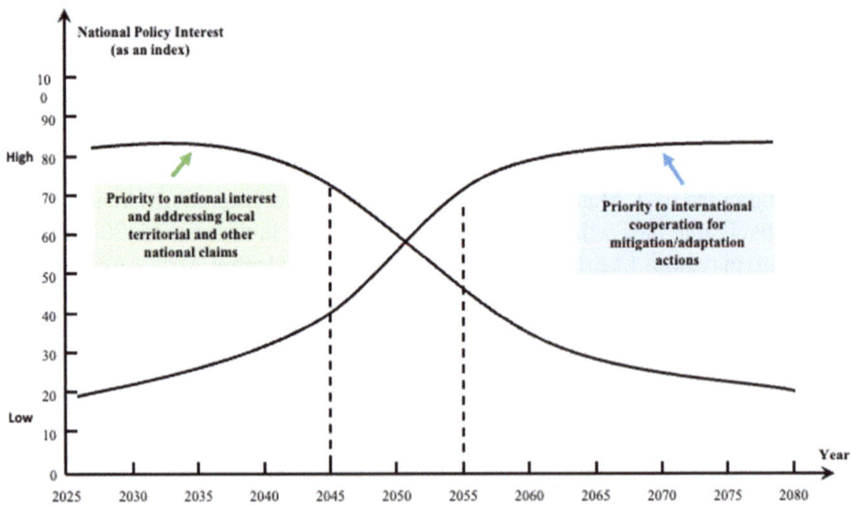

Fig. 10.2 The changing policy priorities towards full decarbonisation and climate change adaptation

Conditionality 3: *As the negative impacts of climate change keep causing bigger and more widespread national and regional disruptions and casualties, the priorities of national governments will shift towards putting at the top of their agendas the facing of climate change and promoting new forms of international cooperation to this end. These new priorities will result, by 2045–50, in a more rigorous and concerted international action for climate change adaptation and full decarbonisation.*

10.2.2 Economic Factors

10.2.2.1 Overall Outlook

The longer-term outlook of the global economy and the international economic order is continuously evolving making any future foresight risky and uncertain. There are many factors that shape the international economic order some of which are shown diagrammatically in Fig. 10.3. On the top half of that Figure are shown those factors that influence and largely shape the national economies. They include the national debt levels, the trading environment (deficits—barriers), the national productivity levels, the degree of global connectedness (i.e., how much "connected" is the national economy to the international—global events and economic developments), major employment disruptions, and others. On the bottom half of the Figure, there are the various "outputs" of national economies in terms of national or global economic and other developments. These refer to economic outputs at both national or world level and include, for example, annual growth levels, competitiveness levels, international trade levels, inflationary trends, and so on. Through the interaction between all these factors and elements, the national economies and through them the global economy and the international economic order is shaped while always evolving and changing.

The world economic order is typically shaped by the policies and actions of a small number of global economic powers such as China and the US as well as others, smaller in scale. Figure 10.4a shows the ten largest economies in the world, in terms of their normal GDP i.e. the market value of all final goods and services produced in the country. The US is no. 1 with a total GDP of 28.78 trillion US dollars followed by China with 18.53 trillion and then Germany with 4.59 trillion and so on. Figure 10.4b, shows the ten largest economies in the world, in 2024, in terms of purchasing power parity (PPP) GDP. This is the sum value of all goods and services produced in the country valued at prices prevailing in the United States in international dollars, i.e., dollars that have the same purchasing power over GDP as the U.S. dollar has in the United States. In terms of GDP/PPP, China is the no. 1 country in the world with 36.29 trillion U.S. dollar worth followed by the US with 28.78 trillion and then India with 14.59 trillion int'l dollars.

With a benign international environment where trade liberalization and free movement of capital and technology takes place freely at global level, the pictures shown in Fig. 10.4a, b are not likely to change radically in the near to medium term future except maybe in the order of ranking. However, given the international trade and technology

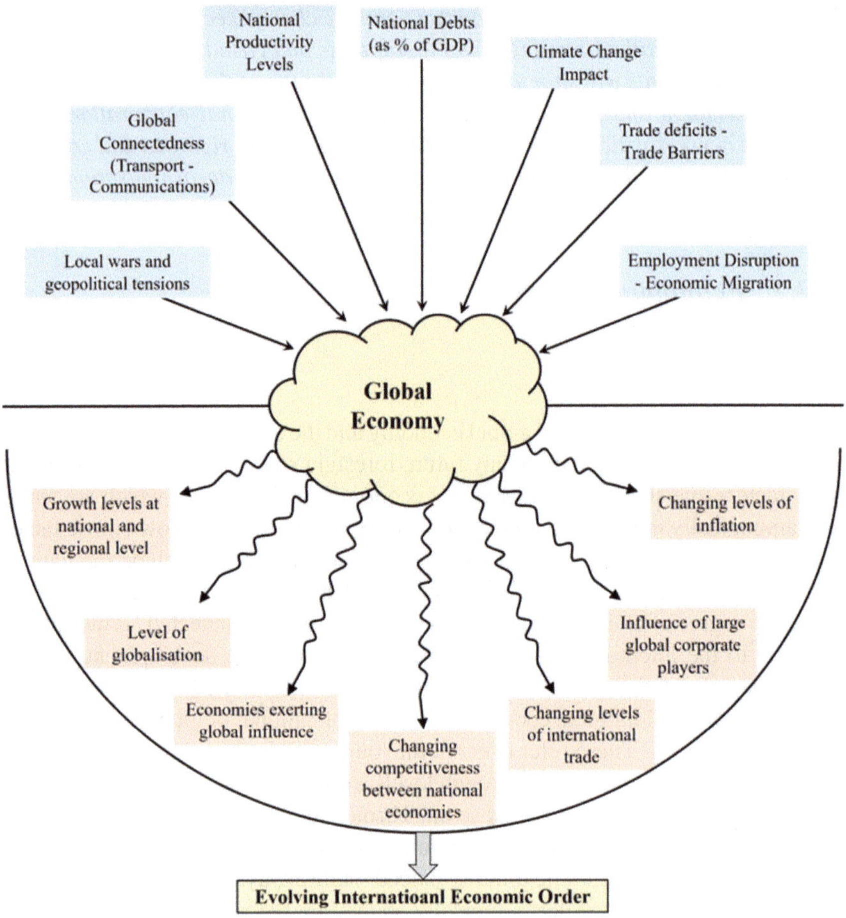

Fig. 10.3 The evolving global economic order: a continuous and changing set of interactions between various influencing factors (top half—straight arrows) and emerging outputs (bottom half—wavy arrows)

limitations that started in the period 2016–2020 of the first Trump administration and continued over his second term in 2025, a more aggressive and restrictive international trade environment is likely to be put in place lasting for an uncertain period. International competition for limited natural resources (e.g., rare earths) and a rise of populism and xenophobia will lead to trade and investment protectionism policies that will create friction and reduced national growth rates. To this, poor governance and weak institutional capacity in many world regions will further harm the growth prospects of the respective countries. To these, one should add the rising costs from the impacts of the expected global warming and climate crisis which can be devastating especially since many of the global economic centres are in coastal areas and subject to rising sea levels and flooding.

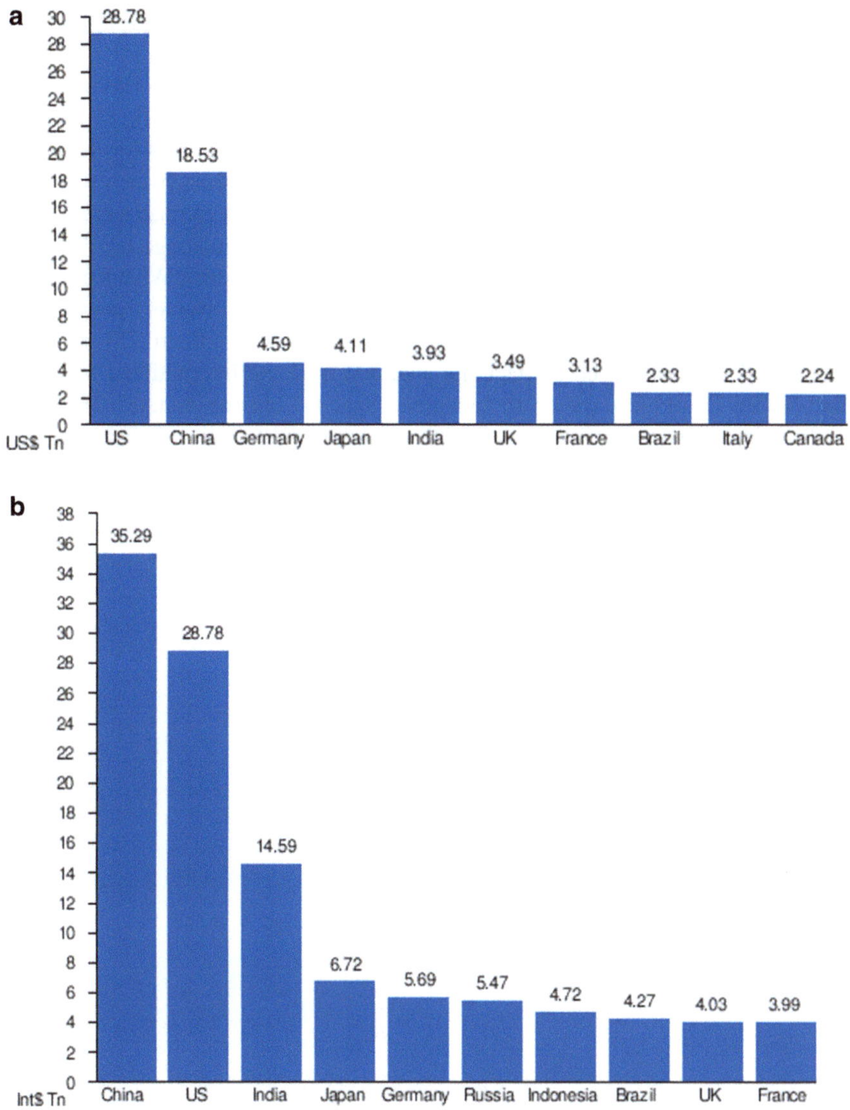

Fig. 10.4 **a** The ten largest economies in the world in 2024 in terms of normal GDP (market value of all final goods and services). *Source* International Monetary Fund (IMF). **b** The ten largest economies in the world in 2024 in terms of purchasing power parity GDP (PPP) in int'l dollars. *Source* International Monetary Fund (IMF). *Notes* An international dollar has the same purchasing power over GDP as the U.S. dollar has in the United States. The figures for China do not include Taiwan and the special administrative regions of Hong Kong and Macau

One factor of major concern is the global cost of climate change adaptation. This is a figure measured in the trillions of US dollars. There are several studies trying to measure the global adaptation costs for the various economic sectors (Brown et al., 2021; COACCH, 2022; IMF, 2022; JRC, 2024; World Bank, 2010). To mention an indicative result, from the IMF study, the global adaptation costs for the public sector can be around 0.25% of the global gross domestic product (which was $105 trillion in 2023) per year in the coming decades. While such estimates can appear manageable for the larger economies, they present a huge challenge for the many poor and vulnerable countries (especially the ones that are islands). What's more, the level of vulnerability to climate change impacts seems to be inversely proportional to the financial and technical capability of the countries to face them. This is shown in Fig. 10.5 which is based on data and analyses from the International Monetary Fund (IMF). This Figure shows the "adaptive capacity" of approximately 100 countries measured against their "exposure index" i.e. their risk to suffer climate related extreme weather impacts.

The scatter of dots (each representing a country) and the resulting regression lines in Fig. 10.5, are quite indicative. The top dotted line is for the advanced economies

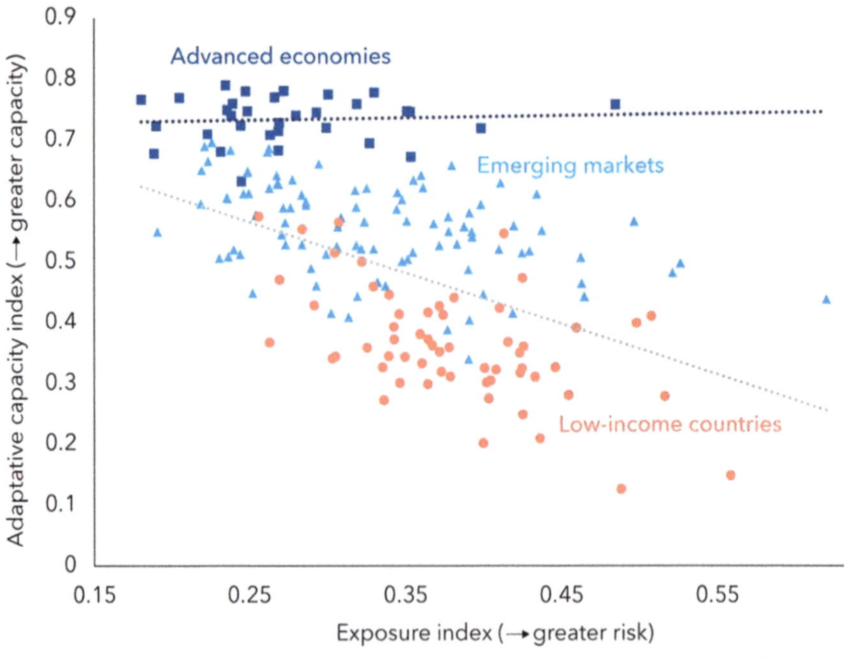

Fig. 10.5 A plot showing the climate change exposure risk (i.e., how exposed a country is to climate change impacts) against their adaptive capacity (i.e., their ability to finance and implement adaptation measures). *Note* Each dot is a country. For the low-income countries (lower dotted line), there is a clear trend showing that the higher the exposure risk the lower the adaptive capacity of the country. *Source* IMF staff calculations based on 2015–2018 data from several sources as explained in IMF (2022)

while the bottom one is for the emerging markets and low-income countries. These lines show a clear trend: the higher the exposure risks the lower the adaptive capacity of the country. The adaptation costs for the lower income and most vulnerable countries will eventually and unavoidably have to be borne by the largest economies on the basis that they are the biggest polluters and largely responsible for the climate change phenomenon. This leads us to our 4th basic hypothesis which is formulated below as conditionality no. 4:

<u>Conditionality 4</u>: *The rising needs for climate change adaptation will eventually be met through a global coalition of countries pulling their resources together to fund the development of the necessary infrastructures. An UN-like, global climate governance organisation will have to be constituted by all the major polluting countries, to coordinate and implement finance and supervision actions for climate adaptation globally with priority to critically affected lower income countries.*

10.2.3 National Debt

Many national debt crises exist (or are likely to develop soon) in most countries, developing or developed. The national debt consists of public debt (i.e. the part owned by the government) and private (owned by households and businesses). For the developed countries, their strong and reliable economic structures and the confidence in their economic performance usually enables them to service their debts viably into the future without many consequences. This is for example the case of Japan whose ratio of public debt to GDP exceeded 250% in 2024. Globally, the total debt (public and private) of all countries together, has reached in 2024 the staggering level of $312 trillion which is approximately 300% of the global GDP. Of this figure, the public debt is one third i.e., nearly 100% of global GDP and the rest is private. The national debts started rising primarily by the global financial crisis of 2008–2009 and worsened substantially during the Covid-19 pandemic.

The most severe and alarming debt crisis is the one in the developing countries that exists since more than a decade now. The total size, including potential liabilities, of the debt of developing countries is estimated, to exceed one trillion US dollars (according to the IMF world economic outlook report of April 2024). According to a 2022 study, the debt service for the more than 100 developing countries that were examined in the study, absorbed an average of 41.5% of their budget revenues and it was 8.4% of their GDP (Martin & Waddock, 2022). In these countries, the debt servicing costs compared to the other national expenditures are on average 2.7 times higher than education expenditures, 4.2 times higher than health expenditures, 11 times higher than social protection expenditures and 54 times higher than the costs spent for climate adaptation. These figures show the devastating effect of national debt for developing countries which see their debt servicing costs far exceeding all social costs.

In 1996, the International Monetary Fund and the World Bank launched the *Heavily Indebted Poor Countries* (HIPC) initiative. In 2005, to accelerate progress toward the UN's sustainable development goals the HIPC was supplemented by the *Multilateral Debt Relief Initiative* (MDRI) which allows countries completing the HIPC initiative to receive a 100% relief on eligible debts by the IMF, the World Bank, and the African Development Fund. There are many more efforts by international organisations or national governments to provide support for heavily indebted developing countries whose detailed mention is out of scope here. However, these efforts are not coordinated and don't have a central focus as they basically reflect the interests and rules of the relief providing entities.

The national debt issue is going to be around for the good part of this century. It is hoped that some sort of global concerted action will be taken to reduce or manage efficiently national debt by preventing corrupt or predatory lending that creates debt crises and strengthen the sustainability of national debts so that the respective economies are free to invest in new technologies and growth promoting activities. Since the issue of national debt is crucial for a healthy economic environment which in turn affects the development and implementation of new mobility solutions and at the same time facilitates the mitigation and adaptation efforts for climate change, we formulate our 5th conditionality as follows:

Conditionality 5: *By 2050 a universal way of lending countries as well as supervising and managing debt repayment in a transparent and sustainable way will have been established under the supervision of an UN-like Global Regulatory Organisation for Debt Issue and Management. This Organisation will monitor, advise, and/or manage national debts to make them sustainable and avert national debt crises.*

10.2.4 International Trade

International trade is the bloodline of international economic growth and cooperation. It is of outmost importance to the future development and well-being of societies with direct and indirect impacts to the future mobility systems, too. There are three distinct periods of international trade in recent history:

A. The period of the "cold war". This period started after the end of the 2nd world war and ended with the dissolution of the Soviet Union in December 1991. This period is known as the "cold war" period because of the rivalry and hostile competition between the two dominant political and economic blocks of the time, the East and the West. International trade was regulated by the *General Agreement on Tariffs and Trade* (GATT) which aimed to reduce tariffs, quotas, and other barriers to trade. The agreement started in 1947 with 23 countries signing it and by 1973 it had 100 signatories. China did not participate in the GATT nor did the Soviet Union. After the dissolution of the Soviet Union, the GATT was succeeded in 1993, by the World Trade Organisation—WTO.

B. The period of increased globalisation that started approximately in 1990 and lasted—in its free and unimpeded form—until around 2016 when the US government, under the first Trump administration, started implementing trade protectionist policies. The period 1990–2016 is characterised by the free and unimpeded international trade and the dispersion of production to places of low labour cost or near the places of consumption and it became known by the word *globalisation.* Globalisation can be said to have lasted until the second Trump administration in the US (i.e. until 2025). This globalisation period is marked by two historical events of global significance. The liberalisation of economic policies in China with its concurrent opening to the world, and the operation of the World Trade Organization (WTO).
C. The current period. This is characterised by the return of international tensions and the application of trade protectionist policies. International trade tends to become more and more restricted and national borders protected by trade tariffs and restrictions. The new "competing" geopolitical blocks are, at the time of writing, not yet clearly defined but it seems certain that the traditional distinctions between "west" and "east" of the past are being demolished. It is difficult to say how long this period will last but, if the past is any indication, its length should be measured in decades and not years.

In the longer term it is anticipated that the divisions that are developing in the current period of international trade, will gradually be eased and even eliminated through top-down agreements or purchases and acquisitions, and a new more homogeneous economic and technological order will arise again globally. This is, obviously, quite critical for the future of mobility and is expressed as our 6th conditionality which is formulated as follows:

Conditionality 6: *The rising rivalries and trade protectionist policies that are currently being initiated by the US government and reciprocated by rival economic blocks, will not last for long on the realization that they are harmful and to the detriment of the interests of all sides. By the end of the 2nd Trump administration or by the early 2030s, international trade will again be free of levies and based on "healthy" competition using international trade rules supervised by the WTO or a new sequel international trade organization.*

10.2.5 Economic Outlook for Each of the Four Regions

The economic outlook for each of the four regions of the world that are included in our futures exercise—i.e., Asia, North America, Europe, and Africa—is of interest as it indicates the ability of those regions to finance and implement new AI-based mobility technologies and systems and at the same time face the climate change crises. The expected economic capacity of these regions is the key to the development and application of the new technologies that will shape the future transport and mobility landscape in their countries. The two key indicators that express the future ability

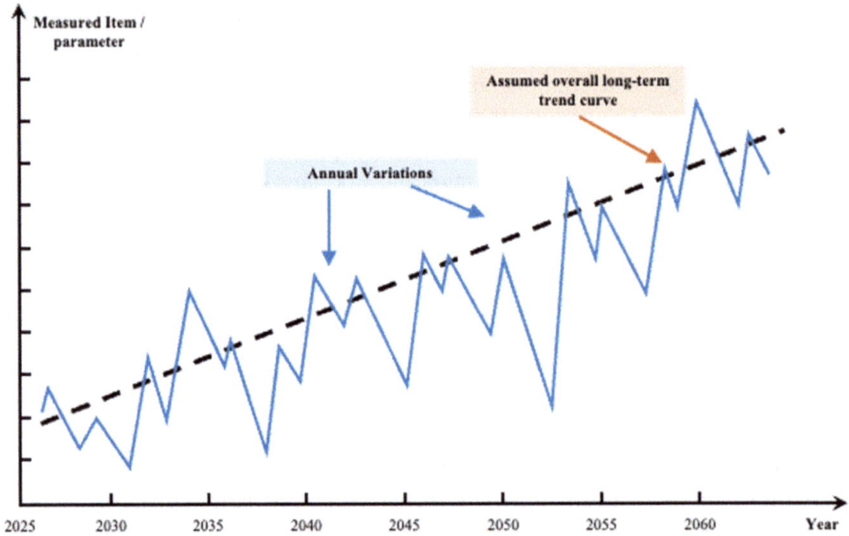

Fig. 10.6 Yearly fluctuations and "median value" curve assumed. The straight line is used here for simplicity, but such overall trend-lines can take the shape of a curve of any kind

of a region to finance climate change mitigation and adaptation measures while at the same time improving the livability of its urban and rural areas through the implementation of new mobility technologies and infrastructures, are taken to be: the GDP per capita and the national debt as a percentage of GDP. Estimating the long-term trends of these indicators is based on past (economic) performance data for each region, their short and medium term forecasted values by reference to existing studies and bibliographic sources and these authors' views and assessments.

In the texts below, a single curve is used to show the expected overall trend in the indicator's values over the years. Such single curves are used as an approximation of the actual trend on the assumption that this is the median or best-fit line to a range of values that this indicator can take over time which would normally produce a fluctuating curve from year to year as it is diagrammatically shown in Fig. 10.6.

10.2.6 Asian Economic Outlook

Most of the current economic studies seem to agree that the Asian economies appear poised to continue their growth for a long time in the future although potentially at a slower pace. This is the gist of several reports issued by prestigious international organisations like the Organisation for Economic Co-operation and Development (OECD, 2023), the Asian Development Bank (ADB, 2025), the World Economic Forum (WEF, 2024), the US National Intelligence Council (NIC, 2021), and others.

The twenty-first century seems to be economically dominated by Asia like the nineteenth century was dominated by Europe and the twentieth century by the United States. A precondition for this situation to materialise is that the continent's governments (especially in the bigger economies of the region, i.e., China, India, Japan, and S. Korea) will pursue policies of inclusive growth, innovation, and good governance and will avoid what economists call the *Middle Income Trap* i.e., a situation where a country develops until its GDP per capita has reached a middle level of income, but then it does not develop further (World Bank, 2024).

Current forecasts point to an increase of Asia's share of global GDP/PPP rising from 32% in 2014 (it was 26% in 2000), to more than 50% in 2050 (ADB, 2024b). Correspondingly, the USA's share of the world GDP (in purchasing power parity—PPP terms), is estimated to be reduced from around 33% in 2014 to 25% in 2050. India could overtake the EU and the US in terms of its share of the world's GDP (in PPP terms) by 2044 and 2049 respectively. The increasing global competition and international trade protectionism as well as the disruptions due to local geopolitical tensions and conflicts may cause temporary economic slowdowns or even reversals, but the "Asian century scenario", will most probably continue to be valid until the end of the century.

The resulting economic expansion in Asian economies will create a virtuous circle of growing purchasing power for their citizens, which will boost domestic demand and bolster the production of high-quality goods and services including high quality transport and mobility networks. Asia is likely to be the first region to employ and implement state-of-the-art transport technologies and AI-empowered mobility services. It is also likely to continue to undergo massive urbanisation, creating industrial and technological mega-clusters where daily mobility and transport needs will require development of new infrastructures for ground and air transport services and develop operational models highly depended on transformative technologies such as AI and its related applications.

A possible timeline, for the economic indicators mentioned earlier, for Asian countries, is shown in Fig. 10.7. The national trends shown in this Figure are given under the hypotheses in conditionalities 2, and 4 (and of course conditionality 1).

10.2.7 North American Economic Outlook

The North America region, i.e. the US and Canada, faces challenges such as an aging population and strong global competition from other world regions especially from Asia. Its strengths in innovation, natural resources, and adaptability are likely to support continued economic growth and development through to 2050' and early '60s but this growth will not be as high or as sustainable as that of Asian countries. The key factors that will shape N. America's future economic outlook are:

a. Technological innovation: This has always been the strong point of the N. American region and is likely to continue in the future though at a lesser pace. Areas

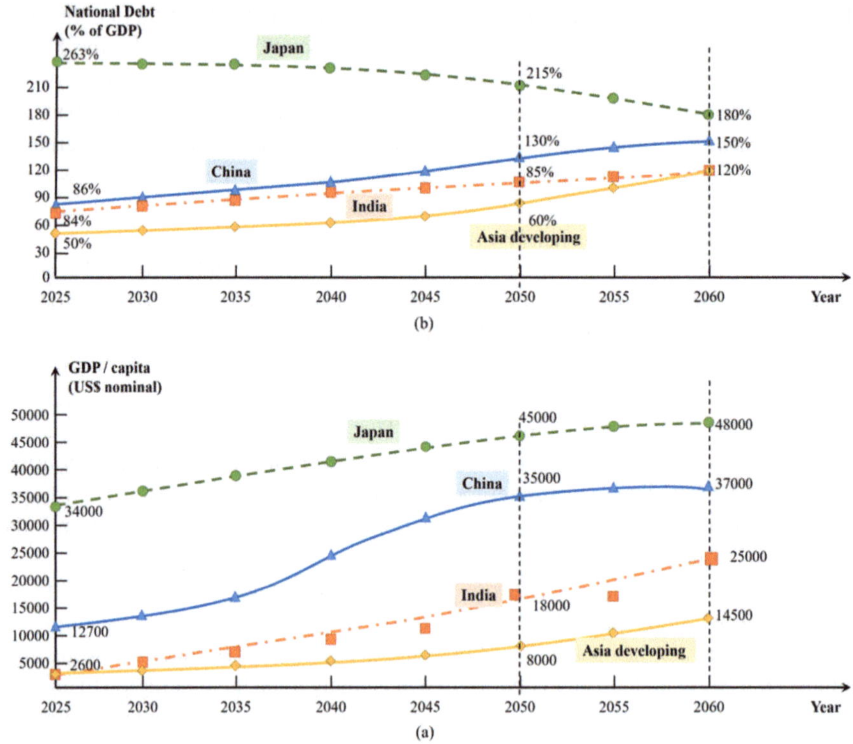

Fig. 10.7 Estimated timelines for Asia's normal GDP/capita and national debt as a percent of normal GDP till 2060 (average values per country or category of countries). *Note* (a) Data extracted, using AI, from: a) the World Bank (World Bank, 2024), b) the IMF (IMF, 2022), c) the Asian Development Bank (ADB, 2025), and d) the McKinsey Global Institute (McKinsey, 2024), reports. Assumed average overall long-term annual growth rates: China 3–4%, India 4–5%, Japan 0.5–1%, Asia developing countries 4–5%. (b) Japan's GDP per capita growth is expected to remain modest and its national debt as a percent of GDP to be reduced due to aging population and low birth rates. (c) India is expected to see robust growth in GDP per capita, driven by its young population and economic reforms and a relatively low ratio of debt to GDP. (d) For developing Asian countries, the figures are an approximate average between the figures for Vietnam and Thailand. Their GDP per capita growth is expected to be high due to industrialization and foreign investment. (e) The yearly fluctuations are not shown in this Figure as explained in Fig. 10.6

like artificial intelligence, biotechnology, and services will be at the forefront of N. American innovation.

b. Demographic change: North America has an aging population which impacts labour markets and economic growth. Legal immigration and the integration of younger more diverse populations especially scientists from other areas of the world, are likely to continue in the longer term in spite the current (2025) policies of the US government that are in the opposite direction i.e. discouraging scientific migration. The need for policies encouraging legal immigration is expected to prevail again as this will mitigate the present labour problems in the US and Canada.

c. Diversity and internal competition: The system of governance in the US promotes diversity and internal competition making the overall outcome positive for the economic development of the area . According to Forbes, economically, the US is not a single country but a collection of seven countries and three quasi-independent city-states (based on the administrative system of governance, as reported in *NewGeography.com* August 29, 2024). Each of these separate entities has its own economic policies, resources and legislation and this enables competition and cooperation as needed to move forward at an overall positive way even in spite federal policies to the opposite.

North America's, ability to remain a major economic world player will be determined by the continuation of its strong innovation capacity, its natural resource wealth and its strong institutions. It will also be determined by three more shaping factors: (a) the quick and efficient adaptation to climate change, (b) the adaptation of its urban infrastructures to the pressures of growing urban populations, and (c) the international geopolitical challenges and the international trade situation as they will evolve in the next decades.

A possible timeline for the North American economic indicators that are examined here, is shown in Fig. 10.8.

10.2.8 European Economic Outlook

The main characteristic of Europe's economic development is that it is based on a more humane form of capitalism. The total social expenditures in the 27 EU-member countries are almost 50% of the social expenditures of all other countries in the world together. As a result, Europe enjoys a high level of social security and quality of life compared to most other countries in the world. The region has lowest mortality rates of infants, higher healthy life expectancy, reduced working hours, high levels of health care coverage, and pursues climate neutrality goals and objectives that are of the most comprehensive and strict, globally. The European economy is currently (2025) going through a phase of slow average annual growth rates (0.5–1,5%). This is 3 times lower than that of the US (+ 2.5%) and 4–10 times lower than that of China (+ 6%). These slow growth rates are partly due to the abrupt change in the energy sources that occurred in 2022 but also due to other factors such as a reduction in population and its aging, bureaucracy and slow decision making (mainly at EU level), etc.

The future economic development of the group of the EU-27 member countries, will depend on developments in the following key areas:

1. *Energy security and provision*: This will remain perhaps the most pressing and demanding problem for Europe in the face of regional geopolitical "eruptions". A massive effort to develop strong and reliable Renewable Energy Sources (RES) that was accelerated after the war in Ukraine, is giving good results. According to the European Environment Agency (EEA) the RES in the EU are expected to

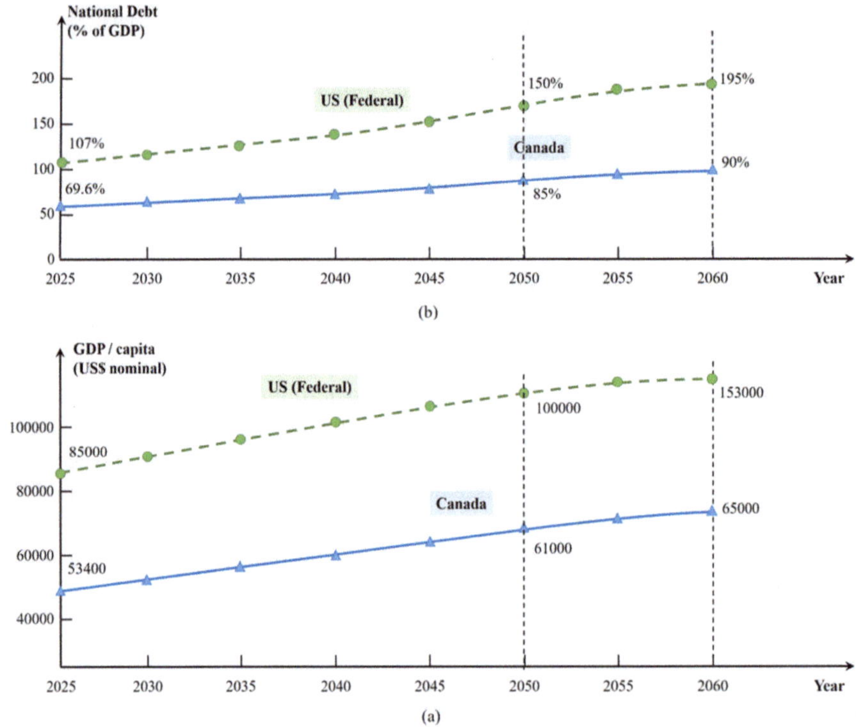

Fig. 10.8 Estimated timelines for North America's GDP/capita and national debt indicators till 2060. *Note* (a) Data extracted, using AI, from various sources: a) The OECD (OECD, 2023), b) US Congressional budget office (www.cbo.gov/data/budget-economic-data), c) US Federal Reserve Bank (FOMC projections materials accessible version, June 2025), d) Penn Warton University of Pennsylvania Budget model (budgetmodel.warton.upenn.edu), e) International Monetary Fund (IMF, 2022), reports. (b) Assumed average overall long-term annual growth rates: US (federal) 1.8%, Canada 0,8%. (c) The federal US debt as % of GDP is projected to reach in 2060 almost as high as the national debt of Japan today. (d) The national trends shown are given under the hypotheses in conditionalities 1, and 4. The yearly fluctuations are not shown in this Figure as explained in Fig. 10.6

rise to more than 40% of the total energy mix by 2030 and close to 90% by 2050 (from 20% in 2020). Nuclear energy, is included in the RES.

2. *Climate neutrality*: The EU, in spite its recent slowdown and rethinking of certain climate goals and policies (see Chaps. 2 and 3) is likely to continue to lead the way, globally, in climate neutrality legislation and implementation. The 27 EU member countries will be transitioning towards an economy with zero greenhouse gas emissions throughout the period 2030–2050 driven by legally binding targets under the European Climate law and national legislation based on the National Energy and Climate Plans (NECPs). This transition will require considerable amounts of investment but at the same time it will create new economic opportunities in green technologies, renewable energy, and sustainable practices that

can then be exported globally against competition from China, the US, and other countries. The outcome of this competition will largely depend on the level of technological innovation that can be developed in Europe.

3. *Technological innovation*: Advances in technology, particularly in digitalization, artificial intelligence, and biotechnology, will play a crucial role in driving economic growth in Europe. Increased security concerns may reduce investments in research and technological innovation, but it is possible that this will not be allowed to happen. The need to increase innovation in the EU is particularly stressed in the recommendations of the Draghi and Letta reports (see below) as otherwise Europe would be left behind its main global competitors.
4. *Demographic changes*: If European governments do not foster policies to support higher birth rates and at the same time regulate immigration and the integration of diverse populations, they will face a disproportionally aging population by 2050. This will severely impact its labour markets and economic growth.
5. *Economic integration and trade*: Continued economic integration within the European Union and strong trade relationships with global partners will be vital. To this end the EU must foster a balancing act by promoting trade agreements with third countries while at the same time cautiously protecting its single market and economic stability and growth against undue foreign competition.
6. *Security independence*. After the war in Ukraine, the EU and other European countries are reconsidering their security (which traditionally relied on the Atlantic treaty organisation and the US). Until January 2025, when the 2nd Trump administration took office in the US, the European debate on a self-financed and common European security structure had started but was not leading to any concrete action. After January 2025, all this has changed and most leading European countries, have moved towards a common European security arrangement independent of the US.

An overall economic assessment for the EU (and the rest of the free trade European countries) is that the region will register a rather modest and horizontal, i.e., for all EU-member countries, growth of around 1–2% per year. If this happens and considering that the expected growth of the emerging economies will rise more rapidly, the share of the EU-27 in the global GDP will decline as we move toward the 2040s and 2050s. The current new international trade environment that brings reduced globalisation, increased national trade protectionism, and a fierce global competition will obviously affect Europe in many ways (positive and negative). If the EU succeeds in re-organising itself to stand up to its global competitors, it will reap positive results. If it fails, it will suffer further decline and modest growth.

The Mario Draghi report of September 2024 made several strong recommendations for the future of the EU which, if implemented, are expected to propel the EU to becoming a strong competitor to the US and China (Draghi, 2024). According to this report, the EU will need to promote most rigorously its technological innovation, productivity and economic self-sufficiency with a more coherent and decisive governance with less bureaucracy and a common defence structure. The total financing to fund the necessary investments according to the Draghi report is an annual 750–800 billion Euros which is about 4–5% of the EU's GDP. For this level of funding, the report suggests the creation of a special development fund funded by loans. Similar recommendations were made in another report by Enrico Letta (Letta, 2024). The key messages of these two reports can be summed up in the phrases "*unite, defend, grow*" and "*do this or die*". In other words, there is an urgent need for the EU to stand up to the harsh global realities and reorganise, to increase the integration of all 27 national economies with simpler decision-making procedures, increased spending on research and innovation, increased investment in climate change and development of new technologies and systems.

Strengthening the European competitiveness in research and technological innovation will also strengthen and expand Europe's already significant achievements in the development of smart mobility systems and services, smart cities and sustainable urban development. However, all the above remain to be seen. If we stay with the current European (EU) economic trends and prospects, a possible timeline for Europe's economic development can be depicted as shown in Fig. 10.9.

10.2.9 African Economic Outlook

The African continent and the prospects for its 55 countries, have been the focus of several independent studies (BCG, 2022; Cilliers et al., 2011; MGI, 2023; WBG, 2023). The average annual growth for Africa for the 30-year period 1990–2019 was a mere 1% but by all counts Africa has a remarkable economic and developmental potential which will unfold during this century probably making this continent the new global economic powerhouse of the next century. The critical element that will determine Africa's future is whether its many countries will manage to educate and retain, for their own economies, the young population that will result from the current baby-boom that exists in most African countries. If the current birth trends persist, by 2050 about 40% of all people on earth under 18 will have been born and raised in Africa whose total population will nearly double by then to 2.5 billion people (BCG, 2022). Africa will also be home to an increasing number of megacities (densely populated areas with 10 million or more residents). By 2050, African urban areas are expected to attract an additional population of 1 billion residents raising the number of "megacities" from the current three (Cairo, Kinshasa, and Lagos) to 14 (MGI, 2023).

A huge working-age young African population can be a disruptive force if there are insufficient jobs around leading to unrest and migration. So, the crucial factor that will determine the African success story will be giving ample opportunity to the African youth to work and catalyze economic growth in Africa rather than abandoning it to migrate to other countries or continents or pushing it to extremes. So far, the signs are that given the opportunity many African young entrepreneurs will spring up in everything from fintech to clean energy. Already, over the past 20 years, a profound structural shift to services has been going on in leading African countries as people left work in the fields to take jobs in trade and other services in cities. According

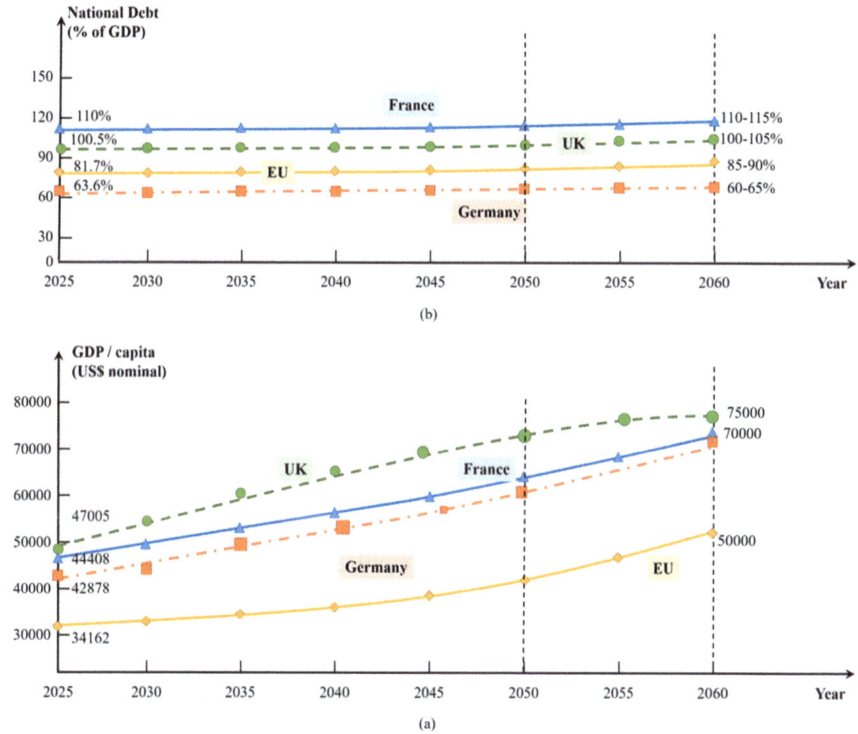

Fig. 10.9 Estimated timelines for GDP/capita and national debt for the EU (27) and three European countries, till 2060. *Note* (a) Approximate average values for all EU member countries. (b) Data extracted, using AI, from various sources: a) Eurostat, b) OECD, c) World Bank, d) Trading Economics (tradingeconomics.com), e) Statista (statista.com), f) Macrotrends (macrotrends.net), reports. (c) Assumed average long-term annual growth rates: EU (27): 1.2%, France: 1.3%, Germany: 1.1%, UK: 1.5%. (d) All European countries are projected to have low annual growth rates, of around 1–2%, according to current estimates. This may change, at least for the EU member countries, if the Letta and Draghi reports are adopted and implemented. In this case the national debt figures may also increase. (e) The national trends shown, are given under the hypotheses in conditionalities 1, 2, and 4. The yearly fluctuations are not shown in this Figure as explained in Fig. 10.6

to the African Development Bank, about 22% of working-age Africans start small businesses, as compared to 18% in Latin America and 13% in Asia. In 2024, Africa will account for eleven of the world's 20 fastest-growing economies (ADB, 2024a).

Despite the positive signs, growth will remain uneven across the African continent. The central part—the sub-Saharan Africa (SSA)—is largely a poor region with some 40 diverse countries half of which have a fragile conflict-affected economy and are characterized by a small population, limited human capital, and a confined land area. The overall economic performance of this region is low (growth rates generally are in the range of 0–1%) with high national debts endemic economic crises, many epidemics, and mounting local conflicts. East Africa has similar problems but with somewhat higher average growth rates while Western Africa has the highest average growth rates, exceeding 3%, over the last decade (WBG, 2023).

For the period under consideration, the African economies are expected to continue to grow and strengthen their internal economic collaboration helped by institutions like the recently constituted *African Continental Free Trade Area* (AfCFTA) or the *African Union* and others. The East African region will probably be the last to implement the novel mobility technologies with the Western one following a bit earlier. South Africa and Nigeria are the most likely countries to be the first in the continent to implement automated shared mobility and other novel mobility services as well as to fully decarbonise the sector, by 2060 or thereabouts. The real story of Africa will unfold towards the end of the century when the continent is expected to take off economically, socially and technologically. Towards the same time, we expect the successful conclusion of the decarbonisation efforts for most African countries and this is because of the high costs involved. According to a recent report for the decarbonisation of the transport sector in Africa (NASAC, 2024), the total cost of implementing climate change mitigation measures in all sectors for the 53 African countries that submitted Nationally Determined Contribution (NDC) reports to the UN (see also Chap. 2), is as high as USD 2.8 trillion. For the foreseeable future, the lack of a well-established regulatory framework and relevant policy incentives discourages private sector investment in the continent's transport decarbonisation.

It is also difficult to give even a rough estimate of the potential timeline of Africa's economic development. The continent is too big and too diverse in its constituent entities (countries) while its full development potential is estimated to take place beyond 2050 and towards the end of the century. Some estimated prospects as regards the penetration of renewables in the energy mix and the degree of climate change adaptation are given in Fig. 10.11 in later sections.

10.3 Technological Factors

By 2050, some of the key technologies that are currently new, or still on their research and development stages, will have matured and applied in at least the more economically advanced countries transforming human mobility and human life. The most influential, and transformative technologies that have the potential to play that role and define human mobility by 2050 and beyond, are the following:

1. Battery electric vehicles.
2. Vehicles using hydrogen (in direct combustion or in fuel cells).
3. Cooperative connected and driverless vehicles (automation levels 4 or 5).
4. Shared automated mobility.
5. Flying vehicles (vertical take-off and landing).
6. Artificial Intelligence (software and services).
7. Advanced freight transport and urban distribution technologies.

A brief description of each of these, as well as their technology readiness and estimated penetration levels by 2050+ is given in summary form in Table 10.1.

Table 10.1 Estimates of technology readiness and expected penetration levels for key future mobility technologies—2050 horizon

Name of technology	TRL		Penetration level (2050)	Comments
	Current	Expected 2050		
Battery electric mobility	7/8	9	Developed economies: $\geq 80\%$ for passenger and $\sim 50\%$ for freight vehicles (urban and interurban) Less developed: half of above	On the assumption that BEVs will continue to be improved technologically with better and more lasting batteries for whom the supply of necessary raw materials will be secured, the charging infrastructures will steadily expand, and the pace of energy transition to renewable sources will continue
Hydrogen-based mobility	4/5	7	Developed economies: Low but starting to expand Less developed: Very low	Full implementation of safe, low-cost, hydrogen systems, operating under suitable regulatory legislation, are expected to start being implemented and used at the time horizon of 2050. Full application and sizeable penetrations of hydrogen mobility will be seen after 2050 during the second half of the century
Cooperative, connected and automated mobility—CCAM (automation level 4 or 5)	7	9	Developed economies: 40 to 50% of the total vehicle fleets on the road Less developed: 10–20%	These estimated levels of penetration, by 2050, will be reached gradually starting perhaps around 2035—'40 by application of CCAM in specialised niches of urban mobility such as: Buses on dedicated busways/Low to medium-speed passenger shuttles, for trips in well designated areas/ Sidewalk package delivery robots/Freight urban delivery light-vehicles/ Automated ride hailing vehicles ("robotaxis")/Automated shared vehicles in mobility as a service packages

(continued)

Table 10.1 (continued)

Name of technology	TRL		Penetration level (2050)	Comments
	Current	Expected 2050		
Shared automated mobility	7/8	9	Developed economies: almost all automated vehicles will be used for shared services (estimated 40–50% of all vehicles on urban roads) Less developed: Shared transport will be highly used but not with automated vehicles	The final penetration of the shared automated mobility market will depend on the type of business models that will be used, the extent of complementing the Public Transport network through First and Last Mile services, also for mobility on-demand services, and the financial feasibility of all these
Flying vehicles (Urban Air Mobility—UAM)	4/5	8	Expected to be applied in $\sim$ 100 metropolitan areas globally—mainly in developed countries	Battery electric flying vehicles are expected to have a low penetration mainly because of the difficulties in developing networks of urban vertiport infrastructures and the difficulties in managing the airspace above "living" cities. The serious safety and environmental issues raised must also be solved. Because of the limited size and carrying capacity of the air vehicles the UAM will be suitable primarily for air taxi operations or parcel deliveries

(continued)

Table 10.1 (continued)

Name of technology	TRL		Penetration level (2050)	Comments
	Current	Expected 2050		
Artificial Intelligence mobility applications	5/6	9	Developed economies: High (>90%) penetration to all aspects of mobility Less developed: Lower penetration (30–40% of all possible use cases)	The full penetration of AI applications for passenger and freight transport operations and planning will depend on the success of the regulatory and legislative frameworks that will be imposed by governments, the competition conditions of the relevant stakeholders (distribution of "power" and dominance of the market). Also, on the limits in the capacity of the semiconductor industry to supply appropriate chips, and the availability of electrical power that will be needed for AI development and deployment. Finally, on the pricing models that will be used and the economics of AI based mobility applications
Freight transport and urban distribution				
Platooning with semi or fully automated trucks	7/8	9	High penetration in intercity road transport	The technology of truck platoons is already quite mature and is expected to expand fully in the 2030s
Fully automated trucks (level 4 or 5) running between hubs and distribution centres	6/7	9	Developed countries: high penetration. Rest of the world: medium to low	Wide implementation, in the roads of many developed countries as early as 2030 with full maturity expected by 2040

(continued)

Table 10.1 (continued)

Name of technology	TRL		Penetration level (2050)	Comments
	Current	Expected 2050		
AI enabled planning and management of intermodal freight transport	7/8	9	High penetration globally	Through cloud-based AI Digital Platforms and Intermodal Capacity Brokers, offering one-stop-shop services finding the fastest, cheapest, environmentally sustainable routes, and the best suppliers of transport services
Automated logistics hubs	7/8	9	Developed economies: High penetration Developing economies: low penetration	The technologies that will enable a fully automated operation of a logistics hub are currently on their way to technical maturity and need to be integrated into an overall automated logistics chain system. Fully automated logistics hubs are expected to be in full implementation by 2030 or thereabouts. Complete automation of freight transportation between distribution centres is anticipated to occur by 2045–2050
Automated local delivery light vehicles	8	9	Developed economies: High penetration. Developing economies: low penetration	Such vehicles can be on the roads currently, but their safety and certification legislation and standards are likely to delay their implementation until 2030 and their wider use and maturity until 2035–'40
Side-walk package delivery robo-vehicles	8	9	High penetration globally	Such vehicles can be on the roads currently, but their safety and certification legislation are likely to delay their implementation until 2030

(continued)

Table 10.1 (continued)

Name of technology	TRL		Penetration level (2050)	Comments
	Current	Expected 2050		
Use of clean fuels (urban or interurban freight transport)	7/8	8/9	Electric-LCVs $\geq$ 50% by 2040 and $\geq$ 80% by 2050 in urban areas e-HCVs ~ 30–40%	Electric light commercial vehicles (e-LCVs) will dominate urban distribution networks with China leading the way Electric heavy commercial vehicles (e-HCVs) slow to emerge (currently below 1% of total) and of low penetration by 2050
Moving passengers and freight underground (Hyperloop)	5	7	Hydrogen-trucks ~ 40/50%	Green hydrogen trucks (h-trucks) over long distances and in regions with limited charging infrastructure, medium–high penetration by 2050
			Real-world trials exist but significant barriers too	Commercial implementation of Hyperloop type of technologies is still not certain in 2050. Their possible realization to any meaningful extend is placed beyond 2060 and possibly by the end of the century

Source For the current situation: (Martínez-Plumed et al., 2022; O'Hern & St. Louis, 2023). For the expected in 2050: the analysis in previous sections and authors' estimates

Note TRL 1: Basic principles observed/TRL 2: Technology concept formulated/TRL 3: Experimental proof of concept/TRL 4: Technology validated in a lab/ TRL 5: Technology validated in a relevant environment/TRL 6: Technology demonstrated in a relevant environment /TRL 7: System prototype demonstration in an operational environment/TRL 8: System complete and qualified/TRL 9: Actual system proven in operational environment

10.3.1 Battery Electric Vehicles

Battery Electric Vehicles (BEV), or New Energy Vehicles (NET) as they are called in China, are growing in numbers. The 2024 level of global annual sales of electric vehicles (EVs) of about 15 million units per year is expected to increase to about 50 million per year in 2030 (McKinsey, 2024). This will be approximately 50% of the expected total car sales in that year. The concerns that have been projected by many early adopters of electric mobility about price and range have temporarily slowed the growth of electric vehicle sales, but ultimately electric mobility is expected to prevail and by 2050 the total electric vehicle population will have reached penetration levels of 80–100% in most developed economies. In the lesser developed ones these penetration levels will be much lower, approximately half of the figures for developed economies. Currently the only country that has reached the 100% level is Norway. In the EU, the trend to electrification will be further strengthened by the recent decision of several European countries to prohibit the sale of internal combustion engine (ICE) vehicles by 2040 or 2045. Chinese brands are currently accounting for about 50% of global BEV sales with the Chinese BYD firm overtaking Tesla in global EV sales in the fourth quarter of 2023.

The final levels of penetration of electric mobility will be determined by the scale of development of the infrastructures for EV charging. According to some estimates, to meet the energy demand of global BEVs, some 135 million charging points will be needed globally by 2030 which is five times the 25 million charging points installed by 2023 (of which about 15 million are public chargers). Considering the many decades over which the gasoline filling stations were developed to the level that we know them today, the BEV charging infrastructure (in homes, offices, shopping centres, highways and other places), will probably reach its maturity by 2040 or thereabouts. Of critical importance for electric mobility is the extent to which it will be based on clean electricity for the charging of batteries. Clean renewable energy, together with the necessary energy storage infrastructures, is a prerequisite so that electric mobility can be a true decarbonization option. These infrastructures are expected to gradually be developed by 2050 when they will have reached a high level of penetration in the medium to high economic output countries. These estimates are based on the current slow pace of progress in EV sales, the slow progress of decarbonisation and renewable energy production in key countries where planned decommissioning of energy generating plants using fossil fuels are postponed or stopped and new policies favoring fossil fuels are introduced, and also on the expected negative influence of (existing or yet unknown) geopolitical events (conditionality 2).

For freight transport, the future of electric mobility looks less optimistic with the current (2024) share of BEVs in the freight vehicle fleets is low, about 5%. By 2040, 60–80% of small and medium-sized trucks in Europe and China may well be electric due to supportive regulations but in other developed regions, such as the United States, the estimated share is lower probably below 30-40% of the total fleet. For the rest of the world, that share is expected to be even lower e.g. between 10 and 30% (McKinsey, 2024). To reach these shares, battery electric trucks would require either significant improvement in battery energy density or reduced payloads compared with traditional diesel trucks.

In conclusion, though electric mobility is here to stay, the way to full electrification will be long and costly even for the advanced economically countries. China and the EU are expected to be the exemption to this general rule. For the lesser developed countries BEV penetration levels are likely to be low even by 2050+. These estimates are made even though it is taken for granted that BEVs will continue to be improved technologically with better and more lasting batteries, for whom the supply of necessary raw materials will be secured, the charging infrastructures will steadily expand, the pace of energy transition to renewable sources will continue, and the announced bans on ICE vehicle sales will remain in place.

10.3.2 Vehicles Using Hydrogen

Hydrogen can be used as a fuel either in engines with direct combustion or in fuel cells that produce electricity which is then used to energise an electric motor. Most of today's national energy policies foresee increased production and use of hydrogen, including its use as vehicle fuel. However, this hydrogen needs to be "green" i.e. produced by electrolysis of water using renewable electricity (solar or wind power), or nuclear (then called "yellow"). Currently, over 99% of hydrogen is produced from fossil sources (methane for example) by a process known as steam reforming. So, its production emits significant quantities of CO_2 and is called for this, "grey" hydrogen. Hydrogen production by electrolysis leads to CO_2 emissions almost twice as high as those of the conventional methane reforming process if the electricity used is not green. For the time being the hydrogen used in the electricity mix of almost all countries producing hydrogen, is largely carbon-based.

The use of hydrogen in the field of transport meets with many difficulties that relate to the transportation of the fuel to filling stations and its storage there, as well as its storage in the vehicle (where the use of high-pressure tanks of 700 bars poses significant problems not yet solved). In terms of energy efficiency over the entire vehicle life cycle, a hydrogen vehicle is significantly less efficient than a BEV (25% vs. 80% efficiency level). The main advantages of using hydrogen as fuel come from the short time for recharging the vehicle and the long ranges achieved for one filling. This is particularly useful for vehicles that must run long distances in areas where re-charging of batteries is difficult, or for airplanes and ships. Today hydrogen-powered vehicles are used mainly for heavy- or semi-heavy vehicles (trucks, buses, vans, garbage collection vehicles) and occasionally cars (e.g. taxicabs in France relying on the Toyota Mirai hydrogen vehicle), and on rare occasions trains. For air transport, especially wide-body aircraft, the use of hydrogen on fuel cells and electric motors is possible but hard to see by 2050 as this possibility must await major advances in the development of suitable propulsion systems, the management of the extreme cryogenics required for liquid hydrogen, the lightening of storage tanks, and in achieving an acceptable overall safety level for aircraft and airport infrastructures (Fontecave et al., 2024). A similar argumentation holds for maritime transport too, though the feasibility of implementation there, is of a more closer time horizon.

The technical difficulties regarding the production and distribution of green (or yellow) hydrogen, are expected to be overcome eventually but the full implementation of safe and low-cost hydrogen systems in transportation, operating under suitable regulatory legislation, should be expected towards the end of the century and not towards the 2050s. For now, the expected explosion of the battery-electric vehicle market, worldwide, suggests that BEVs have won the race with the hydrogen vehicles and will dominate the clean transportation market for light duty vehicles (cars and light vans/trucks) well into the 2050s and ‘60s.

10.3.3 Cooperative, Connected and Automated Mobility (Automation Levels 4 and 5)

Cooperative, connected, and automated vehicles (CCAV), or Intelligent Connected Vehicles (ICV) in China, are automated vehicles that can exchange data and information with their environment or other vehicles and perform their own driving automatically in cooperation with them. Connectivity and automation technologies are already well developed and tested in hundreds of real-life applications connecting infrastructure to vehicle and vice versa (I2V/V2I), vehicle to vehicle (V2V), and vehicle to vulnerable road users (V2P). The most used communication technology today is the 5G cellular system for information and data transfer combined with LiDAR and radar technology. The automated vehicle functionalities up to level 3 of the SAE classification (SAE J3016)[2] are currently well advanced, mature, and commercially available. Automation levels 4 and 5, i.e. the highest levels of automation, are gradually being tested in real life applications with the so called “robotaxis” being the most well-known example. The regulatory and legislative frameworks for the operation of CCAVs are now being developed at national and international levels.

Overall, the advent of connected and automated mobility represents a technological transition with a significance equivalent to that of the emergence of the Internal Combustion Engine (ICE) in the 1900s. Just as the ICE technology in the 1920s and ‘30s was still in its beginnings and it needed a whole century to mature, so is the automated—CCAV/ICV—technology still at its beginnings in the 2020s and ‘30s. It is expected to fully mature towards the end of the century but by the 2050s it should be well on its way to market sustainability and dominance. So, we are currently in the first stages of implementing automation technology and it will take quite a lot of effort and time for this technology to mature.

Deployment will be gradual, limited by technological capabilities and benefit/ cost constraints, so at the beginning CCAVs will start operating in niche markets

[2] See https://www.sae.org/standards/content/j3016_201806/.

and confined areas only. This is beginning to take place now in the most economically advanced regions but fully and universally implemented *cooperative, connected and automated mobility* (CCAM) systems will take more time to come, most likely beyond the 2040s and even '50s.

10.3.4 Shared Automated Mobility

Shared mobility is the shared use of a vehicle, bicycle, or other mode of transport that enables users to have short-term access to the transportation mode concerned on an "as-needed" basis. Shared mobility includes services like:

- Fixed public transport services,
- Car sharing,
- Carpooling,
- Bike sharing,
- Scooter sharing,
- Demand responsive transport (DRT) services (on-demand rides),
- Ridesharing,
- Micro transit,
- First/ last mile services,
- Courier network services.

A *Shared Automated Vehicle* (SAV) is an automated vehicle that operates under shared mode, moving people or cargo. The importance of SAVs and shared mobility comes from the realisation that the use of CCAV/ICVs by individual owners like today's cars, is expected to increase rather than reduce the vehicle miles travelled on the roads. This is due to the comfort, safety, privacy and capability to fully utilise the travel time in a more productive or enjoyable way which these vehicles will offer to their passengers. To reduce the number of vehicle-miles on the roads, automated vehicles will have to be used through intelligent sharing schemes powered by AI. The shared mobility industry changed dramatically in the first decade of the twenty-first century as regional and multinational tech companies began competing with traditional taxi companies by offering ride-hailing services. Overall, shared CCAV/ICVs and CCAM technologies are here to stay and will gradually be the dominant form of mobility in our urban areas by 2050 and beyond.

Shared mobility with traditional vehicles is here today in increasing numbers. The e-hailing market accounts today for a large part of consumer spending in shared mobility globally and in Chinese cities in particular where the e-hailing mobility (provided by hundreds of companies) accounts for more than 70% of daily trips. In a recent survey covering the 10 largest Chinese urban areas, 90% of the consumers asked, stated that they use e-hailing services at least once per week. On the two biggest US e-hailing platforms (Uber & Lyft) more than 40 million e-hailing trips are being booked every day. According to a consumer survey by McKinsey's centre for future mobility, more than 60% of people would share their ride with a stranger if doing so would add less than 15% to their travel time while reducing their cost (Heineke et al., 2021).

According to McKinsey (2024) the shared automated vehicle industry's revenue could grow to between $610 billion and $2.3 trillion as early as 2040. The final size of the shared automated mobility market will depend on several factors and mainly, the:

a. Type of business models that will be formulated and used (today the automated technology providers, the ride-hailing companies, and the carmakers are the main entities likely to be activated in this market).
b. Extent of complementing the Public Transport network through First & Last Mile and on-demand services.
c. Technological progress of the automated vehicles in terms of safety and reliability.
d. Financial feasibility for deploying shared automated vehicles at scale considering the significant capital investment, and the considerable outlays of expenses before attaining profitability.

Our estimate is that with the full use of AI, by 2050, shared automated mobility will be widely popular and used in developed economies while in the developing ones shared mobility will also be high but mostly with traditional vehicles.

10.3.5 Flying Vehicles (Vertical Take-Off and Landing)

Urban air mobility (UAM) using automated or human driven flying vehicles, for passenger and freight transport, is a transport mode that emerged as a viable alternative to congested surface networks in the mid-2010s when it attracted increased attention due to the unstoppable urban road network congestion. It involves electric aircraft taking off and landing vertically and piloted remotely or with a pilot on board. This form of mobility is defended as a safe, secure and sustainable transportation mode for passengers and cargo in urban environments with commercial operations expected to start in European, US, and Asian cities between 2025 and 2030. The first types of services expected to be the delivery of goods by drones or transport of passengers by piloted air-taxis. The aim is to create a safe, environmentally friendly and efficient urban air transport system and to relieve congested ground infrastructures. According to the European Union's Aviation Safety Agency (EASA), UAM operation in European cities will save 15 to 40 min on average on a standard city travel time and more than 70% time saving for emergency and medical delivery vehicles. It will also have a positive economic impact with the creation of an estimated 90,000 jobs in the EU as soon as by 2030 (https://www.easa.europa.eu/en/what-is-uam). UAM is seen as the 'third dimension' in urban mobility making the airspace the new frontier of development for mobility services. Advances in electric propulsion, autonomous flight technology and 5G or 6G communication networks will spawn this form of mobility offering (among others) on-demand air taxi services, scheduled airport shuttles and short intercity flights.

UAM must be integrated into existing urban mobility systems. This will require suitable ground infrastructures for take-off and landing (*vertiports*), and corresponding interfaces with the other ground infrastructures. It will also need a solid regulatory framework focused on safety. Vertiports must be developed, giving safe and comfortable modal transfer from ground to air transport services (Fig. 10.10). This may be difficult to achieve in an environmentally friendly way as they must be fully integrated with the infrastructure of the other modes of transport. The needs for suitable infrastructure |(vertiports) could limit growth of UAM at least in the foreseeable future, pushing it towards the second half of the century. The comprehensive regulations that need to be in place for the registration, certification and safe operation and deployment of Vertical Take-off and Landing (VTOL) vehicles is another limiting factor. Currently each country or region issues such regulations as a response to requests by interested operators but a mature and universally applicable VTOL-specific regulatory environment as well as standards for the drones will need at least one more decade to develop. Early initiatives have already started. For example, in 2023 the US Federal Aviation Administration (FAA) announced the steps needed to achieve a competitive, scaled-up air taxi market by 2028 in the US. In the EU the European Union's Aviation Safety Agency (EASA) is developing rules for the certification of new passenger VTOLs[3] while in China, some companies (e.g., AutoFlight and EHang) have already received airworthiness certification (Alcock, 2024; Cheng, 2023).

By 2050, it is expected that the "third dimension" of urban mobility (i.e. air mobility) will have matured technologically and legislatively but its penetration to urban areas around the world will still be low for several reasons, among which, the:

a. Difficulties to integrate the networks of urban vertiport infrastructures into the rest of the urban fabric.
b. Difficulties in managing the airspace above a "living" city.
c. Serious safety and environmental issues raised (potential accidents will delay deployment considerably).
d. Limited size and carrying capacity of the air vehicles that makes them suitable primarily for small air taxi operations or parcel deliveries.

The need to avoid traffic congestion on the urban road networks and the need for adding resilience to the transport system to meet climate change crises, will—at the end—keep this form of urban mobility developing and expanding. According to earlier estimates, by 2050 close to 100 000 passenger drones could be in service

[3] The European Union's Aviation Safety Agency (EASA) published, in March 2022, specifications for the construction of vertiports the so called 'U-space package' and also specifications for UAM air traffic control that came into force on 26 January 2023. The rules for the operation of air taxis in cities were published in June 2022. In April 2024 the European Commission published an Opinion (No. 03/2023), paving the way for the launch of air taxi services and other innovative drone uses in Europe. This legislation does not cover automated, i.e. pilotless personal air vehicles (PAV) in relation to which the development of certification regulations is still pending.

Fig. 10.10 Image of a future vertiport for London, UK. *Source* Pascall + Watson and Swanson Aviation Consulting (EA Maven). https://www.pascalls.co.uk/projects/work/future-flight/waterloo-vertiport/

worldwide. By the same year approximately 100 metropolitan areas (with over a million inhabitants) may have implemented UAM to provide a variety of passenger transportation services, with roughly 5000–6000 passenger drones in operation (Berger, 2018). Potential early adopters of UAM include China, the US, the EU, Japan, and the UAE.

10.3.6 Artificial Intelligence

AI applications for mobility related systems are becoming the next frontier in the transport and mobility domain. Starting with already well-identified opportunities on the generative AI value-chain, further end-user applications are continuously being offered. Systems tailored for specific applications built with customized AI models and using techniques such as *fine-tuning* and *retrieval-augmented generation* have a huge role and potential to play as presented in more detail in Chaps. 5, 6, and 7. The

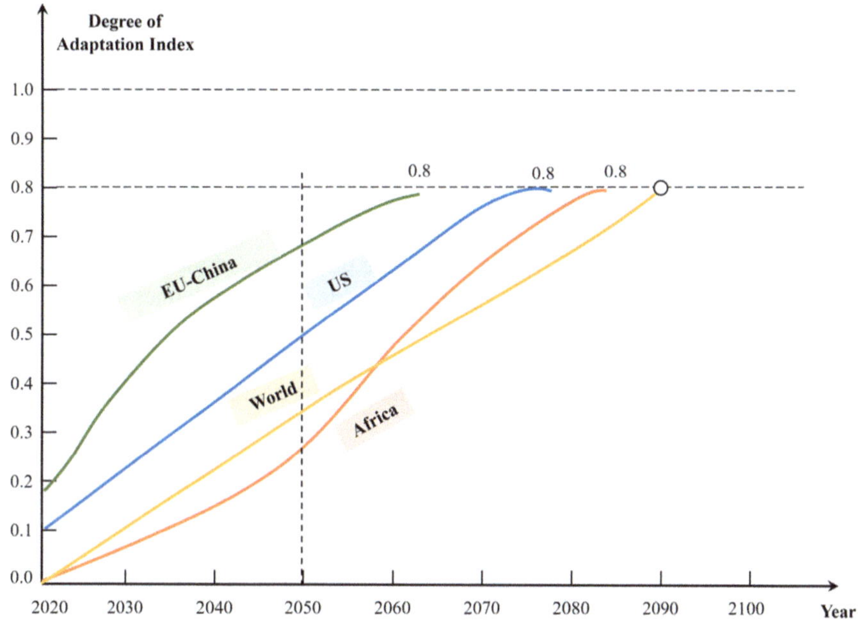

Fig. 10.11 Estimated degree of adaptation to climate change in each of the four regions considered. *Notes* (a) Data and relevant figures about the degree of adaptation achieved in each country or region, do not exist currently. Therefore, the starting figures, i.e. the estimated adaptation degree indices for 2020, are estimates which have been roughly aligned to the IMF data presented of Fig. 10.5 about the adaptive capacity of countries. (b) It is assumed that no country or region will ever be possible to reach full adaptation (i.e. adaptation degree index 1.0) as this will require enormous amounts of financial resources and because the unmitigated climate change impacts will be many and varied, and it will be practically impossible to develop infrastructures that are fully resilient against all of them. The 0.8 level is considered as a satisfactory degree of adaptation (indicating an almost 80% completion of the totality of possible adaptation measures). (c) The EU and China are shown as following the same curve as we estimate that they will follow an adaptation timeline that is the speediest in global terms. This will be the result of strong political commitment, comprehensive policy formulations for climate change adaptation and strong implementation and financing practices

AI revolution in the field of transport and mobility will be based on two different types of software development models. The first, consists of the so called "*frontier foundation*" models, which are the more sophisticated and advanced AI models that can perform a variety of tasks by using expansive artificial neural networks and enormous sets of data. For example, the GPT models by *OpenAI*, the *Gemini* by Google, the *Claude* by Anthropic, the *Command* by Cohere, and the models by *DeepSeek* and others are *frontier* models. Building a frontier foundation model normally requires access to massive amounts of data, capital, and infrastructure. The second type is the *specialized software* models that are enabled by AI frontier foundation models and fulfill a wide variety of specific needs and tasks (AI apps). The industry that develops these apps is, normally, fragmented because of the huge variety of needs and applications. New competitors in this field continually emerge and outcompete established ones.

The full unfolding of the AI potential in the transport and mobility sector is expected to have been established by 2050 provided that several key factors are established in the formative years 2025–2035, namely:

a. A comprehensive regulatory and legislative framework (see also "Legal—Regulatory Factors").
b. A balanced distribution of power among the developers of the frontier foundation models with a "healthy" competition among them that defines an open and competitive AI market.
c. Existence of adequate capacity of the semiconductor industry to supply AI chips.
d. Existence of adequate electrical power that will be needed for AI development and deployment, and
e. Affordable pricing models and sustainable economics for all new AI based mobility applications.

There is no doubt that AI-enabled opportunities in the field of transport and mobility will be of a transformative, revolutionary, nature. They can change entirely the way we move and/or transport our goods and will become multibillion (or even, multitrillion) revenue generating tools, something that will further spring their development and expansion. Our estimate is that AI applications in the transport and mobility sector will develop at a rapid pace and produce significant change as soon as 2030. They will then continue with significant potential for growth and dynamism up to 2050 and beyond.

10.3.7 Freight Transport and Urban Distribution Technologies

The freight transport and urban distribution sector has the potential to employ several key new technologies which are currently being researched and tested but which will require more time to mature and become everyday practice. The ones presenting the most realistic perspectives are:

1. Platooning with semi- or fully- automated trucks.
2. Fully automated trucks of level 4 or 5 operating between freight hubs.
3. AI enabled fully automated planning and management of intermodal freight trips.
4. Automated logistics hubs (warehouse automation and digitization technologies).
5. Automated local delivery light vehicles.
6. Side-walk package delivery robo-vehicles.
7. Use of clean fuels (urban or interurban freight transport).
8. Underground high-speed vehicles (Hyperloop type).

10.3.7.1 Platoons of Semi- or Fully-Automated Trucks

For several years now all major truck manufacturers in the US, Europe, Japan, and China have been testing and developing platooning technology. Truck platooning research using several technologies and integrating V2V communication has produced various advanced driver-assistance systems already currently in use (including cooperative adaptive cruise control, lane departure-assist and active brake-assist). These are already on sight on major motorways in Europe, Australia and the US.

The technology of semi-automated truck platoons is expected to expand rapidly in the long-haul road freight business in the next few years. The most relevant cases of implementation of such technologies are primarily to be seen in the US (East–West long haul freight transport corridors), Europe (North–South, East–West corridors), Australia, and China (the Eastern North–South corridors).

10.3.7.2 Fully Automated Trucks (Level 4 or 5)

Level 4 automated truck technologies, which enable automated driving on specific routes without requiring the driver's intervention, are the focus of research since 2015 or even before. Level 4 automated trucks, running between freight consolidation/distribution hubs (see below Sect. 10.3.7.4), are expected to materialize, in terms of wider implementation as early as 2030 with full maturity (level 5) expected by 2040 and beyond.

10.3.7.3 AI Enabled Fully Automated Planning and Management of Intermodal Freight Transport

Full and unimpeded use of intermodal freight transport services has been a target of transport policies since the early '80s. Current technologies, enabled by new AI digital tools and 5G or 6G connectivity combined with new automation and robotics technologies, make this goal closer to materialise in the medium turn i.e. by early or mid-2030's. Fully automated planning and management of intermodal freight transport will be done by professional teams powered by strong AI tools. These will be the *Intermodal Capacity Brokers* (ICBs) which will be specialised firms that offer one-stop-shop services in finding the fastest, cheapest and most sustainable routes and the best suppliers of transport services in the market. Finding the best routes, getting price lists and other data, will be assessed and imported automatically by AI to the digital intermodal freight platforms that offer the user easy access to the intermodal market with a transparent view on available capacities in large networks of trusted carriers and transport operators. They will also provide real-time data for better planning and reducing manual work, offer ETA (Estimated Time of Arrival) calculations and cancellations automatically, and provide issuance of the invoices and handling the financial transactions. They will enable on-line and real-time communication and collaboration between all players in the freight transportation chain from the very beginning of a freight haul (i.e. the offer) till its final stage (invoice issuance and handling).

10.3.7.4 Automated Logistics Hubs

Automated Logistics Hubs (ALB) will contain warehouses, trucking and shipping services, staff offices and distribution areas, and will perform functions such as semi manufactured goods assembling, labelling, sorting, and others. It will form the crucial interface to enable automated hub-to-hub freight transport operations in anticipation of fully automated transport between hubs. These automated transport and logistics chains will form the crucial interfaces to the future freight and logistics systems serving the megacities. The key elements of an ALB are the fully automated, access control, coupling technologies, yard tractors, loading/unloading operations, charging (invoice and payment), and remote monitoring and control of all these functions. An ALB is split in two sections, one for serving as trailer yard where preloaded trailers will be placed ready to be transported, and an inner yard with all the warehouse operations mentioned in the previous paragraph. The technologies that will enable the fully automated operation of such hubs are currently available and many such hubs operate already in a fully or semi-automatic way. For a detailed description of such hubs, see (Porche Consulting, 2022).

The road to full technical maturity and their integration into an overall system that will materialise the fully automated logistics chains by use of AI is not very long and fully automated logistics hubs are expected to be in implementation and wider use by early 2030s or just after.

10.3.7.5 Automated Local Delivery Light Vehicles

The rapid growth in e-commerce and the need for flexible urban deliveries plus the driver shortage, are making the use of automated local delivery light vehicles cost effective and sustainable. These vehicles are simpler than those carrying passengers, can be of small size (same or smaller than small cars) and don't need passenger protections or safety-of-life design standards for their payloads. Such vehicles are on experimentally on the roads currently, but their safety and certification legislation and standards are likely to delay their full implementation until 2030 with their wider use and maturity foreseen by 2035.

10.3.7.6 Sidewalk Package Delivery Robo-Vehicles

These are very small vehicles, with limited protection, carrying a single payload with very low speed. They move on the sidewalks and are not interacting with other road vehicles, but only with pedestrians. They have lower safety requirements and operate under local regulation and standards. They are economically viable for local delivery of packages in high density areas. Such vehicles can be on the roads currently, but their safety and certification legislation are likely to delay their implementation until 2030.

10.3.7.7 Use of Clean Fuels

According to the most recent International Energy Agency's annual electric vehicle (EV) outlook report, the sales of electric light commercial vehicles (LCV) worldwide in 2022 were 310,000 vehicles (nearly double relative to 2021). This figure is more significant if one considers that the overall LCV sales declined that year by more than 10% due to the COVID pandemic (IEA, 2023). At global level, the electric LCV sales share, is currently around 4% but is increasing, a trend that shows that the electric LCV market is catching up with that of other electric cars. If these trends continue, electric LCVs will gather a substantial percentage of the total number of LCVs as high as 50% by approximately 2040. China leads in terms of overall LCV sales with over 130,000 units sold in 2023, and nearly 15% of the LCVs sold being electric. In terms of global market share, Korea leads the way with 27% of LCV sales being electric (36,000 vehicles) in 2022.

For heavy freight vehicles (trucks) the use of battery electric motors is currently very limited. Worldwide, only 60,000 electric medium- and heavy-duty trucks were sold in 2022, representing about 1.2% of the global truck sales. Of these, only 2000 electric trucks were sold in the European Union with most of the rest sold in Asia. China dominates in production and sales of electric (and fuel cell) trucks (and buses) with 52,000 electric medium- and heavy-duty trucks sold in 2023. This number represents 4% of the total truck sales in China and about 85% of the global e-truck sales. As regards fuel cell electric freight vehicles (FCEFVs), currently only 10% of all FCEVs sold are trucks (IEA, 2023) with China, again, dominating the heavy-duty fuel cell vehicle sales. Except for China, cumulative electric medium- and heavy-duty truck sales, to date, number in the hundreds in most countries with just under 2000 electric trucks sold across the entire European Union in 2022. Global sales of medium- and heavy-duty electric trucks are low (below 1% of total) in regional and long-haul operations. The average declared range of electric trucks does not exceed 300 kms, which compared to the range of electric cars is almost half (IEA, 2023).

The other clean fuel for freight transport that is being researched currently is Hydrogen i.e., trucks with combustion engines that run on hydrogen. These are tested by almost all major truck manufacturers, but their full commercial launch is expected towards 2035. Trucks that run on green hydrogen will represent a significant step forward, but as already mentioned in previous sections the full maturity and penetration of hydrogen-based mobility is expected to take place well into the 2050s. Hydrogen trucks will be especially suitable over long distances and in regions where there is limited charging infrastructure for batteries.

10.3.7.8 Moving Freight Underground (Hyperloop Types of Vehicles)

The idea of moving freight in fixed guideway vehicles moving in underground tunnels was born some 20 years ago with the *Hyperloop* concept which involved very high-speed vehicles running on magnetic induction, or similar high tech supporting system, in narrow-width tunnels. The same type of vehicles and supporting infrastructures can be used for the movement of freight as well as passengers. Proponents of the hyperloop technology claimed that hyperloop systems would be cheaper, cleaner and faster than current long-distance transport and they would generate more energy than they consume. The concept became widely known in 2013, when Elon Musk's hyperloop company released a white paper entitled *Hyperloop Alpha*, which proposed the concept of transporting people in magnetically levitated pods (low capacity, 28–40 people) propelled by linear induction motors at speeds of up to 760 mph (1220 km/h) within an extreme low-pressure (100 pascals) tube environment. The same system would also operate with freight pods which would carry freight containers (see photo). So far, companies working on the technology include Hyperloop One, HTT and The Boring Company. The Hyperloop concept is being in an advanced planning stage, but it is being implemented only in one country so far (Saudi Arabia).

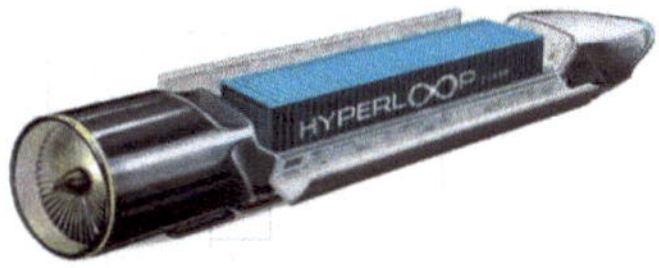

Perhaps the most authoritative assessment of the Hyperloop technology was the one made in 2017 by the UK Science Advisory Council of the UK Department for Transport and a later one by the UK's Transport Research Laboratory (Walker, 2018). According to this later report, whilst Hyperloop offers key benefits in relation to speed and energy, other factors raise doubts over its viability as an alternative transport mode. The claims related to low costs were found unsubstantiated by the evidence and the cost of construction may undermine any economic benefit. According to the same report only real-world trials will prove the validity and viability of the concept as, in addition to the economic and engineering issues, for passenger Hyperloops there are significant barriers to overcome regarding human factors including aspects such as physical and mental passenger comfort, noise, vibration, motion sickness, accelerating and decelerating forces, etc. (Walker, 2018).

We can only conclude that commercial implementation of Hyperloop type of technologies is far from certain and certainly not short term. Their possible realization to any meaningful extend is placed well beyond 2050 i.e., by the end of the century.

10.3.8 Technology Readiness and Penetration Assessment

Table 10.1 shows estimates of the *Technology Readiness Levels*—TRL (a scale used to assess the maturity of a particular technology) for the transport technologies examined above and summarises our expectations for their penetration/implementation potential, by the year 2050.

10.4 Environmental and Climate Change Factors

10.4.1 The Current (Alarming) Trends

Environmental and climate factors include issues regarding the quality of the environment in terms of air or noise pollution, the degree of decarbonisation achieved, the degree of adaptation of national or local infrastructure to climate change impacts, the governmental policies for the protection of the environment (including issues like waste disposal and noise control), creating a cyclical economy, and other. The current environmental situation in most countries of the world—especially the developing world—is characterised by a continuous degradation. Climate change and related extreme weather phenomena impact the coastal and inland areas alike causing severe damage to infrastructures, disruption of economic and social life, and many deaths. Although ambitious policies have been announced by many world governments as early as 2020 for climate change mitigation, their implementation is delayed or postponed due to various reasons some of which have been discussed in earlier sections. As a result, the UN's Environment Programme (UNEP) is sounding the alarm repeatedly. In the COP29 meeting that took place in Baku, Azerbaijan in December 2024 the inactivity and lack of progress towards decarbonisation was amply evident in the statements of UN officials and scientists. The title of the UN's *Emissions Gap Report* of 2023 is quite indicative of its desperation: "*Broken record—Temperatures hit new highs, yet world fails to cut emissions (again)*" (UNEP, 2023).

Besides the necessary mitigation measures aimed at withholding climate change within "benign" limits, world governments should already be adapting their vital infrastructures (among them the transport and mobility related ones) to be resilient to climate change impacts. Adaptation to climate change requires two forms of response: dealing with the long-term effects on the infrastructure such as rising sea levels and developing resilience to avoid interruptions of basic services during acute and extreme weather events. In the EU region, as already was detailed in Chap. 3, the European Environment Agency (EEA) estimated the minimum cost of not adapting to climate change at €100 billion per year in the early period (the 2020–2030s) and €250 billion per year in the 2050s and beyond, for the whole of the EU. To the same objectives, the European Economic Area (EEA) countries have developed and operate, since 2012, the European Climate Adaptation Platform, *Climate-ADAPT* (http://climate-adapt.eea.europa.eu/). This platform provides resources to support adaptation policies and decision-making and produce a toolset for adaptation planning, a database of projects and case studies, and information on adaptation actions at all administrative levels. The recently published *European Climate Risk Assessment* (EUCRA) report stressed the need to scale up adaptation measures and actions across all policy sectors and governance levels to address escalating climate risks like extreme heat, drought, wildfires and flooding (EEA, 2024). In China, climate change adaptation efforts are high on the government's agenda as explained in the Chinese *National Strategy for Climate Change Adaptation 2035* (see Chap. 3). Adaptation in China is named as "ecological civilization", or "ecological security" and its goals are "enhancing sustainable harmony between humans and the natural environment".

Based on the estimates by the International Monetary Fund (IMF) on the *adaptive capacity* of more than 100 countries worldwide, shown in Fig. 10.5, we have drawn the curves in Fig. 10.11. This Figure shows the estimated timelines (trends in time) of the potential *adaptation degree* achieved, i.e. the level of climate change adaptation that is expected to be achieved at a given time expressed as an index from 0 (no adaptation) to 1 (full adaptation). As shown in Fig. 10.11, none of the 4 regions examined is expected to reach an 80% adaptation level (index 0.80) by 2050. The 0.8 index is expected to be reached by 2060+ for EU and China and this should be considered as a satisfactory level of adaptation. An 80% adaptation level is indicating an almost 80% completion of the totality of possible adaptation measures for all infrastructures in the transport and mobility sector for the majority of the expected extreme weather events. Achieving more than an 80% level of adaptation is considered virtually impossible both technically as well as financially because the costs would rise exponentially. The 80% level is expected to be reached by the US in 2080–2090, on account of its current policies as regards climate change that will delay actions for many years or even decades, while for Africa and the rest of the world beyond 2090.

Whatever the governmental intentions are, the reality is that successive external events of geopolitical and/or economic nature will invariably delay or postpone the application of the announced mitigation or adaptation policies. Our basic assumption, expressed as *Conditionality 3*, is that when governments and their societies realise the full impacts and implications of the climate crisis, they will gradually give priority to actions implementing mitigation and adaptation policies and to international cooperation to that effect. However, this is expected to happen gradually and sometimes toward the middle of the century probably the decade 2040–2050. So far, the evidence is overwhelmingly pointing to the fact that humanity is neither doing enough to stop climate change nor is it adapting its infrastructures to withstand the adverse climate crisis impacts. This is hopefully expected to start taking place systematically, by 2040 and thereon. For many cases this change of policies may come too late but there seems to be very little that can be done to change the priorities and focus of world politics currently!

10.4.2 2050—a Full Blown Climate Crisis

If the current trends and indications continue unchanged, as it now seems to be the case, the climate change phenomenon that has already started showing its ugly face will, by 2050, be transformed into a full-blown crisis. Extreme weather-related phenomena (typhoons, tornados, extreme heavy rainfall, extreme heat waves, desertification and similar) will be hitting most countries of the world with a frequency of a few years or in some areas even yearly. Global warming will be endemic with the average overall increase in global temperatures above the + 2 °C UN limit, and close to + 3 °C as compared to pre-industrial levels. Urban and rural areas will experience longer and more frequent extreme heat waves or flooding. The extended heatwaves will result in many negative impacts in many sectors including of course the mobility sector. The most widely felt impact will be the manyfold increase of energy demands for cooling. It is very unlikely that these cooling needs will be met adequately through passive cooling measures (planting of trees, increased insulation and ventilation of buildings, etc.), so most people will opt for mechanical methods of cooling (air conditioners). The resulting high consumption of energy for the air conditioners as well as their use of freon for cooling will give rise to further global warming. The mitigation–adaptation trade-offs of using air conditioners are analysed in (Colelli et al., 2023).

Besides heatwaves, larger and more frequent flooding from sea rise or terrestrial rains will result in putting, permanently in the first case or frequently in the second, under water large parts of the urban and rural areas. The impacts on infrastructures and systems will differ by geographic location, with the low-income countries suffering the most. In all cases there will be immense negative consequences for people's lives and livelihoods, as well as for the national or local economies, infrastructures and public health systems (IPCC, 2014; WMO, 2024; WRI, 2024a, 2024b). The low-income countries will be suffering the most and the forecasted (in 2021) situation in which about 50 major cities around the world will be in need to construct extensive defences against sea rise or lose, under the sea, significant parts of their populated areas and infrastructures, will have come true and largely so for low-income coastal or island countries (Strauss et al., 2021). The climate crisis will be impacting all urban areas of the world making it painfully visible to governments and societies that they will have to pay a high price for the delays and non-implementation of the needed mitigation and adaptation actions of the past decades. As it was described in (Wong et al., 2024), based on data from 996 cities around the world with populations greater than 500,000 people, the impacts from the climate crises can be categorised by income levels. For low-income countries climate related extreme events would take a greater toll on public health and social and economic terms than for higher income countries.

It is hoped that our conditionality 3 will prevail and global cooperation and financing will be enacted fast enough to make it possible to implement the necessary climate adaptation measures that will mitigate or even eliminate the doomsday situations described above. The governments of the most developed and polluting countries will hopefully commit the necessary resources to climate change mitigation and adaptation actions helping also the less developed nations both financially and technically to face the impacts of climate change phenomena.

10.4.3 Energy Related Issues

The energy sector merits special mention both because it is the major force behind economic and social development in human societies at all levels, but also because the energy producing sector is (currently) one of the most polluting sectors of human activities. The International Energy Agency's (IEA) flagship report *World Energy Outlook* that is published every year, one of the most authoritative global sources of energy analysis and projections, identifies and explores the trends in energy demand and supply, as well as what they mean for energy security, emissions and economic development. The World Energy Outlook report for 2024, identified a steady rise of clean energy technologies such as solar, wind, geothermal and others in most parts of the world but primarily in the EU and China. Earlier versions of the same report, for the years 2022 and 2023, were very optimistic and put forward expected trends for the global energy producing system that by 2030 foresaw:

- Almost 10 times as many electric cars on the road worldwide as compared to 2023.

- Solar photovoltaic panels generating more electricity than the entire US power system did in 2023.
- A share of renewables in the global electricity mix nearing 50%.
- Heat pumps and other electric heating systems outselling fossil fuel boilers globally; and
- Three times as much investment going into new offshore wind projects than into new coal- and gas-fired power plants.

All this, in 2025, sounds too optimistic! Although the transition to clean energy seems to become a generally accepted principle with governments, companies and investors having aligned for many years behind clean energy policies, the developments after the recent US elections, make the achievement of the IEA reports' estimates highly unlikely. Following the economic development outlook for the main regions of the world, as analysed in the previous sections, we can estimate the time for achieving 100% clean energy by region as shown in Fig. 10.12. As shown in that Figure, Africa is the region expected to reach 100% renewables the latest (circa 2080+) and the EU the earliest (circa 2040 to '45). China is expected to reach the 100% level ten years later than the EU (circa 2055–2060). India and the US even later, circa 2060 or '65 and for the rest of the world, the 100% mark is expected to be reached by the year 2080+.

10.4.4 Legal—Regulatory Factors

Within the overall legal frame of a country or region, the so-called regulatory factors are those related to the regulations that specify how a specific technology is developed, certified, licenced, and used. They include, for example, the technical standards that are put forward regarding the operation of a specific technology or mobility system, the international conventions that specify procedures and national certification obligations, the market mechanisms that apply, and so on. Legal and regulatory factors increase public trust in a new technology and seek to address mainly the safety concerns of the users. In this respect they are crucial in fostering public trust to a new technology through imposing rigorous testing and transparent legal and regulatory safeguards for safety. Such legal and regulatory safeguards are currently gradually being introduced in many countries regarding employment, consumer laws, health and safety, the application of Artificial Intelligence technologies, trade regulations and many other aspects of human activities.

There are many areas of the future mobility landscape that need legal and/or regulatory intervention before they can be fully developed and flourish. All new technologies that were presented in previous sections will need to be fully regulated to develop the necessary public trust to them and several initiatives to this effect already exist. For example, for automated guided vehicles within enclosed private areas there are norms such as the Machinery Directive 2006/42/EC of the European Commission that regulate their operation. Several EU member countries have their

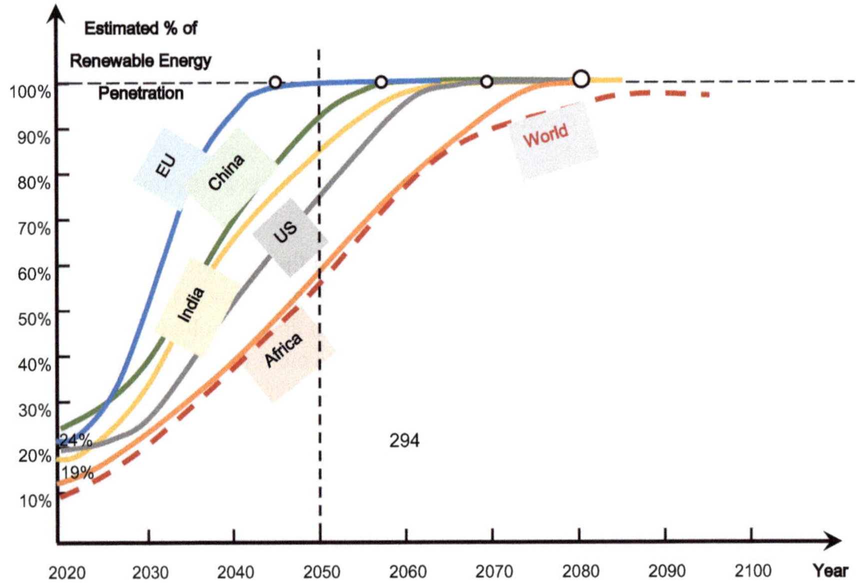

Fig. 10.12 Estimated curves of the % penetration of renewables in the energy mix, by region. *Note* For 2020, the figures are based on the "Renewables 2022—Analysis and forecast to 2027" as well as the "Renewables 2024" (9th October 2024) reports of the International Energy Agency (IEA). The estimated trends shown in this figure have considered: the China's 14th Five-Year Plan for National Economic and Social Development (https://doi.org/10.22617/BRF210192-2), the REPowerEU plan of the EU, and the new US policies for climate change after the elections of November 2024

own regimes for regulation of automated vehicles in accordance with the Vienna Convention and the Geneva Convection on Road traffic. In 2021, Germany was the first European country to introduce a regulatory framework that allows for SAE level 4 and 5 vehicles to circulate on its road network. In July 2022, a European-wide harmonized type approval that allows for the approval of up to 1500 automated vehicles per manufacturer each year, was introduced as part of a larger update on the EU's General Safety Regulation (GSR).

Clear and comprehensive legal frameworks are not only needed to ensure safety and define liability in case of accidents, but also to provide a plain, transparent, and sustainable competition environment for the industrial community who will strive to introduce, promote, and maintain a new technology. Creating the most appropriate legal and regulatory environment for each new technology is always a major task that requires time and involvement of all relevant stakeholders at national and international level. Harmonising such legislation and/or regulatory environment between different countries within a region, or even internationally, is also necessary. This is a tedious and time-consuming work, but it is of vital importance in ensuring the safety of use as well as the viability of the required investments for the new technologies.

Artificial Intelligence (AI), Automated Mobility (AM), and Urban Air Mobility (UAM) are three new technologies for whom a comprehensive and globally harmonised legal and regulatory environment is not yet in place. Without it, public safety and trust in these new technologies will be low and so will be the chances of an early adoption and sustainable operation. For AM and UAM the legal framework is still in the making a task that is undertaken by all countries that aspire to use these new technologies competitively and at global level. In the case of UAM (electric aircraft taking off and landing vertically), as well as in the case of automated mobility (AM) a comprehensive legal framework that will cover all aspects of their operations and ensure safety and security, is even less complete. There are only some temporary legal provisions in a small number of countries that are put in place mainly to provide a legal basis for the demos and test trials of these technologies that take place in well specified areas in these countries.

In the case of AI, the first comprehensive legal framework to be put in place was the *AI act* of the EU (EC, 2024). This is a Council Regulation that regulates the production and use of AI in the EU. It was put in force in August 2024 and provides many rules for the production and use of AI applications. In its 12 Chapters and 13 Annexes, it distinguishes the AI applications in four categories according to the risk estimated to arise from their use and tries to safeguard human rights (of users) and to ensure that the whole AI business is performed with transparency. Of interest, are the AI systems that are prohibited by the *AI act* as having unacceptable risk. According to Chapter II, Article 5 of this Act, the following AI systems are prohibited to users located in the EU, and to third country users if the AI system's output is used in the EU:

a. Systems deploying subliminal, manipulative, or deceptive techniques to distort behaviour and impair informed decision-making, causing significant harm.
b. Systems exploiting vulnerabilities related to age, disability, or socio-economic circumstances to distort behaviour, causing significant harm.
c. Biometric categorisation systems inferring sensitive attributes (race, political opinions, trade union membership, religious or philosophical beliefs, sex life, or sexual orientation), except labelling or filtering of lawfully acquired biometric datasets or when law enforcement categorises biometric data.
d. Social scoring systems, i.e., evaluating or classifying individuals or groups of individuals based on social behaviour or personal traits, causing detrimental or unfavourable treatment of those people.
e. Systems assessing the risk of an individual committing criminal offenses solely based on profiling or personality traits, except when used to augment human assessments based on objective, verifiable facts directly linked to criminal activity.
f. Systems compiling facial recognition databases by untargeted scraping of facial images from the internet or CCTV footage.
g. Systems inferring emotions in workplaces or educational institutions, except for medical or safety reasons.

h. Systems providing 'real-time' remote biometric identification (RBI) in publicly accessible spaces for law enforcement, except when:

- searching for missing persons, abduction victims, and people who have been victims of human traffic or sexually exploited.
- preventing substantial and imminent threat to life, or foreseeable terrorist attack; or
- identifying suspects in serious crimes (e.g., murder, rape, armed robbery, narcotic and illegal weapons trafficking, organised crime, and environmental crime, etc.).

This is the first of similar legislative frameworks that are expected to be formulated within the next few years by interested countries. Our estimate is that for the world to achieve the necessary horizontal (in global or regional scale) harmonisation of rules and regulations concerning AI and to apply them, it will take at least one decade more i.e. until 2035 or even 2040. Nevertheless, an early adoption of AI related legal frameworks at national level can be foreseen in the short term (e.g., by 2027 or 2030) because there is a competitive advantage to being an early adopter and possibly leader in a new technology. Any comprehensive attempt to create or complete the legal framework regulating the new technology landscapes in mobility should be coupled with measures to safeguard the proper functioning of the relevant markets.

10.5 2050—An Urban Mobility Vision

10.5.1 General Outlook

The global mobility outlook for 2050 will be one of diversity. It will be characterised by *"islands" of transformational change* and *"seas" of transition.* The islands of transformational change will be the countries or urban areas where the influencing factors discussed earlier will have created the conditions for most of the novel mobility technologies and concepts to be implemented and flourish. Protected from climate change adverse impacts and consequences through strong adaptation and high resilience measures, these areas will have mastered the necessary funds and political will as well as social backing to implement radical, "revolutionary", changes to their mobility systems as compared to today's standards. Private car ownership and use will have been reduced to almost zero in favour of well-planned and affordable mobility services offered by mobility providing companies using shared or other forms of public transport services. Automated transport vehicles of all kinds will be transporting groups of travellers as well as freight, in AI controlled trips in AI controlled transport networks. They will move people from A to B with maximum efficiency and environmental sustainability while nearly 100% of all transport will have been decarbonised on a life cycle level. Unfortunately, these *"islands"* of transformational change will be realised in a small number of urban areas in the form of smart digital megacities in countries with advanced economies who will be eager to demonstrate the new mobility technologies to the rest of the world. Our estimate is that by 2050 there will be no more than 10 such smart megacities globally with most of them located in Asia.

For all other areas (urban or interurban) mobility systems and services will be in a transitional phase moving from a level close to their current conventional transport services toward the AI-controlled, clean, and automated mobility level of the transformational "islands". The degree of adaptation to climate change of these areas will differ and their resilience and protection from climate change extreme events will be varied. The degree of progress within these transitional "seas" will range and a multitude of different mobility landscapes will be seen across the globe ranging from cases of well organised and smoothly running transport networks but using mainly conventional and partly non-conventional transport vehicles and systems, to chaotic and congested situations still running on conventional vehicles and transport systems and technologies very much like we know today. AI will be recognised and used to differing degrees as the relevant tool of choice but the level of development of the necessary infrastructures for data collection and communication will result in differing degrees of AI adoption and implementation. So, the progress and level of advancement in mobility related areas will not be a universal or uniformly seen situation in 2050. It will most probably be a mixture of mobility "scenes" and situations that will result from the differences in developmental and economic growth potential in each case (city, country or geographic region) as well as from the different degree of adaptation to climate change.

We further elaborate on this vision of future mobility focussing first on urban areas and then on interurban and rural ones.

10.5.2 Three Types of Urban Areas

The great "battlegrounds" of change in 2050 will be the urban areas where more than 75% of the earth's population will be living. The most populous urban areas will be emerging in Africa and more generally the developing world. According to a *Futures Platform* forecast (Futures, 2024), by 2100 the world's 20 largest megacities will be in the developing world, with the majority in Africa.[4] At the projected + 2 °C, or more, global warming levels by 2050 and in the absence of substantial adaptation, the large urban populations will face long heat waves, severe flooding episodes, skyrocketing energy demand (for air conditioning, etc.), a shifting risk for insect-borne diseases all of them, sometimes, simultaneously. Mobility services and infrastructures will

[4] It is worth mentioning here, that according to one of the scenarios in (Futures, 2024), by 2040 the once-dominant economic centres of the developed world will have been steadily losing their competitive advantage to the emerging megacities of the developing world. By 2050 most of the developed megacities will have outpaced the world's largest national economies, in terms of economic output, emerging as global centres of economic activity and political power with some of them even declaring themselves as independent states by 2060!

be hard hit under these circumstances. It is useful to distinguish different levels of urban areas according to their degree of infrastructural development, their level of adaptation to climate change, and their implementation/utilisation of the new mobility technologies. The following three categories of urban areas, called *urban scenes or levels*, are used in our futures thinking exercise:

Urban scene—Level 1: Higher income and more technologically advanced countries will have managed—by 2050—to adapt their urban infrastructures to face the adversities of climate change and radically transform their urban mobility to use automated vehicles moving on land, air or underground. These areas include the ultra-developed cases mentioned before as *"islands" of transformational change* but also other highly advanced cases. Land automated vehicles, which by 2050 will have overcome most of their initial technical and regulatory problems, will be extensively used in these urban areas. AI technologies for travel demand management, traffic management, and other mobility applications will have been highly developed and used to create new and more reliable mobility services that are both efficient and socially accepted. Highly developed infrastructures in all sectors, following the smart city model, will have been developed within the concept of a balanced urban growth and all urban infrastructures will conform to sustainability norms. They will make a solid background for the implementation of all new mobility technologies that will by then have been developed and matured as described in previous sections.

Self-driving cars and shared mobility schemes will be integrated into the urban lifestyle, reducing considerably the total number of vehicles on the urban roads. These urban areas will be pioneers in embracing the circular economy, drastically reducing waste and urban pollution making rubbish or garbage a most valuable resource. Everything will be treated as a resource, and recycling will be made the cornerstone of urban life. The characteristic elements for this type of "smart" urban areas can be summarised as being the following:

- Comfortable, accessible, and high-quality living environment that is green and safe with clean air that invites people to spend time outdoors.
- Healthy lifestyles with sufficient activities at all society levels.
- Human scale urban planning that has achieved to a considerable degree that daily employment and other needs are located at nearby places of residence.
- Full use of renewable energy, most of which is produced within the city boundaries in full compliance to the "smart city" concept.
- Electric mobility applied almost universally for passenger as well as freight vehicles.
- Development of a medium level of infrastructures for supply of hydrogen to niche vehicles.
- Fully decarbonised transport systems as well as decarbonised practices in other sectors (buildings, industry).
- Considerable adaptation against extreme weather events (high resilience).
- Car ownership reduced to very low levels in favour of high-level public transport and shared mobility services.

- Public transport services operating on a fully integrated way between modes and infrastructures providing customers with door-to-door trip making capability.
- Shared mobility services offered everywhere, easy to use and fully operating under AI software (making optimised combinations of passengers and routes).
- Mobility-as-a-Service packages are offered at reasonable cost and high-quality levels by specialised mobility broker companies some of which are car manufacturers that gradually transformed themselves into mobility service providers.
- Many other "smart city" services are implemented offering to the urban resident services with high degree of automation, electronic documentation, remote service provision, and so on (for a smart city list of functions see, Bhatnagar, 2024).

Urban scene—Level 2: Urban transport infrastructures growth and sustainability is still incomplete and uneven. The urban fabric is characterised by uneven development with areas of urban decay and other areas which are well developed and grown. The implementation of new mobility and AI technologies is not uniform but done on a piecemeal basis with the necessary infrastructures still lagging those of level 1 urban areas and causing bottlenecks and increased vulnerability to climate crises. There are several elements that can be termed "highly developed" and using the new technologies mentioned in earlier sections, but they are not integrated adequately with the rest of the mobility systems and services which are of the conventional type.

The following are some, indicative, mobility and other characteristic elements of this type of urban areas:

- Comfortable, accessible, and good-quality living environment on parts of the urban area.
- Uneven development and inequalities exist in the social fabric.
- Renewable energy is being used widely but not totally and is produced outside the city.
- Electric mobility applied almost universally in passenger vehicles but still not so for freight ones.
- Full decarbonisation is still not there but with close cooperation between all stakeholders and firm application of the decarbonisation policies that have been put forward, it can be attained in the short term.
- Adaptation against climate crises attained only in the more affluent parts of the city (medium level resilience).
- Car ownership reduced considerably but not drastically.
- Public transport services have only partly developed the necessary infrastructures for seamless integration of modes (transfer hubs) while trying to provide customers with efficient door-to-door trip making capability.
- Mobility-as-a-Service and shared mobility is offered but not at a universal basis nor at high levels of quality. As a result, it is not attractive enough to fully reduce car ownership and use.

Urban scene—Level 3: Economic or political instability and reduced development at national or regional level, result in poor urban infrastructures and mobility services. Sudden increases in the urban population that occurred during previous decades (internal or external migration) has resulted in large urban slams in the periphery of these cities and the resulting poor social and physical infrastructures are initiating high crime rates and low liveability. A trend for outward emigration and depopulation towards other urban areas in the same country or abroad is continuously taking place but its size is compensated by equal or stronger immigration flows due to climate change crises. Mobility is served through mostly traditional means with slow introduction and operation of new mobility technologies and services.

Level 3 urban areas face a multitude of challenges such as congestion caused by the influx of rural populations and high birth rates, relatively small progress in decarbonisation and adaptation, and low levels of "liveability" due to high levels of pollution, spread of diseases, water shortages and escalating crime rates. The urban societies are highly divided, unequal and suppressed. The characteristic elements for this type of urban areas can be summarised as follows:

- Low to medium level quality of life with high inequalities.
- Large areas of slums with low-income people and new immigrants.
- Limited liveability.
- Limited outdoor activities.
- Medium to low accessibility to efficient mobility services.
- Conventional public transport services of low service quality only occasionally using AI at a limited scale mainly due to low capabilities for data collection and analysis.
- Low adaptation to climate change with large parts of the urban area vulnerable to flooding and other climate related impacts.
- Renewable energy and decarbonization policies of low to medium level of penetration.
- Electric mobility applied for public transport and other niche markets but still not generally applied in private cars and light trucks.
- Mobility services of no uniform quality with service levels varying by city sector.
- Decarbonisation not yet complete especially in housing and the industrial sector with relatively limited use of renewable energy.

10.5.3 Penetration and Development of Novel Transport Technologies

Automated ground and air transport technologies will have reached, by 2050, high levels of development. They will, by then, have been established as the new "norm" in urban and interurban transport but they will have mainly been applied mainly in Level 1 urban areas. Electric mobility will reach high levels of maturity, and a dense network of charging stations will exist in most countries. Hydrogen-based mobility (as technology) will have been making inroads with safe and secure storage and delivery networks but still lacking in filling stations. Internal combustion engines will have made large improvements in terms of emissions and performance, but their production and use is very likely to have been banned in many countries.

Table 10.2 shows the estimated level of technological development as well as the estimated penetration rates for the horizon of 2050+ for all the mobility technologies that were presented in the previous section, by geographic region.

10.5.4 Degree of Adaptation of Transport Infrastructures

The development of transport infrastructures that are more resilient to climate change is a necessary precondition for the future climate-friendly cooperative, connected and automated mobility operations (CCAM). By 2050, China and the EU are expected to have raised the resilience of their transport infrastructures to a considerable extent approximately at the level of 60% while they are expected to reach the 80% level in more than a decade later i.e., in 2060+ (see Fig. 10.12). It is reminded that an x% level of adaptation is meant here as the state in which approximately x% of all infrastructures are protected against a high level of the expected major climate change events. For the US this figure, is expected to have reached by 2050 the level of approximately 45%. For the lesser developed countries, the level of 60% adaptation is likely to be reached some 20–30 years later, i.e. towards 2070 or 2080 and on the assumption that international adaptation aid will be secured with the participation of the economically more advanced economies. As shown in Fig. 10.12, none of the countries are expected to have their infrastructures fully adapted (i.e., to a degree of 100%). A maximum level of 80% adaptation is assumed to be achieved for all countries as a "ceiling" beyond which it will not be feasible to go for financial as well as organisational and technical limitation reasons.

As an indication of what is meant by an 80% level of adaptation, for transport infrastructures, we can mention the following indicative features:

- Bridges that can resist flooding of a 1 in 200 years frequency level.
- Roads, railway lines, and airports (by the sea) that can function adequately and with no disruption with a rise in sea levels of + 1 m.
- Transport services that are capable of withstanding major floods.
- Urban areas that can function adequately in extreme heat events or that can refunction adequately within 24–48 h after a cataclysmic rain and flooding event of a 1 in 100 years frequency level.
- Power lines that are all, or to more than 90%, underground.
- Battery charging stations functioning at all weather conditions and 100% powered by renewable electricity.

Table 10.2 Estimated technology development and level of penetration of the new technologies in the four regions, by 2050+

Name of technology	Development level of the technology by 2050* and TRL	Estimated penetration levels** (%)			
		Asia (China, Japan, S. Korea)	North America (US, Canada)	EU	Africa
Electric mobility	100%/9	80	60	90	30
Hydrogen-based mobility	75%/7	40	40	30	20
Automated veh. up to level 3	100%/9	100	100	100	80
Automated vehicles level 4 and 5	90%/9	45	40	40	20
Urban Air Mobility (VTOL vehicles for urban transport)	80%/8	60	50	50	30
Underground freight (Hyperloop)	70%/7	20	40	30	0
New business and organisational models for shared mobility (***)	85%/9	80	80	90	30
AI applications in human mobility and freight	85%/9	80	85	90	40
Freight transport and urban distribution					
Platooning with semi or fully automated trucks	90%/9	90	90	90	70
Fully automated trucks (level 4 or 5) running between hubs and distribution centres	70%/9	70	70	70	30
AI enabled planning and management of intermodal freight transport	80%/9	80	80	80	40
Automated logistics hubs	90%/9	80	85	80	30
Automated local delivery light vehicles	90%/9	80	80	80	20
Side-walk package delivery robo-vehicles	90%/9	80	80	80	20

(continued)

Table 10.2 (continued)

Name of technology	Development level of the technology by 2050* and TRL	Estimated penetration levels** (%)			
		Asia (China, Japan, S. Korea)	North America (US, Canada)	EU	Africa
Use of clean fuels (urban or interurban freight transport)	90%/9	90	70	90	30
Moving freight underground (Hyperloop type)	70%/7	10	20	10	0

Source Table 10.1 (for TRLs), and authors' estimates
Notes *Expressed as a percentage with Level 0% indicating no development, 50% indicating halfway to full development and maturity, and 100% full development and maturity. The TRL's are also shown as an added measure of technology maturity and are taken from Table 10.1. **Given as a percentage from level 0 (no penetration), to 50% (technology applied to an average 50% of the total potential capacity of applications), to 100% (full application of the technology). The penetration figures for each region are estimates of an average overall penetration level that reflects the differentiated ability of the countries in the corresponding region to apply the new technology. ***Car manufacturers are assumed to have gradually transformed themselves into *mobility service providers* that provide both the automated vehicles and a variety of services using them in the mobility ecosystem.

10.5.5 2050 Urban Mobility Scenes

The anticipated 2050 situation of urban mobility in the three categories of urban areas (Levels 1, 2, 3) is described in Tables 10.3 and 10.4. This is done in terms of some *key operational features* of the urban mobility system which are the following:

As regards the *infrastructures*:

- *Types of networks and transport modes and V2X infrastructures*
- *Degree of adaptation to climate change.*

As regards the *services:*

- *Shared Mobility offered as a service*
- *Accessibility, convenience and affordability of the urban mobility system*
- *Personalised travel advice*
- *Smart travel demand management*
- *Cooperative connected and automated mobility*
- *Smart traffic management*
- *Car ownership and car use levels*
- *Freight transport and urban delivery services.*

The lists of key characteristics in each of the above elements, are not (and cannot be) exhaustive. They are meant to be indicative only, aimed at giving the reader a general overall picture of the expected situation for each of the three levels of urban scenes mentioned earlier.

Table 10.3 Urban mobility 2050—**Infrastructures**

	Urban scene—level 1	Urban scene—level 2	Urban scene—level 3
Transport modes and networks—V2X infrastructure  	• Seamlessly interconnected modal networks with transfer hubs evenly dispersed in space • Poly-centric types of cities and transport networks • V2X connectivity infrastructure installed throughout the city • Urban air transport infrastructure for airborne automated freight vehicles and taxis • Hubs for smooth transitions with regional transport networks • Integrated systems of public transport offering door-to-door services with last mile use of automated vehicles • Logistics and freight transport networks seamlessly integrated into the urban fabric with optimal location of freight distribution centres • Automated light trucks and sidewalk robots, for all urban freight deliveries	• Partly interconnected modal networks with limited coordination of schedules and services • Low to medium quality infrastructure for interchange between urban and interurban/ regional networks • Limited infrastructure for V2X connectivity (mainly V2I) • Urban air network (vertiports) limited to specific land uses (e.g. airports, railway stations, large private companies' offices) • Integrated systems of public transport infrastructures partly offering door-to-door services • Infrastructure for automated vehicles operation in phase of development • Logistics and freight transport network infrastructure not well integrated into the urban fabric	• Poorly interconnected modal transport networks with poor modal transfer infrastructure • V2I connectivity in development in parts of the city • Urban air network (vertiports) not yet developed • Transitions from urban to regional transport networks difficult with poorly located and developed hubs • Public transport networks operate with conventional vehicles and suboptimal modal interconnection • Infrastructure for last-mile services offering door-to-door service not yet in place • Logistics and freight transport networks developed with less-than-optimal location of freight distribution centres and use of conventional vehicles and freight hub infrastructure

(continued)

Table 10.3 (continued)

	Urban scene—level 1	Urban scene—level 2	Urban scene—level 3
Degree of adaptation of transport infrastructures to climate change 	• All transport infrastructure at sea-level protected against sea rise • Flood control and flood protection installations for all critical transport infrastructure • Resilience plans and infrastructure developed for transport and mobility operation in case of extreme weather events • Emergency evacuation routes and plans prepared and in operational readiness • Installation of a foolproof system for information dissemination in cases of extreme weather events and emergencies	• Sea-level transport infrastructure protected only at critical areas (major ports—airports) • Flood protection measures taken at a select number of critical transport infrastructure • Alternative plans for transport and mobility operation in case of extreme weather events planned but not in operational readiness	• Sea-level transport infrastructure not protected against sea rise • Limited adaptation of infrastructure at areas of special interest such as government buildings and hospitals • Flood control and flood protection measures not taken except occasionally and only at very critical infrastructure • Alternative plans for transport and mobility operation in case of extreme weather events do not exist

Table 10.4 Urban mobility 2050—**Mobility and freight transport services**

	Urban scene—level 1	Urban scene—level 2	Urban scene—level 3
Mobility offered as a Service (MaaS) and Shared Mobility  	• Offer of high-quality mobility service packages using automated or human driven electric vehicles, containing a wide range of alternative routes and modes of transport and using AI to suit different personal preferences and lifestyles • Promotion of vehicle-sharing using automated, or human driven electric or hydrogen fuelled vehicles • Ride hailing well developed • The offered MaaS packages provide enjoyable and convenient travel experience able to compete with the privacy and security of the private car • Flexibility and multiple choices offered by different MaaS price packages • The use of AI ensures that the combination of people sharing the same vehicle (in shared mobility packages) have similarity and compatibility in their social, professional, and personal, characteristics • Urban Air Mobility (UAM) is offered as part of MaaS packages	• Limited number of mobility service packages utilising automated as well as conventional vehicles with partial use of AI • Car or van sharing (automated or not) is offered on limited scale • Ride hailing well developed • Public transport door-to-door services partially developed and in experimental stage • MaaS packages are on offer but with small variety and scope due to still relatively underdeveloped mode interchange hubs, modest use of AI, and still relatively high car ownership and use • Small scale use of automated vehicles mainly for public transport and last mile services • Urban Air Mobility (UAM) services at low level of development, not well integrated with the rest of the urban transport services and aimed mainly for accessibility to and from specific land uses	• Mobility as a Service (MaaS) packages not yet offered • Wide use of shared transport services but mainly using conventional human driven vehicles and operated more like the traditional paratransit • Small scale use of automated vehicles mainly on exclusive rights of way and in well designated and controlled areas (e.g. within university campuses) • Shared mobility using AI and automated vehicles in the stage of planning or used on a demo basis

(continued)

Table 10.4 (continued)

	Urban scene—level 1	Urban scene—level 2	Urban scene—level 3
Accessibility, convenience and affordability of the urban mobility system*  	• Cooperative Connected and Automated Mobility - CCAM services available and easily accessible everywhere • Transport hubs provide easy transfers and use of intermodal transport • Easy to use, reliable, and user-friendly electronic reservation and ticketing platforms for all mobility services • Contactless payment of fares in all public transport systems using all means of payment including face recognition • Public transport operators have evolved into mobility service providers offering MaaS packages and door-to-door service using automated vehicles for last-mile service • Fares (or mobility fees) varying according to levels of competition and/or subsidisation* • MaaS packages and restrictions to car ownership make the use of private cars unattractive and undesirable	• CCAM services available but of limited access and high cost • Intermodal transport hubs provide adequate modal transfer services, but not yet available in all parts of the urban area • MaaS packages not fully developed • Electronic reservation and ticket issuance systems available for all transport modes • Payment of fares in all public transport systems using all means of payment but mainly credit or debit cards • Fares (or mobility fees) varying according to levels of competition and/or subsidisation*	• Intermodal transfer hubs of low quality and not in sufficient numbers • Electronic reservation and ticket issuance systems available for certain modes of transport. Ticket collection through mostly conventional means (as today) • Public transport fares affordable, through subsidisation (*) • CCAM used only experimentally in a limited number of clearly designated areas

(continued)

Table 10.4 (continued)

	Urban scene—level 1	Urban scene—level 2	Urban scene—level 3
Personalised travel advice 	• Personalised travel advice is available ubiquitously based on AI platforms using real-time data and adjusted to personal needs and preferences at the time of travel • Smart adjustments to travel advice during the trip, based on real time data • Advice based on cross-modal information and optimisation of carbon footprints and travel times • Advice accessible through a multitude of electronic and conventional means (mobile phones apps, social media, VMS, internet, etc.)	• Personalised travel advice available, via AI apps, using partly real-time and partly historical data • Adjustments to travel advice in real time (during the trip) • Cross-modal information given but largely based on historical data • Advice accessible through electronic and conventional means	• Personalised travel advice occasionally available based on pre-existing, data and information • Travel information services exist but mostly based on historical data • Advice accessible through electronic and conventional means

(continued)

Table 10.4 (continued)

	Urban scene—level 1	Urban scene—level 2	Urban scene—level 3
Smart travel demand management  	• AI assisted travel demand management, based on real-time information and data as well as short-term predictions, by use of AI apps or web-based platforms • Information to the traveller accessible through mobile phones and other electronic means including use of V2X connectivity • AI-assisted ability to balance the travel advice given to individuals with the systems managing traffic flows on the ground • Offer of two or more travel alternatives to choose from • Calculation of carbon footprints for each travel choice, automatic	• Travel demand management advice based partly on real-time data and partly on historical ones • AI use partly based on analysis of historical travel records and occasionally on real time data • Information accessible through mobile phones, and other conventional means • Limited ability to relate the advice given to travellers with the traffic flows on the ground • Ai-assisted short-term predictions possible depending on data availability • Offer of two or more alternatives for travel choices • Calculation of carbon footprints for each travel choice optional	• AI assisted travel demand management not in operation • Advise on travel options available occasionally but mainly connected to public transport operations based on historical data • AI-assisted ability occasional and only if proper data available • No feedback of the advice given to individual travellers to the traffic control systems on the ground • Calculation of carbon footprints occasional

(continued)

Table 10.4 (continued)

	Urban scene—level 1	Urban scene—level 2	Urban scene—level 3
Cooperative connected and automated mobility  	• Cooperative connected and automated mobility (CCAM) in full operation mixed with human driven vehicles • Robotaxis on the ground or air, are the main ride hailing vehicles • Convenient and easy to use systems of last-mile service, using automated vehicles, available by most public transport operators • Seamless transfer hubs, for intermodal trips mainly between public transport and other modes • Safe, convenient, accessible and competitive mobility packages offered providing door-to-door services	• Cooperative connected and automated mobility (CCAM) in its initial stages with vehicles on designated routes and for special types of transport, mainly public transport • Automated vehicles only in full or partial separation from human driven vehicles • Last mile public transport passenger services introduced on an experimental basis using human driven and occasionally automated vehicles • Robotaxis on the ground, are often used together with human driven taxis or ride hailing vehicles	• Cooperative connected and automated mobility (CCAM) not yet in application except on rare cases on well designated routes or areas and for special types of transport • Last mile passenger services, not yet available • Robotaxis and other ride automated hailing vehicles not available

(continued)

Table 10.4 (continued)

	Urban scene—level 1	Urban scene—level 2	Urban scene—level 3
Smart traffic management  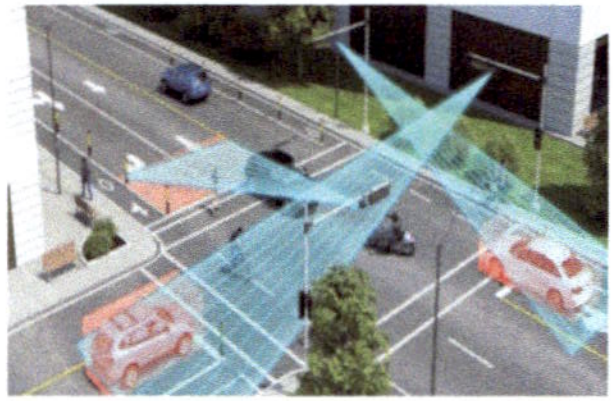 	• Optimisation of traffic for urban road networks through network-wide monitoring and analysis using AI and real time data • V2X connectivity enables high levels of safety and smart coordination of traffic lights • Reliable short-term predictions of traffic situations • Reliable AI-based management of mixed traffic flows (automated and human driven vehicles) • Reliable traffic monitoring and accident recognition video data from cameras with clear colour imaging technology, even in extremely low light conditions, or at night • All traffic management modules and control centres are operated through AI-driven software and real time data from many sources • Use of real-time traffic and incident data to enhance dispatch of emergency teams, or to guide traffic in case of accidents or other disruptions • Urban Air Mobility operating with a network of vertiports and well designated air traffic control centres • Ground and air traffic control coordinated by AI assisted software capable of providing resilience to transport services in case of emergencies	• Traffic management based on real-time cross modal information and data using conventional software • AI assisted short-term predictions of traffic situations based on historical data • Management of mixed traffic flows (automated and human driven vehicles) is done by human controllers (supervisors) • Partial V2X connectivity enables dynamic but suboptimal coordination of traffic lights • Use of real-time traffic and incident data to enhance dispatch of emergency teams at the site of incidents • Partial installation of equipment for traffic violations monitoring and detection dispersed throughout the network	• Coordination of traffic lights and overall traffic management, based on conventional models and algorithms (partly using real-time data) • Short-term predictions of traffic situations based on historical data • V2I connectivity not used for intelligent coordination of traffic lights • Partial installation of equipment for traffic violations monitoring and detection (mainly at major intersections)

(continued)

Table 10.4 (continued)

	Urban scene—level 1	Urban scene—level 2	Urban scene—level 3
Car ownership and car use 	• Car ownership, at an all-time low following a downward trend from previous decades • Daily trip making primarily through shared and automated vehicles on demand • High use of public transport and shared transport services all operating under AI control and using electric vehicles • Mobility as a Service packages provide privacy and convenience comparable to those of the private car • Parking spaces are freed for other urban uses • Purchase of new internal combustion vehicles is prohibited or restricted in favour of electric or hydrogen	• Car ownership low but still exists • Shared mobility is well developed but mainly by conventional vehicles • Use of automated vehicles exists on a limited scale • Mobility as a Service exists but of limited attraction due to yet incomplete networks and limited use of AI • Alternatives to car use, primarily through ride hailing services • Use of internal combustion vehicles restricted in favour of electric vehicles but still permitted in some cases • Public transport vehicles are electric with limited use of AI in schedules setting and operation	• Car ownership and use still quite high • Automated vehicles are not used to any substantial extent • High use of conventional public transport services • Mobility as a Service, offers limited alternatives to car ownership, but privacy and convenience remain considerably lower than those of the private car • High use of ride hailing services of conventional type

(continued)

Table 10.4 (continued)

	Urban scene—level 1	Urban scene—level 2	Urban scene—level 3
Freight transport and urban delivery services  	• Platoons of automated trucks run regularly on the intercity roads • Automated delivery vans and trucks perform deliveries in urban areas controlled by central control centres and linking urban freight consolidation hubs with local freight delivery points • Automated sidewalk package delivery robots, carrying single payload (packages or other), deliver the packages from freight delivery points to the final user • Local package delivery electric vehicles used (that run on the road and not the sidewalk) • All logistics operations performed by AI-driven software that ensures optimised warehouse management, loading/unloading operations, minimisation of empty runs, reduction of delivery mileage, and other • AI-assisted freight distribution management in full use	• Platoons of automated trucks run regularly on the intercity roads • Human driven delivery vans and trucks perform daily deliveries in urban areas • Limited use of automated vehicles for urban freight and last mile deliveries usually in confined and well designated areas • Automated sidewalk package delivery robots, or other automated delivery vehicles not used or used occasionally • Limited use of AI in logistics and urban deliveries management	• Platoons of automated trucks are used occasionally on the intercity roads • Conventional delivery vans and trucks perform daily deliveries in urban areas • Local, urban freight delivery centres used for local deliveries by human driven vehicles • Automated sidewalk package delivery robots, or other automated delivery vehicles not used • AI-driven software for logistics operations, not used • Advanced logistics systems and services hampered by limited data and analysis capabilities • No use of automated vehicles (light trucks) for urban freight transport and/or deliveries

*The financing of public transport services in the future will become a hot subject of debate and deliberation that will develop as the systems become more and more sophisticated and using expensive new tools and technologies (AI, automation, connectivity, etc.). Private enterprises offering mobility services will be using infrastructure developed and paid for by the local or central government at a cost which inevitably will have to be transferred to the users. With the increased costs of developing all the sophisticated infrastructure that will enable the cooperative, connected, and automated mobility (CCAM) services the financial viability of these services will be put in question. It is therefore realistic to expect that on an increasing number of urban areas, of all three levels, the future (2050 or even earlier) may bring increased intervention of local or central governments . This will result in an increasing need for public subsidisation of certain mobility services (especially those related to public transport operation). As the level of such subsidisation increases many urban areas will inevitably adopt the system of **full coverage** of public transport costs along the examples of today's "free public transport services" that are offered in several urban areas (mainly in Europe with the most well-known case being that of Luxembourg).

10.6 2050—Interurban and Rural Mobility Vision

10.6.1 General Outlook

As already mentioned, all current projections point to an increase of urbanisation that will raise urban populations to a total of more than 75% of the global population as early as 2040 or '45. Despite these trends, rural areas will continue to play a significant role by hosting the remaining 25% of global population and eventually acting as an initiator of the opposite movement to urbanisation. The current trends of increasing urbanisation may result at first stage in an increasing desertification and economic stagnation of the interurban and rural areas but by 2040 and beyond, these trends are likely to reverse. Interurban and rural areas are expected to witness a renaissance towards 2050 and may evolve into thriving communities capable of forming attractive alternatives to city life. As the increased urban populations will feel more and more the stress of living on congested, skyscraper-dominated cityscapes and as urban socio-economic inequalities grow, pressures will increase towards enhancing infrastructures in the rural regions to act as an alternative. Remote working (telework) which after the COVID pandemic has gained traction will solidify its place as a lasting trend as early as by 2030 and this will provide an added impetus for taking residence in interurban and rural areas. At the same time, an efficient and ever-expanding intercity high-speed train network will provide easy and affordable connections between rural and urban areas further reducing the "isolation" feelings of the former.

As early as 2035 or '40, systems and infrastructures that provide adequate high-quality electric mobility and full digital connectivity will have been implemented in most low-density rural areas in developed economies and in several developing ones. Innovative mobility solutions tailored to the specific needs of each rural community moving away from traditional urban approaches will be planned and implemented. Contrary to urban populations, rural ones will continue having a sizeable level of car ownership and use (automated or human driven). The emphasis will be on public transport on demand focusing on forging adequately accessible communities in low population density regions. Public transport on demand based on automated buses will offer a basic network of rural public transport services embracing concepts like the "*30-min rural community*" of today which aims to enable rural residents to access essential services and amenities within a 30-min journey. A more sophisticated form of the public transport on demand concept would be the introduction in rural areas of mobility service packages in the frame of a *Rural mobility as a Service*—RmaaS system.

Demand-responsive transport (DRT) systems operating with AI (to collect data and analyse mobility needs in real-time), will be optimising routes and merging the flexibility of automated on-demand vehicles (robo-taxis) while keeping the basic structure of a fixed public transportation system. In the 2050s, aerial rural mobility systems using passenger drones will most probably be included in the mobility offer to rural populations. Air rural mobility platforms will be thriving in the 2050s and

'60s offering easy access to flexible public transportation services in rural areas. In regions with very low density, AI-powered bottom-up mobility initiatives based on volunteering will also be developed.

Sharing a car, will be more popular in rural areas than in urban ones with carpooling being the main type of sharing used. This rural shared mobility will be using an expanded system of strategically placed benches in public locations. A glimpse of such an advanced system of rural shared mobility is provided by the "Mitfahrbänkle" (car-sharing bench) system used currently in Germany. Of similar effectiveness and novelty will be the rural freight delivery system that will connect rural populations to a lean and flexible system that includes package and small goods delivery. By using automated robot-like small vehicles, as well as drones delivering urgently needed goods or medical supplies, freight and package delivery will be accomplished in a most efficient way. The inevitable diversity of demands will result in a combination of solutions tailored to the needs of each rural community with the help of AI and the active involvement of citizens (a citizen-centred approach).

All these novel transportation options will greatly increase the attractiveness of rural areas in developed economies so that, by 2050 or even earlier, the trends towards more urbanisation are expected to stop and even be reversed. Interurban and rural areas in these countries will therefore experience increased attention by governments and a rising allocation of funding.

10.6.2 Rural Mobility Scenarios and Policy Recommendations

The above description of optimal interurban and rural mobility services will not of course be a uniform development globally. Like in the case of urban mobility, there will be many variations depending on the status of economic and social development of the country or region. In the lesser developed economies, the rural mobility scene will be less developed and the time necessary to reach the above optimum, much longer. A more encompassing analysis of future rural mobility in developed and less developed economies can be seen in a relevant study by *Scenario Management International*—ScMI (Kuhle & Günther, 2024). This study developed eight scenarios (numbered from no. 1 to no. 8). They range from a doomsday scenario, no. 1 ("rural frustration"), to an "ecological transition" scenario no. 4, and to a high-end "revolutionary" one (scenarios 6 and 7) in which most mobility services are based on automated systems and digital technologies like the ones described above.

On the policy level, a recent study by the International Transport Forum (ITF) has contributed several recommendations related to the necessary policy and administrative measures that would assist in materialising the types of future mobility described above for rural areas. These policy recommendations are summarised below (ITF, 2021):

- Formulate a countrywide strategic accessibility policy and implement Sustainable Regional Mobility Plans (SRMP).
- Adopt a whole-of-government approach for rural public services and the local economy to ensure that rural mobility policies are embedded within broader policies towards public service provision, regional economic development, land-use planning, digital connectivity and climate policies.
- Promote and support hybrid car and ridesharing to develop in rural areas.
- Increase central government funding for shared and active travel in rural areas.
- Promote mobility hubs to connect local services to the core network.
- Support the development of national or regional Rural mobility as a Service (RmaaS) scheme.
- Make regulations more flexible to allow for the development of innovative, cost-effective mobility solutions.
- Combine public budgets for mobility to achieve cost savings.
- Fund pilot schemes to test innovative concepts.
- Prioritise financial support for innovative services according to high impact levels, rather than high-tech levels.
- Use innovative financing approaches to increase funding pools and viability of individual services.
- Develop a *Rural Mobility Technical Assistance Programme* building local capacity and providing access to centralised expertise, including legal and technical support, to accompany not only local authorities but also community-based bottom-up initiatives.

In summary, circa 2050 at least for the developed economies, the countryside will be rediscovered by young and old people alike as a favourable place to live with a high quality of life. Digitalization will be used extensively to reduce the everyday mobility needs while on demand automated vehicles will provide physical mobility and freight delivery services. The 2050 rural mobility system will be integrating and prioritising on-demand public transportation with AI powered personalised and shared mobility services and including collaborative and volunteer-based options supported and safeguarded by appropriate regulation. Strong, data-driven, AI platforms will handle and manage all these services addressing rural mobility in a most efficient way.

References

Aalto, E., & Kuosa, T. (2024). Four scenarios on the future of megacities. *Futures Platform*, authors, March 5, 2024. https://www.futuresplatform.com/blog/scenarios-future-megacities-urbanisation

ADB. (2024a). Africa's macroeconomic performance and outlook (MEO). *African Development Bank*, February 2024.

ADB. (2024b). ASIA 2050—Realizing the Asian century. *Asian Development Bank*, May 2024.

ADB. (2025). Asian Development Outlook. *Asian Development Bank*, reports on Asian countries development issued and updated continuously. https://www.adb.org/outlook/editions
Alcock C. (2024). AutoFlight delivers first prosperity eVTOL and certifies freighter. *AIN Media Group*, April 4, 2024.
ARUP. (2019). 2050 Scenarios: Four plausible futures. *An ARUP Foresight report*, November 19, 2019. https://www.arup.com/insights/2050-scenarios-four-plausible-futures/
BCG. (2022). Six megatrends that are changing Africa—And how to navigate them. *Boston Consulting Group*, report by P. Dupoux, Q. Dhanani, S. Rafiq, T. Oyekan, & M. Hassoun, August 29, 2022. https://www.bcg.com/publications/2022/six-megatrends-that-are-changing-africa
Berger. (2018). Urban air mobility: The rise of a new mode of transportation. *Roland Berger Focus*, November 2018.
Bhatnagar, S. (2024). Smart cities—Concept, pillars, and challenges. In S. Parkinson, A. Nikitas, & M. Vallati (Eds.), *Deception in autonomous transport systems. Wireless networks.* Springer. https://doi.org/10.1007/978-3-031-55044-7_3
Brown, S., Jenkins, K., Goodwin, P., Lincke, D., Vafeidis Ath, Tol, R. S. J., Jenkins, R., Warren, R., Nichols, R. J., Jevrejeva, Sv., Sanchez Arcilla, A., & Haigh, I. D. (2021). Global costs of protecting against sea-level rise at 1.5 to 4.0 °C. *Climatic Change, 167*, 4. https://doi.org/10.1007/s10584-021-03130-z
Canina, M., Bruno, C., & Monestier, E. (2022). Futures thinking. In *The Palgrave encyclopaedia of the possible.* Palgrave Macmillan. https://doi.org/10.1007/978-3-319-98390-5_272-1
Cheng, E. (2023). China gives Ehang the first industry approval for fully autonomous, passenger-carrying air taxis. *CNBC Newsletter*, November 8, 2023.
Cilliers, J., Hughes, B. B., & Moyer, J. (2011). African futures 2050: The next forty years. *Institute for Security Studies* and the *Pardee Center for International Futures*, Monograph, January 2011. https://doi.org/10.2139/ssrn.2690239
COACCH. (2022). CO-designing the assessment of climate change costs, deliverable D5.7 climate change impacts & policy synthesis. *EU research project COACCH* (Grant agreement no. 776479). In https://www.coacch.eu/. Accessed October 17, 2024.
Colelli, F. P., Wing, I. S., & Cian, E. D. (2023). Air-conditioning adoption and electricity demand highlight climate change mitigation–adaptation trade-offs. *Scientific Reports, 13*, article no. 4413, 17 March 2023. https://doi.org/10.1038/s41598-023-31469-z
Cordova-Pozo, K., & Rouwette, E. A. J. A. (2023). Types of scenario planning and their effectiveness: A review of reviews. *Futures, 149*, May 2023. https://doi.org/10.1016/j.futures.2023.103153
Draghi M. (2024). The future of European competitiveness: Part A, A competitiveness strategy for Europe. September 2024. https://commission.europa.eu/document/97e481fd-2dc3-412d-be4c-f152a8232961_en
EC. (2024). The artificial intelligence act. *European Commission*, Regulation (EU) 2024/1689, Official Journal of the European Union, 12 July 2024.
EEA. (2024). European climate risk assessment—Executive summary. *European Environment Agency*, report 01/2024. ISBN 978-92-9480-627-7. https://doi.org/10.2800/204249
Fontecave, M., Candel, S., & Poinsot, T. (2024). Hydrogen today and tomorrow. French Academy of Sciences (Académie des sciences), 26 August 2024. https://hal.science/hal-04678172v1
Heineke, K., Kloss, B., Möller, T., & Wiemuth, Ch. (2021). Shared mobility: Where it stands and where it's going. *McKinsey Center for Future Mobility*, August 2021.
IEA. (2023). Global EV outlook 2023: Catching up with climate ambitions. *International Energy Agency*, April 2023. www.iea.org
IMF. (2022). Poor and vulnerable countries need support to adapt to climate change. *International Monetary Fund* blog by K. Georgieva, V. Gaspar, & C. Pazarbasioglu, March 23, 2022. https://www.imf.org/en/Blogs/Articles/2022/03/23/blog032322-poor-and-vulnerable-countris-need-support-to-adapt-to-climate-change
IPCC. (2014). Fifth assessment report—AR5. Intergovernmental Panel on Climate Change (IPCC), October 2014.

ITF. (2021). Innovations for better rural mobility. *International Transport Forum—ITF*, Research report, OECD Publishing.

JRC. (2024). Assessing climate risks in Europe: a quantitative and consistent long-term scientific analysis to support EU climate policies. Project TRACE (Territorial Risk Assessment of Climate in Regions of Europe) European Commission, Joint Research Center part of the JRC PESETA V programme. First results August 2024. https://joint-research-centre.ec.europa.eu/scientific-activities-z/peseta-climate-change-projects/jrc-peseta-v_en

Kuhle, J.-P., & Günther, J. (2024). The future of mobility in rural areas 2040. *Scenario Management International (ScMI)*. https://www.scmi.de/en/future-rural-mobility

Letta, E. (2024). Much more than a market: Speed, security, solidarity, empowering the Single Market to deliver a sustainable future and prosperity for all EU Citizens. Report, April 2024 (commissioned by the Belgian and the Spanish governments with the support of the European Council). https://www.consilium.europa.eu/media/ny3j24sm/much-more-than-a-market-report-by-enrico-letta.pdf

Martin, M., & Waddock, D. (2022). A Nordic initiative to resolve the new debt crisis. *Debt Relief International Organisation* for the "Norwegian Church Aid", September 2022. dfi@dri.org.uk

Martínez-Plumed, F., Caballero, F., Castellano-Falcón, D., Fernández-Llorca, D., Gómez, E., Hupont-Torres, I., Merino, L., Monserrat, C., & Hernández-Orallo, J. (2022). AI Watch: Revisiting technology readiness levels for relevant artificial intelligence technologies. Joint Research Center, Publications Office of the European Union, Luxembourg. ISBN 978-92-76-52328-4. https://doi.org/10.2760/495140

McKinsey. (2024). The next big arenas of competition. *McKinsey Global Institute*, October 2024. mckinsey.com/mgi

MGI. (2023). Reimagining economic growth in Africa: Turning diversity into opportunity. *McKinsey Global Institute*, Report by M. Kuyoro, A. Leke, O. White, L. Woetzel, K. Jayaram, & K. Hicks, June 5, 2023. https://www.mckinsey.com/mgi/our-research/reimagining-economic-growth-in-africa-turning-diversity-into-opportunity

NASAC. (2024). *Decarbonisation of transport in Africa: Opportunities, challenges and policy options*. Network of African Science Academies (NASAC) and InterAcademy Partnership (IAP), March 2024. ISBN 978-9914-9636-2-5.

NIC. (2021). Global trends 2040—Economics. *US National Intelligence Council*, March 2021. https://www.dni.gov/index.php/gt2040-home/gt2040-structural-forces/economics

OECD. (2023). Economic outlook for Southeast Asia, China and India 2023. *OECD report*, 31 March 2023. https://doi.org/10.1787/f677c529-en

O'Hern, S., & St. Louis, R. (2023). Technology readiness and intentions to use conditionally automated vehicles. *Transportation Research Part F: Traffic Psychology and Behaviour, 94*, 1–8. https://doi.org/10.1016/j.trf.2023.02.001

Porche Consulting. (2022). The autonomous logistics hub of the future. www.porsche-consulting.com

Strauss, B. H., Kulp, S. A., Rasmussen, D. J., & Levermann, A. (2021). Unprecedented threats to cities from multi-century sea level rise. *Environmental Research Letters, 16*(11). https://doi.org/10.1088/1748-9326/ac2e6b

UK. (2023). Net-Zero society: Scenarios and pathways—How could societal changes affect the path to net zero? *UK government Office for Science* in collaboration with FORESIGHT. April 2023.

UK. (2024). The futures toolkit. *UK government Office for Science* in collaboration with SAMI Consulting. June 2024.

UNEP. (2023). Emissions gap report 2023: Broken record—Temperatures hit new highs, yet world fails to cut emissions (again). *UN Environment programme*, Nairobi. ISBN: 978-92-807-4098-1. https://doi.org/10.59117/20.500.11822/43922

van Notten, P. W. F. (2005). *Writing on the wall: Scenario development in times of discontinuity* [Doctoral Thesis, Maastricht University]. Thela Thesis & Dissertation.com, pp. 211. https://doi.org/10.26481/dis.20050408pn

Walker R. (2018). Hyperloop: Cutting through the hype. *Transportation Research Laboratory* (TRL), June 2018.

WBCSD. (2014). Futures thinking: A guide to using futures thinking to help drive corporate resilience and transformational innovation. *The World Business Council for Sustainable Development* (WBCSD), Switzerland, March 2014. https://docs.wbcsd.org/2017/01/Future_Thinking/1_Futures_Thinking.pdf

WBG. (2023). The World Bank in Africa. *The World Bank*, Oct 05, 2023. https://www.worldbank.org/en/region/afr/overview

WEF. (2024). Why we can expect the return of the 'Asian Century' amid weak global economic forecasts. *World Economic Forum*, Opinion, 8, January 2024. https://www.weforum.org/agenda/2024/01/asian-century-weak-global-economic-forecasts/

WMO. (2024). WMO global annual to decadal climate update (2024–2028). World Meteorological Organization, 5 June 2024.

Wong, T., Campos, R., Mackres, E., Staedicke, S., & Doust, M. (2024). What would cities look like with 3 degrees C of warming vs. 1.5? Far more hazardous and vastly unequal. *World Resources Institute* report, 17 September 2024. www.wri.org/insights

World Bank. (2010). Economics of adaptation to climate change: synthesis report, *The World Bank*. https://openknowledge.worldbank.org/handle/10986/12750

World Bank. (2024). The middle-income trap—World Development Report 2024. The *World Bank Group*, flagship report, Washington DC, 2024.

WRI. (2024a). City-scale, city-relevant climate hazard indicators under 1.5 °C, 2.0 °C, and 3.0 °C of global warming. World Resources Institute, Technical Note, 19 September 2024. https://doi.org/10.46830/writn.23.00154

WRI. (2024b). Cities will suffer health and infrastructure crisis under 3 °C vs. 1.5 °C warming, with low-income cities worst impacted. World Resources Institute, Technical Note, September 19, 2024.

[illegible]

Epilogue, a Call for Action

In the face of climate change challenges, armed with the tools supplied by modern advanced mobility and information technologies, modern societies must urgently secure the necessary resources and master the political will to enable a safe, efficient, and resilient transition to the anticipated more collective, intelligent, and automated mobility systems of the future. Adopting a "do nothing" approach would only aggravate the problems of congestion, delays, and reduced mobility in the future to an unsustainable degree so this is not an option. The first and foremost message of this book is that besides the aggravating mobility problems modern societies sooner or later will also have to confront the looming climate crisis or face severe and unbearable consequences. In the face of these consequences any of the novel mobility technologies that are now being put forward for implementation or are emerging through research and innovation, will not be possible to be implemented or, if they do, they will render themselves unusable for large periods of time. The survival of our future urban areas calls for finding ways to integrate and operate the new mobility technologies and systems within an operating environment that is climate friendly, resilient, and fully intelligent to offer operational efficiency in line with real user needs.

The second key message is that the making of such an intelligent, human centred, and climate resistant mobility system should start right now if we are to reap its maturity and full benefits by 2050 and beyond. Unfortunately, at present, the world seems to move on a "business as usual" path absorbed by short term economic, political, and geopolitical concerns with many countries or regions largely ignoring the larger scale and longer-term climate related consequences. We have argued that these relaxed attitudes of "doing little or nothing" will "forcefully" change when the urgency and severity of the climate crisis phenomena start hitting hard on national economies and societies with an ever-increasing frequency and strength. It will be an abrupt awakening that will change priorities and put mitigation and adaptation measures first, but until then many governments happily move-on with business as usual or, in a few yet very remarkable cases, they even reverse their climate action

G. A. Giannopoulos and Y. Li, *Human Mobility, Artificial Intelligence and Climate Change*, Lecture Notes in Mobility, https://doi.org/10.1007/978-3-032-08171-1

course and move to the opposite direction. For them, there will come a time—possibly by 2035–2040 or thereabouts—when the hard realities of climate induced catastrophes will hit hard and make the need for urgent action sink in. Then all the governments that are not willing now to accept and face the threat, will change course and align with the rest of the world in taking climate mitigation and adaptation actions seriously, including in the transport and mobility sector.

A third key message is that new transport technologies alone will not be enough to deliver optimally sustainable mobility and transport systems. Societies will need to go beyond adopting and implementing technological breakthroughs to also change attitudes and organisational structures to enable ways to "absorb" the new technologies by enabling them to function in the most efficient way within a suitable social, legal, and administrative environment. This is where AI will play a most important and decisive role besides all its other applications and roles. It will take bold political, financial, and organisational decision-making aided and supported by strong Artificial Intelligence capabilities to create the new mobility landscape of the future. A landscape that is user centric, multimodal, efficient, and adequately resilient to climate change. By 2050, it is expected that such bold decisions and commitment of resources will only be possible to be achieved by a handful of countries and urban areas whose number is not expected to exceed 10 or 12 megacities globally.

In trying to outline the most likely mobility context and operations by 2050 +, we have considered the above preconditions and opted for a pragmatic approach i.e. that the mobility landscape in 2050 + will be the result of a balancing interplay between the following main influencing factors:

a. Ability to install, implement and operate (especially in urban areas) the physical and administrating/supervising infrastructures necessary for the functioning and smooth operation of the novel transport and mobility technologies including cooperative, connected and automated mobility.
b. Making all urban and rural infrastructures resilient to climate related extreme events at an 80% level i.e. covering at least 80% of the total number of vulnerable infrastructure for most of the expected extreme weather events.
c. Ability to optimally operate the new mobility systems while at the same time adapting user behaviour to utilise harmoniously these systems and exploit the installed capacities with the help of AI, and
d. Develop and implement internationally harmonised technical standards and operational rules as well as legal and regulatory frameworks to enable a socially accepted use of AI as an enabler of smart mobility systems especially the cooperative, connected, and automated ones.

The degree to which future human societies will combine successfully the above influencing factors while safeguarding the lifestyles and wellbeing of their citizens and at the same time mastering the economic, political, and geopolitical challenges that will always exist, will determine the resulting future mobility landscapes. Our estimate is that the future mobility landscape across the various countries and urban or rural areas, for 2050 and beyond, will be one of diversity. There will be a few *"islands" of transformational change* where—as already mentioned—the influencing factors would have been combined successfully creating favourable conditions for the sustainable implementation of all novel mobility technologies and concepts. Such "islands" will be characterised by very low private car ownership and use in favour of well-planned and affordable mobility services offered by mobility providing companies using shared but well "personalised" forms of automated mobility. Automated transport vehicles of all kinds will be transporting groups of travellers, as well as freight, in AI controlled fully decarbonised trips and AI controlled transport networks.

In contrast to the advanced mobility "islands", the rest of the urban and rural areas in the world, by 2050 +, will be struggling in *transition*. This means that their mobility systems and services will be in a transitional phase between a level close to their current (conventional) state and one that aspires to reach the AI-controlled, clean, and automated mobility level of the transformational "islands". The degree of adaptation to climate change of these areas will also differ and their resilience and protection from climate related extreme events will be uneven and variant. So, there will be a multitude of different mobility landscapes materialising globally by 2050 including cases of well organised and smoothly running transport networks but still using mainly conventional transport vehicles and systems, as well as chaotic and congested situations still running on conventional vehicles and transport systems very much like the ones we see in many countries today. In all cases, AI will be recognised as the most relevant tool of choice for effecting efficient and climate friendly mobility but the level of its development and application will depend on the necessary infrastructure (especially for data collection and communication) that will be built and operate.

In harmony with the above views, a picture of the expected future mobility in 2050 + is presented in Tables 10.3 and 10.4 of Chap. 10, in terms of three types of urban areas termed *Urban scenes* or *Levels*. These types of urban areas are differentiated according to their level of economic development, the capacity to utilise new technologies and mobility systems, and the degree of adaptation to climate change. Besides the distinction between types of urban areas, further insights were sought by looking at the development potential of four key geographic regions i.e., Asia, Europe, North America, and Africa. This potential was examined under the following three attributes considered as the key drivers of change in the transport and mobility sector:

a. Economic and social maturity of the region to finance and implement disruptive technological change.
b. User behavioural attitudes and "maturity" in terms of issues such as abolishing car ownership and use, acceptance of mobility sharing and mobility-on-demand, full respect for the environmental and climate change impacts of trip making, and similar.
c. Degree of adaptation to climate change and prevailing environmental conditions.

The assumed level of development for the above three key drivers in the target year of 2050 was the result of a synthesis of the findings in a larger number of six influencing factors: Political and geopolitical events and disturbances/national economic development prospects/social attitudes and preferences/technological developments/legal and regulatory rule setting prospects/and environmental and climate change mitigation and adaptation policies and prospects. For the advanced economies where all new transport technologies and new mobility business models are most likely to be implemented first, the picture for urban and interurban/rural mobility in 2050 + can be quite radically different from today. It will consist of automated vehicles being used freely on the roads, in the air, and eventually on the water. These will be fully AI-driven and adapted to individual mobility needs with services offered by several competing mobility service providers offering highly time- and energy-efficient and fully decarbonised trip making options using renewable energy. The detailed characteristics of the future mobility situation that are likely to be found in both advanced economies as well as for the lesser developed economies, are presented in Tables 10.3 and 10.4 of Chap. 10 of this book. As we move beyond 2050 and towards the end of the century, more and more urban mobility systems of the lesser-developed urban areas group should be expected to enter the advance mobility category.

The overriding concern in all these developments, will be the environmental protection and the proper handling of all necessary mitigation or adaptation actions for climate change. The full use of AI, the need to develop resilience against climate change, and the capacity of the existing socio-political systems to support innovation and change, will be critically enabling factors. Equally important will be the timing of the necessary enabling decisions. Crucial decisions and commitments must be taken "now" i.e., until 2030 or 2035 the latest. The developing climate change crisis and the extent to which its impacts will "awaken" the decision makers (especially in the leading and more advanced economies) will determine the final outcome as regards the shape of human mobility in the future. The world must unite behind the United Nations, or other similar future organisations, for proper coordination and cooperation actions in achieving commonly defined plans for sustainable development which will include future mobility systems. The current Sustainable Development Goals list by the UN that is aimed to promote a sustainable and equitable future for all, provides a good example of what such coordination actions can entail. The decisions to be made and the actions to be taken by individuals, businesses and governments over the next decade (2025–2035) will determine our trajectory till 2050 and beyond.

Humanity—all of humanity—must understand the dangers of climate change and work together to make technology shape a better world with better mobility for all.

MIX
Papier aus verantwortungsvollen Quellen
Paper from responsible sources
FSC® C105338

If you have any concerns about our products,
you can contact us on
ProductSafety@springernature.com

In case Publisher is established outside the EU,
the EU authorized representative is:
Springer Nature Customer Service Center GmbH
Europaplatz 3, 69115 Heidelberg, Germany

Printed by Libri Plureos GmbH
in Hamburg, Germany